Pi to One Million Digits

(A most unusual notebook where you write between the pi)

Archi Medes

Berhampore Press

Wellington,

New Zealand

BerhamporePress@gmail.com

ISBN-13:

978-1974036691

ISBN-10:

1974036693

3.14159265358979323846264338327950288419716939937510582097494459230781640628620899862803482534211706798214808651328230664709384460955058223172535940812848111745028410270193852110555964462294895493038196442881097566593344612847564823378678316527120190914564856692346034861045432664821339360726024914127372458700660631558817488152092096282925409171536436789259036001133053054882046652138414695194151160943305727036575959195309218611738193261179310511854807446237996274956735188575272489122793818301194912983367336244065664308602139494639522473719070217986094370277053921717629317675238467481846766940513200056812714526356082778577134275778960917363717872146844090122495343014654958537105079227968925892354201995611212902196086403441815981362977477130996051870721134999999983729780499510597317328160963185950244594553469083026425223082533446850352619311881710100031378387528865875332083814206171776691473035982534904287554687311595628638823537875937519577818577805321712268066130019278766111959092164201989380952572010654858632788659361533818279682303019520353018529689957736225994138912497217752834791315155748572424541506959508295331168617278558890750983817546374649393192550604009277016711390098488240128583616035637076601047101819429555961989467678374494482553797747268471040475346462080466842590694912933136770289891521047521620569660240580381501935112533824300355876402474964732639141992726042699227967823547816360093417216412199245863150302861829745557067498385054945885869269956909272107975093029553211653449872027559602364806654991198818347977535663698074265425278625518184175746728909777727938000816470600161452491921732172147723501414419735685481613611573525521334757418494684385238239073941433345477624168625189835694855620992192221842725502542568876717904946016534668049886272327917860857843838279679766814541009538837863609506800642251252051173929848960841284886269456042419652850222106611863067442786220391949450471237137869609563643719172874677646575739624138908658326459581339047802759009946576407895126946839835259570982582262052248940772671947826848260147699090264013639443745530506820349625245174939965143142980919065925093722169646151570985838741059788595977297549893016175392846813826868386894277415599185592524595395943104997252468084598727364469584865383673622262609912460805124388439045124413654976278079771569143599770012961608944169486855584806353422072225828488648158456028506016842739452267467678895252138522549954666727823986456596116354886230577456498035593634568174324112515076069479451096596094025228879710893145669136867228748940560101503308617928680920874760917824938589009714909675985261365549781893129784821682998948722658804857564014270477555132379641451523746234364542858444795265867821051141354735739523113427166102135969536231442952484937187110145765403590279934037420073105785390621983874478084784896833214457138687519435064302184531910484810053706146806749192781911979399520614196634287544406437451237181921799983910159195618146751426912397489409071864942319615679452080951465502252316038819301420937621378559566389377870830390697920773467221825625996615014215030680384477345492026054146659252014974428507325186660021324340881907104863317346496514539057962685610055081066587969981635747363840525714591028970641401109712062804390397595156771577004203378699360072305587631763594218731251471205329281918261861258673215791984148488291644706095752706957220917567116722910981690915280173506712748583222871835209353965725121083579151369882091444210067510334671103141267111369908658516398315019701651511685171437657618351556508849099898599823873455283316355076479185358932261854896321329330898570642046752590709154814165498594616371802709819943099244889575712828905923233260972999712084433573265489382391193259746366730583604142813883032038249037589852437441702913276561809377344403070746921120191302033038019762110110044929321516084244485963766983895228

8478312355265821314495768572624334418930396864262434107732269780280731891544110104468232527162010526522721116603966655730925471105578537634668206531098965269186205647693

1257058635662018558100729360659876486117910453348850346113657686753249441668039626579787718556084552965412665408530614344431858676975145661406800700237877659134401712749

470420562230538994561314071127000407854733269939081454664645880797270826683063432858785698305235808933065757406795457163775254202114955761581400250126228594130216471550 9

7925923099079654737612551765675135751782966645477917450112996148903046399471329621073404375189573596145890193897131117904297828564750320319869151402870808599048010941214

722131794764777262241425485454033215718530614228813758504306332175182979866223717215916077166925474873898665494945011465406284336639379003976926567214638530673 6096571209

180763832716641627488880078692560290228472104031721186082041900042296617119637792133757511495950156604963186294726547364252308177036751590673502350728354056704038674351 3

622224771589150495309844489333096340878076932599397805419341447377441842631298608099886874132604721569516239658645730216315981931951673538129741677294786724229246543668

00980676928238280689964004824354037014163149658979409243237896907069779422362508221688957383798623001593776471651228935786015881617557829735233446042815126272037343146 53

19777741603199066554187639792933441952154134189948544473456738316249934191318148092777710386387734317720754565453220777092120190516609628049092636019759882816133231666 36

5286193266863360627356763035447762803504507772355471058595487027908143562401451718062464362679456127531813407833033625423278394497538243720583531147711992606381334677687

9695970309833913077109870408591337464144282277263465947047458784778720192771528073176790770715721344473060570073349243693113835049316312840425121925651798069411352801314

70130478164378851852909285452011658393419656213491434159562586586557055269049652098580338507224264829397285847831630577775606888764462482468579260395352773480304802900 58

760758251047470916439613626760449256274204208320856611906254543372131535958450687724602901618766795240616342522577195429162991930645537991403734043287526288896399587947

572917464263574552540790914513571113694109119393251910760208252026187985318877058429725916778131496990090192116971737278476847268608490033770242429165130050051683233643 5

0389517029893922334517220138128069650117844087451960121228599371623130171144484640903890644954440061986907548516026327505298349187407866808818338510228334508504860825039

302133219715518430635455007668282949304137765527939751754613953984683393638304746119966538581538420568533862186725233402830871123 82789212507712629463229563989898935821 1

674562701021835646220134967151881909730381198004973407239610368540664319395097901906996395524530054505806855019567302292191393391856803449039820595510022635353619204 1994

745538593810234395544959778377902374216172711172364343543947822181852862408514006660443325888569867054315470696574745855033232334210730154594051655379068662733379958511 5

625784322988273723198987571415957811196358330059408730681216028764962867446047746491599505497374256269010490377819868359381465741268049256487985561453723478673303904688 3

8343634655379498641927056387293174872332083760112302991136793862708943879936201629515413371424892830722012690147546684765357616477379467520049075715552781965362132392640

61601363581559074220202031872776052772190055614842555187925303435139844252323415762336106425063904975008656271095359194658975141310348227693062474353632569160781547818 11

528436679570611086153315044521274739245449454236828860613408414863776700961207151249140430272538607648236341433462351897576645216413767969031495019108575984423919862916 4

2193994907236234646844117394032659184044378051333894525742399508296591228508555821572503107125701266830240292952522011872676756220415420516184163484756516999811614101002

9960783869092916030288400269104140792886215078424516709087000699282120660418371806535567252532567532861291042487761825829765157959847035622262934860034158722980534989650

04925485346941469775164932709504934639382432227188515974054702148289711177792376122578873477188196825462981268685817050740272550263329044976277894423621674119186269439 65

06715157795867564823993917604260176338704549901761436412046921823707648878341968968611815581587360629386038101712158552726683008238340465647588040513808016336388742163 71

40643549556186896411228214075330265510042410489678352858829024367090488711819090949453314421828766181031007354770549815968077200947469613436092861484941785017180779306 81

08546900094458995279424398139213505586422196483491512639012803832001097738680662877923971801461343244572640097374257007359210031541508936793008169980536520276007277496 74

58400283624053460372634165542590276018348403068113818551059797056640075094260878857357960373245141467870368809880609716425849759513806930944940151542222194329130217391 2

53835591503100333032511174915696917450271494331515588540392216409722910112903552181576282328318234254832611191280092825256190205263016391147724733148573910777587442538 76

11746578671169414776421441111263583553871361011023267987756410246824032264834641766369806637857681349204530224081972785647198396308781543221166912246415911776732253264 33

56861461865452226812688726844596844241610785401676814208088502800541436131462308210259417375623899420757136275167457318918945628352570441335437585753426986994725470316 56

61399199968262824727064133622217892390317608542894373393561889165125042440400895271983787386480584726895462438823437517885201439560057104811949884239060613695734231559 07

96703461491434478863604103182350736502778590897578272731305048893989009923913503373250855982655867089242612429473670193907727130706869170926462548423240748550366080136 04

66895118400936686095463250021458529309500009071510582362672932645373821049387249966993394246855164832611341461106802674466373343753407642940266829738652209357016263846 48

52851490362932019919968828517183953669134522244470804592396602817156551565666111359823112250628905854914509715755390024393153519090210711945730024388017661503527086260 25

37881797519478061013715004489917210022201335013106016391541589578037117792775225978742891917915522417189585361680594741234193398420218745649256443462392531953135103311 47

63949119950728584306583619353693296992898379149419394060857248639688369032655643642166442576079147108699843157337496488352927693282207629472823815374099615455987982598 91

09371712621828302584811238901196822142945766758071865380650648702613389282299497257453033283896381843944770779402284359883410035838542389735424395647555684095224844554 13

92394100016207693636846776413017819659379971557468541946334893748439129742391433659360410035243477706588867811394986164787471407932638587386247328896456435987746676384 7

94665040741118256583788784548581489629612739984134427260860618724545423606431537101127468097787044640947582803487697589483282412392929605829486191966709189580898332012 10

31843034012849511620353428014412761728583024355983003204202451207287253558119584014918096925339507577840006746552603144616705082768277222353419110263416315714740612385 04

25845988419907611287258059113935689601431668283176323567325417073420817332230462987992804908514094790368878687894930546955703072619009502076433493359106024545086453628 93

54568629585313153371838682656178622736371697577418302398600659148161640494496501173213138957470620884748023653710311508984279927544268532779743113951435741722197597993 59

68525228574526379628961269157235798662057340837576687388426640599099350500081337543245463596750484423528487470144354541957625847356421619813407346854111766883118654489 37

76979566517279662326714810338643913751865946730024434500544995399742372328712494834706044063471606325830649829795510109541836235030309453097335834462839476304775645015 00

85075789495489313939448992161255255977014368589435858775263796255970816776438001254365023714127834679261019955852247172201777237004178084194239487254068015560359983905 48

4

98572354674564239058585021671903139526294455439131663134530893906204678438778505423939052473136201294769187497519101147231528932677253391814660730008902776896311481090022

09724520759167297007850580717186381054967973100167870850694207092232908070383263453452038027860990556900134137182368370991949516489600755049341267876436746384902063964 01

97666855923356546391383631857456981471962108410809618846054560390384553437291414465134749407848844237721751543342603066988317683310011331086904219390310801437843341513 70

92435301367763108491351615642269847507430329716746964066653152703532546711266752246055119958183196376370761799191920357958200759560530234626775794393630746305690108011 49

42714100939136913810725813781357894005599500183542511841721360557275221035268037357265279224173736057511278872181908449006178013889710770822931002797665935838758909395 68

81485602632243937265624727760378908144588378550197028437793624078250527048758164703245812908783952324532378960298416692254896497156069811921865849267704039564812781021 79

91321741630581055459880130048456299765112124153637451500563507012781592671424134210330156616535602473380784302865525722275304999883701534879300806260180962381516136690 33

41111386538510919367393835229345888322550887064507539473952043968079067086806445096986548801682874343786126453815834280753061845485903798217994599681154419742536344399 60

29025100158882721647450068207041937615845471231834600726293395505482395571372568402322682130124767945226448209102356477527230820810635188991526928891084555711266039650 34

39789627825001611015323516051965590421184494990778999200732947690586857787872098290135295661397888486050978608595701773129815531495168146717695976099421003618355913877 78

17698458758104466283998806006162298486169353373865787735983361613384133853684211978938900185295691967804554482858483701170967212535338758621582310133103877668272115726 94

95181795897546939926421979155233857662316762754757035469941489290413018638611943919628388705436777432242768091323654494853667680000010652624854730558615989991401707698 38

54831887501429389089950685453076511680333732226517566220752695179144225280816517166776672793035485154204023817460892328391703275425750867655117859395002793389592057668 27

89677644531840404185540104351348389531201326378369283580827193783126549617459970567450718332065034556644034490453627560011250184335607361222765949278393706478426456763 38

81880756561216896050416113903906396016202215368494109260538768871483798955999911209916464644119185682770045742434340216722764455893301277815868695250694993646101756850 60

16714535431581480105458860564550133203758645485840324029871709348091055621167154684847780394475697980426318099175642280987399876697323769573701580806822904599212366168 90

25962730430679316531149401764737693873514093361833216142802149763399189835484875625298752423873077559555955465196394401821840998412489826236737714672260616336432964063 35

72810707887581640438148501884114318859882769449011932129682715888413386943468285900666408063140777577257056307294004929403024204984165654797367054855804458657202276378 40

46682337985282710578431975354179501134727362577408021347682604502285157979579764746702284099956160156910890384582450267926594205550395879229818526480070683765041836562 09

45554346135134152570065974881916341359556719649654032187271602648593049039787489589066127250794828276938953521753621850796297785146188432719223223810158744450528665238 02

25328438913752738458923844225354726530981715784478342158223270206902872323300538621634798850946954720047952311201504329322662827276321779088400087861480221475376578105 819

70222630971749507212724847947816957296142365859578209083073323356034846531873029302665964501371837542889755797144992465403868179921389346924474198509733462679332107268 68

70768062639919361965044099542167627840914669856925715074315740793805323925239477557441591845821562518192155233709607483329234921034514626437449805596103307994145347784 57

4699992128599999399612281615219314888769388022281083001986016549416542616968586788372609587745676182507275992950893180521872924610867639958916145855058397274209809097817

2932393010676638682404011130402470073508578287246271349463685318154696904669686939254725194139929146524238577625500474852954768147954670070503479995888676950161249722820

4030399546327883069597624936151010243655535223069061294938859901573466102371223547891129254769617600504797492806072126803922691102777226102544149221576504508120677173571

2027180242968106203776578837166909109418074487814049075517820385653909910477594141321543284406250301802757169650820964273484146957263978842560084531214065935809041271135

9200419759851362547961606322887361813673732445060792441176399759746193835845749159880976674470930065463424234606342374746660804317012600520559284936959414340814685298150

5394717890045183575515412522359059068726487863575254191128887737176637486027660634960353679470269232297186832771739323619200777452212624751869833495151019864269887847171

9396649769070825217423365662725928440620430214113719922785269984698847702323823840055655517889087661360130477098438611687052310553149162517283732728676007248172987637569

8163354150746088386636406934704372066886512756882661497307886570156850169186474885416791545965072342877306998537139043002665307839877638503238182155355973235306860430106

7576083890862704984188859513809103042359578249514398859011318583584066747237029714978508414585308578133915627076035639076394731145549583226694570249413983163433237897595

5680856836297253867913275055542524491943589128405045226953812179131914513500993846311774017971512283785460116035955402864405902496466930707769055481028850208085800878115

7738171917417760173307385547580060560143377432990127286772530431825197579167929699650414607066457125888346979796429316229655201687973000356463045793088403274807718115553

3090988702550520768046303460865816539487695196004408482065967379473168086415645650530049881616490578831154345485052660069823093157776500378070466126470602145750579327096

2047825615247145918965223608396645624105195510522357239739512881816405978591427914816542632892004281609136937773722299983327082082969955737727375667615527113922588055201

8988762011416800546873655806334716037342917039079863965229613128017826797172898229360702880690877686605932527463784053976918480820410219447197138692560841624511239806201

1318454124478205011079876071715568315407886543904121087303240201068534194723047666672174986986854707678120512473679247919315085644477537985379973223445612278584329684664

7513336573692387201464723679427870042503255589926884349592876124007558756946413705625140011797133166207153715436006876477318675587148783989081074295309410605969443158477

5397009439883949144323536685392099468796450665339857388878661476294434140104988899316005120767810358861166020296119363968213496075011164983278635316145168457695687109000

2999769841263266502347716728657378579085746646077228341540311441529418804782543876177079043000156698677679576090996936075594965152736349811896413043311662774712338817400

6037317439705406703109676765748695358789670031925866259410510533584384656023391796749267844763708474978333655579007384191473198862713525954625181604342253729962863267496

8240580602964211463864368642247248872834341704415734824818333016405669596686867695634914163842641497453334999948000266998758881593507357815195889900539512085351035722613

7364034336753471410483601754648830040784641674521673719048310967671134434948192626811107399482506073949507350316901973185211955263563258433909982249862406703107683184660

7291248747540316179699411397387765899868554170318847788675929026070043212666179192235209382278788809886335991160819235355570464634911320859189796132791319756490976000139

9623444553501434642686046449586247690943470482932941404111465409239883444351591332010773944111840741076849810663472410482393582740194493566516108846312567852977697346843

030614624180358529331597345830384554103370109167677637427621021370135485445092630719011473184857492331816720721372793556795284439254815609137281284063330393735624200 1604

5664557414588166052166608738748047243391212955877763906969037078828527753894052460758496231574369171131761347838827194168606625721036851321566478001476752310393578606896

11125996028183930954870905907386135191459181951029732787557104972901148717189718004696169777001791391961379141716270701895846921434369676292745910994006008498356842520 19

155937037010110497473394938778859894174330317853487076032219829705797511914405109942358830345463534923498268836240433272674155403016195056806541809394099820206099941 4021

68909007082133072308966211977553066591881411915777836272927461561857103721724710095214236964830864102592887457999322374955191221951903424452307535133806856807354464995 127

2031744871954039761073080602699062580760202927314552520780799141842906388443734996814582733720726639176702011830046481900024130835088465841521489912761065137415394356 572

113903285749187690944137020905170314877734616528798482353382972601361109845148418238081205409961252745808810994869722161285248974255555160763716750548961730168096138 0381

191436114399210638005083214098760459930932485102516829446726066613815174571255975495358023998314698220361338082849935670557552471290274539776214049318201465800802156 6536

067765508783843041343105918046068008345911366408348874080057412725867047922583191274157390809143831384564241509408491339180968402511639919368532255733896695374902662 09

232613188558915808324555719484538756287861288590041060060737465014026278240273469625282171749415823317492396835301361786536737606421667781377399510065895288774276626 3684

1830680190804609849809469763667335662282915132352788806157768278159588669180238940333076441912403412022316368577860357276941541778826435238131905028087018575047046312 933

35375728538660588890458311145077394293520199432197117164223500564404297989208159430716701985746927384865383343614579463417592257389858800169801475742054299580124295810 54

565108310462972829375841611625325625165724980784920998979906200359365099347215829651741357984910471116607915874369865412223483418877229294463351786538567319625598520 2607

294767407261676714557364981210567771689348491766077170527718760119990814411305864557910525684304811440261938402322470939249802933550731845890355397133088446174107959 162

511714864874468611247605428673436709046678468670274091881014249711149657817724279347070216688295610877794405048437528443375108828264771978540006509704033021862556147 3321

177711744133502816088403517814525419643203095760186946490886815452856213469883554445602495566684366029221951248309106053772019802183101032704178386654471812603971906 8846

237085751808003532704718565949947612424811099928867915896904956394762460842406593094862150769031498702067353384834955083636601784877106080980426924713241000946401437 3603

265645184566792456669551001502298330798496079949882497061723674493612262229617908143114146609412341593593095854079139087208322733549572080757165171876599449856937956 2387

555161757543809178052802946420044721539628074636021132942559160025707356281263873310600589106524570802447493754318414940148211999627645310680066311838237616396631809 3144

4671298615527598201451410275600689297502463040173514891945763607893528555053173314164570504996443890936308438744847839616840518452732884032345202470568516465716477139 323

775517294795126132398229602394548579754586517458787713318138752959809412174227300352296508089177705068259248822322154938048371454781647213976820963320508305647920482 0859

204754998573203888763916019952409189389455767687497308569559580106595265030362661597506622508406742889826590751063756356996821151094966974458054728869363102036782325 018

23237084597901115484720876182124778132663304120762165873129708112307581598212486398072124078688781145016558251361789030708608701989758980745664395515741536319319198 1070

5753366337380382721527988493503974800158905194208797113080512339332219034662499171691509485414018710603546037946433790058909577211808044657439628061867178610171567409676

6208029576657705129120990794430463289294730615951043090222143937184956063405618934251305726829146578329334052463502892917547087256484260034962961165413823007731332729830

5001602567240141851520418907011542885799208121984493156999059182011819733500126187728036812481995877070207532406361259313438595542547781961142935163561223496661522614735

3996740515849986035529533292457523888101362023476246690558164389678630976273655047243486430712184943734853006063876445662721866617012381277156213797461498613287441177145

5244470899714452288566294244023018479120547849857452163469644897389206240194351831008828348024924908540307786387516591130287395878709810077271827187452901397283661484214

2871705531796543076504534324600536361472618180969976933486264077435199928686323835088756683595097265574815431940195576850437248001020413749831872259677387154958399718444

9072791419658459300839426370208756353982169620553248032122674989114026785285996734052420310917978999057188219493913207534317079800237365909853755202389116434671855829068

5371189795262623449248339249634244971465684659124891855662958932990903523923333647435203707701010843880032907598342170185542283861617210417603011645918780539367444747205

9985023582891833692922337323999480437108419659473162654825748099482509991833006976569367159689364493348864744213500840700660883597235039532340179582557036016936990988671

1321097988970705172807558551912699306730992507040702455685077867906947661262980822516331363995211709845280926303759224267425755998928927837047444521893632034894155210445

9726188380030067761793138139916205806270165102445886924764924689192461212531027573139084047000714356136231699237169484813255420091453041037135453296620639210547982439212

5172540132314902740585892063217589494345489068463993137570910346332714153162232805522972979538018801628590735729554162788676498274186164218789885741071649069191851162815

2854867941736389066538857642291583425006736124538491606741373401735727799563410433268835695078149313780073623541800706191802673285511919426760912210359874692411728374931

2616339500123599924050845437569850795704622664619000103500490183034153454842833764378111988556318777792537201166718539541835984438305203762819440761594106820716970230228

5152250573126093046898423433152732131361216582808075212631547730604423774753505952287174402666389148817173086436111389069420279088143119448799417154042103412190847094080

2540239329429454938786402305129271190975135360009219711054120966831115163287054230284700731206580326264171161659576132723515666625366727189985341998952368848309993027574

1991646384142707798870887422927705389122717248632202889842512528721782603050099451082478357290569198855467886079462805371227042466543192145281760741482403827835829719930

1017888345674167811398954750448339314689630763396657226727043393216745421824557062524797219978668542798977992339579057581890622525473582205236424850783407110144980478726

6919901864388229323053823185597328697809222535295910173414073348847610055640182423921926950620831838145469839236646136398910121021770959767049083050818547041946643713122

9969235889538493013635657618610606222870559942337163102127845744646398973818856674626087948201864748767272722206267646533809980196688368099415907577685263986514625333631

2450536402610569605513183813174261184420189088853196356986962795036738424313011331753305329802016688817481342988681585577810343231753064784983210629718425184385534427620

1282345707169885305183261796411785796088881503296022907056144762209150947390359466469162353968092013945781758910889319921122600739281491694816152738427362642980982340632

0024402449589445612916704950823581248739179964864113348032475775219708932772262349486015046652681439877051615317026696929704928316285504212898146706195331970269507214437

823047687528028735412616639170824592517001071418085480063692325946201900227808740985977192180515853214739265325155903541020928466592529991435379182531454529059841581 7637

058927906909896911164381187809435371521332261443625314490127454772695739393481546916311624928873574718824071503995009446731954316193855485207665738825139639163576723 1510

055560372633948672082078086537349424401157996675073607111593513319591971209489647175530245313647709420946356969822667377520994516845064362382421185353488798939567318 780

660610788544000550827657030558744854180577889171920788142335113866292966717964346876007704799953788338787034871802184243734211227394025571769081960309201824018842705 7046

092622564178375265263358324240661253311529423457965569502506810018310900411245379015332966156970522379210325706937051090830789479999004999395322153622748476603613677 6979

785673865846709366795885837887956259464648913766521995882869338018360119323685785585581955560421562508836502033220245137621582046181067051953306530606065010548871672 4537

794283133887163139559690583208341689847606560711834713621812324622725884199028614208728495687963932546428534307530110528571382964370999035694888528519040295604734613 1138

263878897551788560424998748316382804046848618938189590542039889872650697620201995548412650005394428203930127481638158530396439925470201672759328574366661644110962566 3373

054092195196751483287348089574777527834422109107311135182804603634719818565557295714474768255285786334934285842311874944000322969069775831590385803935352135886007960 034

209754739229673331064935960181223781285458431760556173386112673478074585067606304822940965304111830667108189303110887172816751957967534718853722930961614320400638132 2465

841111157758358581135018569047815369381377184728147519983505047812977185990847076219746058874232569958288925350419379582060162118423687685114183160683158679946016520 577

405294230536017803133572632670547903384012573059123396018801378254219270947673371919872873852480574212489211834708766296672072723256505651293331260595057777275424712 4164

831283298207236175057467387012820957554430596839555568686118839713552208445285264008125202766555767749596926612604565245684086139238265768583384698499778726706555191 854

468698469478495734622606294219624557085371272776523098955450193037321666491825781546772920052126671434632096378918523232150189761260343736840671941930377468809992968 775

824410478781232662531818459604538535438391144967753128642609252115376732588667226040425234910870269580996475958057946639734190640100363619040420331135793365424263035 6145

700901124480089002080147805660371015412232889146572239314507607167064355682743774396578906797268743847307634645167756210309860409271709095128086309029738504452718289 2749

689212106670081648583395537359191369501531620189088748421079870689911480466927065094076204650277252865072890532854856143316081269300569378541786109696920253886503457 71

831766868859236814884752764984688219497397297077371871884004143231276365048145311228509900207424092558592529261030210673681543470152523487863516439762358604191941296 9769

040526483234700991115424260127343802208933109686367898694977994001260164227609260823493041180643829138347354679725399262338791582998486459271734059225620749105308531 537

182911681637219395188700957788181586850464507699343940987433514431626330317247747486897918209239480833143970840673084079589358108966564775859905563769525232653614424 7802

308268118310377358870892406130313364773710116282146146616794040905186152603600925219472188909181073358719641421444786548995285823439470500798303885388608310357193060 0277

119455802191194289992722353458707566246926177663178855144350218287026685610665003531050216318206017609217984684936863161293729518730789726373537171502563787335797718 08

184878458866504335824377004147710414934927438457587107159731559439426412570270965125108115548247939403597681188117282472158250109496096625393395380922195591918188552 6780

6214992317276316321833989693807561685591175299845013206712939240414459386239880938124045219148483164621014738918251010909677386906640415897361047643650006807710565671848

6281496371118832192445663945814491486165500495676982690308911185687986929470513524816091743243015383684707292898982846022237301452655679898627767968091469798378268764311

598832109043715611299766521539635464420869197567370005738764978437686287681792497469384274652563163230055513041742273416464551278127845777724575203865437542828256714128

8583454443513256205446424101103795546419058116862305964476958705407214198521210673433241075676757581845699069304604752277016700568454396923404171108988899341635058515788

7353430815520811772071880379104046983069578685473937656433631979786803671873079693924236321448450354776315670255390065423117920153464977929066241508328858395290542637687

6689688050333172278001858850697362324038947004718976193473443084374437599250341788079722358591342458131440498477017323616947197657153531977549971627856631190469126091825

9124989036765417697990362375528652637573376352696934435440047306719886890196814742876779086697968852250163694985673021752313252926537589641517147955953878427849986645630

2878831962099830494519874396369070682762657485810439112232618794059941554063270131989895703761105323606298674803779153767511583043208498720920280929752649812569163425000

522908872646925284666104665392171482080130502298052637836426959733707053922789153510568839381132497570713310295044303467159894487868471164383280506925077662745001220035

2620370946602341464899839025258883014867816219677519458316771876275720050543979441245990077115205154619930509838698254284640725554092740313257163264079293418334214709041

25425335232480219322770753555546795871638358750181593387174236061551171013123525633485820365146141870049205704372018261733194715700867578539336078622739558185797587258744

1025420771054753612940474601000940954449596628814869159038990718659805636171376922272907641977551777201042764969496110562205925024202177042696221549587264539892276976603

1052498085575947163107587013320886146326641259114863388122028444069416948826152957762532501987035987067438046982194205638125583343642194923227593722128905642094308235254

408411086454536940496927149400331978286131818618881111840257865928757426384450059944229568586460481033015388911499486935436030221810943466764000022362550573631294626296

0961987605642599639461386923308371962659547392346241345977957485246478379807956931986508159776753505539189911513352522987361127791827485420086895396583594219633315028695

6119201229888988700607999279541118826902307891310760361763477948943203210277335941690865007193280401716384064498787175375678118532132840821657110754952829497493621460821

5583205687232185574065161096274874375098092230211609982633033915469494644491004515280925089745074896760324090768983652940657920198315265410658136823791984090645712468948

4702093577611931399802468134052003947819498662026240089021501661638135383815150377350229660746279529103840686855690701575166241929872444827194293310048548244545807188976

33003232525821581280327467962002814762431828622171054352898348208273451680186131719593324711074662285087106661177034653528395776259977446721857158161264114327179434788

5990892808486694914139097716736900277758502686646540565950394867841110790116104008572744562938425494167594605487117235946429105850909950214958793112196135908315882620682

3321561530868337308381732793281969838750870834838804638847844188400318471269745437093732983624028751979208023218787448828728437273780178270080587824107493575148899789117

3974612932035108143270325140903048746226294234432757126008664250833318768865075642927160552528954492153765175149219636718104943531785838345386525565644065725136357506435

3236508936790431702597878177190314867963840828810209461490079715137717099061954969640070867667102330048672631475510537231757114322317411411680622864206388906210192355223

5467116621374996932693217370431059872250394565749246169782609702533594750209138366737728944386964000281103440260847128990007468077684440887113413525033678773167977093727

7868216611786534423173226463784769787514433209534000165069213054647689098505020301504488083426184520873053097318949291642532293361243151430657826407028389840984160295030

9241897120971601649265613413433422298827909921786042679812457285345801338260995877178113102167340256562744007296834066198480676615805021691833723680399027931606420436812

0799003162644491461902194582296909921227885539487835383056468648816555622943156731282743908264506116289428035016613366978240517701552196265227254558507386405852998303791

8035043287670380925216790757120406123759632768567484507915114731344000183257034492090971243580944790046249431345502890068064870429353403743603262582053579011839564908935

4345101342969617545249573960621490288728932792520696535386396443225388327522499605986974759882329916263545973324445163755334377492928990581175786355555626937426910947117

0021654117182197505198317871371060510637955585889055688528879890847509157646390746936198815078146852621332524738376511929901561091897779220087057933964638274906806987691

6819749236562422608715417610043060890437797667851966189140414492527048088197149880154205778700652159400928977601330756847966992955433656139847738060394368895887646605498

3871478968482805384701730871117761159663505039979343869339119789887109156541709133082607647406305711411098839388095481437828474528838368079418884342666222070438722887413

9478010177213922819119923654055163958934742639538248296090369002883593277458550608013179884071624465639979482757836501955142215513392819782269842786383916797150912624105

4872570092407004548848569295044811073808799654748156891393538094347455697212891982717702076661360248958146811913361412125878389557735719498631721084439890142394849665925

1731388171602663261931065366535041473070804414939169363262373767770958503132559900957627319573086480424677012123270205337426670531424482081681303063973787366424836772539

8374876909806021827857862165127385635132901489035098832706172589325753639939790557291751600976154590447716922658063151110280384360173747421524760851520990161585823125715

9073342173657626714239047827958728150509563309280266845893764964977023297364131906098274063353108979246424213458374090116939196425045912881340349881063540088759682005440

8364386516617880557608956896727531538081942077332597917278437625661184319891025007491829086475149794003160703845549465385946027452447466812314687943441610993338908992638

4118474252570445725174593257389895651857165759614812660203107976282541655905060424791140169579003383565748692528007430256234194982864679144763227740055294609039401775363

3565547193100017543004750471914489984104001586794617924161001645471655133707407395026044276953855383439755054887109978520540117516974758134492607943368954378322117245068

7344231989878844128542064742809735625807066983106979935260693392135685881391214807354728463227784908087002467776303605551232386656295178853719670346347012229395816067092

5091532174890308408865160611190114984434123501246469280288059961342835118847154497712784733617662850621697787177438243625657117794500644777183702219991066950216567576440

4499794076503799995484500271066598781360380231412683690578319046079276529727769404361302305178708054651154246939526512710105292707030667302444712597393995051462840476743

1363739978259184541176413327906460636584152927019030276017339474866960348694976541752429306040727005059039503148522921392575594850788679779252539317651564161971684443524

3697944473559642606333910551268260615957262170366985064732812667245219890605498802807828814297963366967441248059821921463395657457221022986775997467381260693670691340815

5941201611596019023775352555630060624798326124988128819293734347686268921923977783391073310658825681377717232831532908252092733047850724977139448333892552081117560845296

6590553940965568541706001179857293813998258319293679100391844099286575605993598910002969864460974714718470101531283762631146774209145574041815908800064943237855839308530

8283054760767995243573916312218860575496738322431956506554608528812019023636447127037486344217272578795034284863129449163184753475314350413920961087960577309872013524840

7505763719925365047090858251393686346386336804289176710760211115982887553994012007601394703366179371539630613986365549221374159790511908358829009765664730073387931467891

31814651093167615758213514248604422924453041131606527009743300884990346754055186406773426035834096086055337473627609356588531097609942383473822208729246449768456057 9562

51676557408841032173134562773585605235823638953203853402484227337163912397321599544082842166663602329654569470357718487344203422770665383738750616921276801576618109 54200

977083636043611105924091178895403380214262523948929686439808926114635414571535194342850721353453018315875628275733898268898523557799295727645229391567477566676051 08788764

84534936360682780505646228135988858792599409464460417052044700463151379754317371877560398159626475014109066588661621800382669899619655805872086397211769952194667 89857011

79833244060181157565807428418291061519391763005919431443460515404771057005433900018245311773371895585760360718286050635647997900413976180895536366960316219311325 02238517

916720551806592635180362512145759262383693482266589557699466049193811248660909979812857182349400661555219611220720309227764620099931524427358948871057662389469 388944649

50939603304543408421024624010487233287500817491798755438793873814398942380117627008371960530943839400637561164585609431295175977139353960743227924892212670458 08183313764

16581826956210587289244774003594700926866265965142205063007859200248829186083974373235384908396432614700053242354064704208949921025040472678105908364400746638 00208701266

6420945718170294675227854007450855237772089058168391844659282941701828823301497155423523591177481862859296760504820386434310877956289292540563894662194826871 104282816389

3975711757786915430165058602965217459581988878680408110328432739867198621306205559855266036405046282152306154594474489908838908199973874745296981077620148713 4000122535522

2466954093152131153379157980269795557105085074738747507580687653764457825244326380461430428892359348529610582693821034980000405248407084403561167817170512813 3788057056434

50616119330424440798260377951198548694559152051960093041271007277849301555038895360338261929343797081874320949914159593393636811062755729527800425486306005 452383915106899

8913578820019411786535682149118528207852130125518518493711503422159542244511900207393539627400208110465530207932867254740543652717595893500716336076321614 72581540 7642053

020045340183572338292661915308354095120226329165054426123619197051613839357326693760156914429944943744856809775696303129588719161129294681884936338647392 47601226 9641588

4890096571708616059814720446742866420876533479985822090619802173211614230419477754990738738567941189824660913091691772274207233367635032678340586301930 193242996 39720444

517928812285447821195353089891012534297552472763573022628138209180743974867145359077863353016082155991131414420509144729353502223081719366350934686585 8656314855575862447

818620108711889760652969899269328178705576435143382060141077329261063431525337182243385263520217735440715281898137698755157574546939727150488469793619 5004777209705617939

1382898984532742622728864710888327017372325881824465843624958059256033810521560620615571329915608489206434030339526226345145428367869828807425142256745 180618414956468611

16354049718976821542277224794740335715274368194098920501136534001238467142965518673441537416150425632567134302476551252192180357801692403266995417460 8759240 9207004669340

39651017813485783569444076047023254075557764728450751826890418293966113310160131119077398632462778219023650660374041606724962490137433217246454097412 99557052 91424382080

12

7609836482346597388669134991978401310801558134397919485283043673901248208244481412809544377389832005986490915950532285791457688496257866588599917986752055455809900455646

1178755249370124553217170194282884617402736649978475508294228020232901221630102309772151569446427909802190826689868834263071609207914085197695235553488657743425277531197

2474308730436195113961190800302558783876442060850447306312992778889427291897271698905759252446796601897074829609491906487646937027507738664323919190422542902353189233772

9316673608699622803255718530891928440380507103006477684786324319100022392978525537237556621364474009670539439838235764606992465260089090624105904215453927904411152958034

5334500256244101006359530039598864466169595626351878060688513723462707997327233134693971456285542615467650632465676620279245208581347717608521691340946520307673391841147

5041401689241213198268815686645614853802875393311602322925556189410429953356400957864953409351152664540244187759493169305604486864208627572011723195264050230997745676478

3848897346431721598062678767183800524769688408498918508614900343240347674268624595239589035858213500645099817824463608731775437885967767291952611121385919472545140030118

0503437875277664402762618941017576872680428176623860680477885242887430259145247073950546525135339459598789619778911041890292943818567205070964606263541732944649576612651

9534957018600154126239622864138977967333290705673769621564981845068422636903678495559700260798679962610190393312637685569687670292953711625280055431007864087289392257145

1248113577862766490242516199027747109033593330930494838059785662884478744146984149906712376478958226329490467981208998485716357108783119184863025450162092980582920833481

3638405421720056121989353669371336733392464416125223196943471206417375491216357008573694397305979709719726666642267431117762176430686813103518991122713397240368870000996

8629225464650063852886203938005047782769128356033725482557939129852515068299691077542576474883253414121328006267170940090982235296579579978030182824284902214707481111240

1860761341515038756983091865278065889668236252393784527263453042041880250844236319038331838455052236799235775292910692504326144695010986108889991465855188187358252816430

2520939285258077969737620845637482114433988162710031703151334402309526351929588680690821355853680161000213740851154484912685841268695899174149133820578492800698255195740

2018181056412972508360703568510553317878408290000415251186577945396331753853209214972052660783126028196116485809868458752512999740409279768317663991465538610893758 79522

1497173172813151793290443112181587102351874075722210012376872194474720934931232410706508061856237252673254073332487575448296757345001932190219911996079798937338367324257

610393898534927877747398050808001554476406105352202325409443567718794565430406735896491017610775948364540823486130254718476485189575836674399791508512858020607820554462

9917232020282229148869593997299742974711553718589242384938558585954074381048826246487880533042714630119415898963287926783273224561038521970111304665871005000832851773117

7648973523092666123458887310288351562644602367199664455472760831011878838915114934093934475007302585581475619088139875235781233134227986650352272536717123075686104500454

8970360079569827626392344107146584895780241408158405229536937499710665594894459246286619963556350652623405339439142111271810691052290024657423604130093691889255865784668

4612156795542566054160050712766417660568742742003295771606434486060201239821698271723197826816628249938714995449137302051843669076723577400053932662622760323659751718 9259

0180110429038427418550789488743883270306328327996300720069801244365116394086922207453202446241211558043545420642151215850568961573564143130688834431852808539759277 3443

3655384188340303517822946253702015782157373265523185763554098954033236382319219892171177449469403678296185920803403867575834111518824177439145077366384071880489358256868

13

5420116450313576333355509440319236720348651010561049872726472131986543435450409131859513145181276437310438972507004981987052176272494065214619959232142314439776546 7083517

1474936798618655279171582408065106379950018429593879915835017158075988378496225739851212981032637937621832245659423668537679911314010804313973233544909082491049914332584

3298821033984698141715756010829706583065211347076803680695322971990599904451209087275776225351040902392888779424630483280319132710495478599180196967835321464441189260631

5266181674431935508170818754770508026540252941092182648582138575266881555841131985600221351588872103656960875150631875330029421186822218937755460272272912905042922597877

1066787384000061677215463844129237119352182849982435092089180168557279815642185819119749098573057033266764646072875743056537260276898237325974508447964954564803077159815

3955827779139373601717422996027353102768719449444491793978514463159731443535185049141394155732938204854212350817391254974981930871439661513294204591938010623142177419918 4

0601803479498876910515579055548069538785400664533759818628464199052204528033062636956264909108276271159038569950512465299960628554438383303276385998007929228466595035512

11245284087516229060262011857775313747949362055496401073001348853150735487353905602908933526400713274732621960311773433943673385759124508149335736911664541281788171454 02

3054750667136518258284898099512139193995633241336556777098003081910272040997148687418134667006094051021462690280449159646545330107754695413088714165312544813061192407821

1886900560277818242350226961893443525476335735364856193632544177566139817039306328721669057222597452091291726219984440964615826945638023950283712168644656178523556516 41

2771282691868861557271620147493405227694659571219831494338162211400693630743044417328478610177774383797703723179525543410722344551255558999864618387676490397246116795901

8100035098928641204195163551108763204267612979826529425882951141275841262732790798807559751851576841264742209479721843309352972665210015662514552994745127631550917636730

259462132930190402837954246323258550301096706922720227074863419005438302650681214142135057154175057508639907673946335146209082888934938376439399256900604067311422093312 1

9593620298297235116325938677224147791162957278075239505625158160313335938231150051862689053065836812998810866326327198061127154885879809348791291370749823057592909186293

9195014721197586067270092547718025750337307993971345395326461952699965963856549175904583335857991020127132045839032008538788816336376851208372788513117522776960978796 2

1423721625452145912818317982160441113116714069148271709810154577819392023115638719508050246797257924976057726259133285597263712112019057207714091486450740949267180358151

5757151405039761096384675556929897038354731410022380258346876735012977541327953206097115450648421218593649099791776687477448188287063231551586503289816422828823274686610

6592732197907162384642153489852476216789050260998045266483929542357287343977680495774091449538391575565485459058976495198513801007958010783759945775299196700547602252552

034453988712538780171960718164078124847847257912407824544361682345239570689514272269750431873633263011103053423335821609333191218806608268341428910415173247216053355849 9

9322454873077882290525232423486153152097693846104258284971496347534183756200301491570327968530186863157248840152663983568956363465743532178349319982554211730846774529708

5839507616458229630324424328237737450517028560698067889521768198156710781633405266759539424926280756968326107495323390536223090807081455919837355377748742029039018142937

311529334644468151212945097596534306284215319445727118614900017650558177095302468875263250119705209476159416768727784472000192789137251841622857783792284439084301181121 4

9636642465903363419454065718354477191244662125939265662030688852005559912123536371822692253178145879259375044144893398160865790087616502463519704582889548179375668104647

46141051424988702521399368705093723054477341126413548928068410591077166778212383328102621855877513127211793444482014404257450830639447383637939062830089733062413806 14589

41422769474793166571762318247216835067807648757342049155762821758397297513447899069658953254894033561561316740327647246921250575911625152965456854463349811431767 02572956

61844775487469378464233737238981920662048511894378868224807279352022501796545343757274163910791972952950812942922205347717304184477915673991738418311710362524395 71615271

46690058147000026330104526435478659032907332054683388720787354447626479252976901709120078741837367350877133769776834963442524199499513883150748775374338494582597 65560996

55595431804092017849718468549737069621208852437701385375768141663272241263442398215294164537800049250726276515078908507126599703670872669276430837722968598516912 23050374

62744310852934305273078865283977335246017463527703205938179125396915621063637625882937571373840754406468964783100704580613446731271591194608435935825987782835266 53115106

50416232953290477217408355934972375855213804830509000964667608830154061282430874064559443185341375522016630581211103345312074508682433943215904359443031243122747 1385842

03039010607094031523555617276799416002039397509989762933532585557562480899669182986422267750236019325797472674257821111973470940235745722227121252685238429587427 35015636

60093188045493338989741571490544182559738080871565281430102670460284316819230392535297795765862414392701549740879273131051636119137577008929564823323648298263024 60797587

57677453771601024908046243018565241617566556001608591215345562676021926899828553778725831451440826545834844094784631787773747946535801699607794055687011923286080 41130904

62935087182712593466871276669487389982459852778649956916546402945893506496433580982476965165142090986755203808309203230487342703468288751604071546653834619611223 0137594

51579252696743642531927390036038608236450762698827497618723575476762889950752114804852527950845033958570838130476937881321123674281319487950228066320170022460331 98967197

06491637411758548518784840120548446725888514015627250198217190669608126277854859648183696214107217142149863619187747545096503089570994709343378569816744658282679 11940611

95603784539785583924076127634410576675102430755981455278616781594965706255975507430652108530159790807334373607943286675789053348366955548680391343372015649883422 08933999

71641479746938696905480089193067138057171505857307148815649920714086758259602876056459782423770242469805328056632787041926768467116266879463486950464507420219373 94525926

26686135529406247813612062026364981999994984051438682852589563422643287076632993048917234007254717641886853513723326678779217383475414800228033929973579361524127 55829569

27683723123479898944627433045456679006203242051639628258844308543830720149567210646053323853720314324211260742448584509458049408182092763914000854042202355626021 85643489

94145439950410980591817948882628052066441086319001688568155169229486203010738897181007709290590480749092427141018933542818429995988169660993836961644381528877214 08526808

87574882932587358099056707558170179491619061140019085537448827262009366856044755965557476485674008177381703307380305476973609786543859382187220583902344443508867 499866506

04064587434600533182743629617786251808189314436325120510709469081358644051922951293245007883339878842933934243512634336520438581291283434529730865290978330067126 17981303

16794385535726296998740359570458452230856390098913179475948752126397078375944861139451960286751210561638976008880092746115860800207803341591451797070303683519697 7766607637

37853330120241201120469886092093390853657732223924124490515327809509558664594776344822699860748132973026309750288121035177231244650953496536930900186377640940943 49837313

25132186208021480992268550294845466181471555744470966953017769043427203189277060471778452793916047228153437980353967986142437095668322149146543801459382927739339 60327540

480095522318166673803571839327570771420467238386246178039762923771312095807893638414479298025880655221292620936239306373134966401866195108115834711733120258058667276399

276357907806381881306915636627412543125958993611964762610140556350339952314032311381965623632719896183725484533370206256346422395276694356837676136871196292181875457608

61705303159072882870071231366630872275491866139577373054606599743781098764980241401124214277366808275139095931340415582666789510846776118665957660165998178089414985754

76284387856100263796543178313634025135814161151902096499133548733131115022700681930135929595971640197196053625033558479980963488718039111612813595968565478868325856437

617315976200241962155289629279048198221994622694871374624447290934564700285376949588595916067892824910544125159963007813683674902093749157328962700286568293444313423473

2392982591667395034259958689706972673325827359031212887466604514614878503461428277659916080903986525757172630818334944418201935333850712923457743755793440621787113300631

06003324053991693682603746176638565758877580201229366353270267100681261825172914608202541892885935244491070138206211553827793565296914576502048643282865557934707209634

737269214118689546732276775133569019015372366903686538916129168888787640752549349424973342718117889275993159671935475898809792452526236365903632007085444078454479734829

8020820449266706344204375553250505275228337788870408040335319234076856301093477212563908864041310107381785333831603813528082811904083256440184205374679299262203769871

180611226244909092426419858208617511771137890516091403815750033664241560952163281971223350231674226005679412814062172196418427057843289598028823350598282081966662490358

77899403331522748177769528436816300885317696947836905806710648280835980466988410981351586549069333195223943632879239905348109878302745001720654336990661177845543646877

631844464768069142828004551074686645392805399409108754939166095731619715033166968309929466349142798780842257220697148875580637480308862995118473187124777291910070227588

934869394562895158029653721504096031077612898312635899648934102470360366450568728758905140684123812424738638542790828273382797332688550493587430316027474906312957234974

26112215174171531336186224109138695006888358989623492763173164783400774608866555987333821138299287769114954921841920877716060684728746736818861675072210172611038306717

856694812948785048943063086169948798703160515884108282351274153538513365895332948629494495061868514779105804696039069372662670386512905201137810858616188886947957607413

85534585151768051973334433495230120395770739623771316030242887200537320998253008977618973129817881944671731160647231476248457551928732782825127182446807824215216469567

9294098238926284943760248852279003620219386696482215628093605373178040863727268426696421929946819214908701707533361094791381804063287387593848269535583077395761447997270

003472880182785281389503217986345216111066608839314053226944905455527867894417579202440021450780192099804461382547805858048442416404775031536054906591430078158372430123

375115622840158386442708907182848167575271238467824595343344496220100960710513706084618011875431207254913349942476171156333214089346091565615506003173842187015702261031

1916603887064661438897736318780940711527528174689576401581047016965247557740891644568677171585005832699434016772021567677240681283665652641229824394651331973591997094

275938502669557470231813203243716420586141033606524536939160050644953060161267822648942437397166717661231048975031885732165554988342121802846912529086101485527815277625

237504563757694977343368460156077270355096290493924870884062810679436224187047470083688426710225583024035998416459511224852726336326451140173952480861946358407837535568

56223171155209472230654370926067973510005655493812245754837285457117973936157561676416928598052572975223385586113883221711073622658162188424431788574887981090266537934

16

6642169909140565364322493013348679881548866286650523469972355747384248305904236771432787923164224038777643301926001922847783138376325361210253369358126240868666997382759

7736568222790721583247888864236934639616436330873013981421143030600873066616480367898409133592629340230432497492688783164360268101130957071614191283068657732353263965367

7390317661361315965553584999398600565155921936759977179330197446881483711032065036931928945214026509154651843099365534933371834252984336799159394174662239003895276738 13

3306177476295749438687169784537672194935065908757119177208754771071899379608947745126547575018711948707387367858902006173733210756933022163206284320656711920969505857611

7396163232621770894542621460985841023781321581772760222273813349541048100307327510779994899197796388353073444345753297591426376840544226478421606312276964696715647399904

3715903323906560726644116438605404838847161912109008701019130726071044114143241976796828547885524779476481802959736049439700479596040292746299203572099761950140348315380

9477146010563334469988208221205872815107291829712119178764248803546723169165418522567292344291871281632325969654135485895771332083399112887759172261152733790103413620856

1457799239877832508355073019981845902595835598926055329967377049172245493532968330000223018151722657578752405883224908582128008974790932610076257877042865600699617621217

6845478996440705066241710213327486796237430229155358200780141165348065474882306150033920689837947662550365498228053296628621179306284301704924023019857199789488368971 83

0438051821744191476604297524372516834354112170386313794114220952958857980601529387527537990309388716835720957607152219002793792927863036372687658226812419933848081660216

0372215471014300737753779269990695871212892880190520316012858618254944133538207848834653116326504076424283908701210151942319616522684220037112304643006734420647477180 2135

3070124098860353399152667923871101706221865883537381210935179775604425634694999787251125440854522274810914874307259869602040275941178942581281882159952359658979181144077

6533543217575952555361581280011638467203193465072968079907939637149617743121194020212975731251652537680173591015573381537720019524445436200718484756634154074423286210609

9761324348754884743453966598133871746609302053570271952983943271425371155766600025784423031073429551533945060486222764966687624079324353192992639253731076892135352572 32

1080889819339168668278948281170472624501948409700975760920983724090074717973340788141825195842598096241747610138252643955135259311885045636264188300338539652435997416931

3228947198783084276004013680747039040972384739458348961865397905941185993103561684368692194853820557803957738813606795499000851232594425297244866667668346414021899159445

6530942344065066785194841776677947047204195882204329538032631053749488312218039127967844610013972675389219511911783658766252808369005324900459741094706877291232821430463

5337283519953648274325833119144459017809607782883583730111857543659959882745319253105881150263075425714939430244539318701799236081666113054262539958338979429716020703 38

7678150330102801200959972522222808014235710947603519255444349299867678178910455590630159538097618759203589373419789623589311259839025983102671933041892151096891562250696

5911982832345550305908173073519550372166587028805399213857603703537710517802128012956684198414036287272562321442875430221090947272107347413497551419073704331827662617727

5996888826027225247133683353452816692779591328861381766349857728936900965749562287103024362590772412219094300871755692625758065709912016659622436080242870024547362036394

8412559548817272724736534677836472019183039987176270375157246499222894679323226936191776416146187956139566995677830682903165896994307673335082349907906241002025061340573

4430069574547468217569044165154063658468046369262127421107539904218871612617787014258864825775223889184599523376292377915585744549477361295525952226578636462118377598 47

37003479714082069941455807190802135907322692331008317595106590191212947954086036407573587502058902087045796700070552625058114206639074592152733094068236494415908910092.20

29668052332526619891131184201629163107689408472356436680818216865721968826835840278550078280404345371018365109695178233574303050485265373807353107418591770561039739.50626

40355442275156101107261779370634723804990666922161971119425912044508464174638358993823994651739550900085947999013602667426149429006646711506717542217703877450767356374215

47829059110126191575558702389570014051178226469899449179083017954758767601680941001358376135785913569244556477644641786671153919513576961048649224900834467154863830544.77

91433009768048687834818467273375843689272431044740680768527862558516509208826381323362314873333671476452045087662761495038994950480956046098960432912335834885999029452.64

00284994280878624039811814884767301216754161106629995553668193123287425702063738352020086863691311733469731741219153633246745325630871347302792174956227014687325867891.73

45583799643513588009593508775563562488104938529990076751355135277924124292774885658885665132473025147102105753525165118148509027504768455182520963318990685276144351382.13

66215236889057878669943228816028377482035506016029894009119713850179871683637441392759736440170070147637066557035043381211135764150184518214136198234951596010647527125

75935185304332875537783057509567425442684712219618709178560783936144511383335649103256405733898667178123972237519316430617013859539474367843392670986712452211189690840.23

63274114966012434830989299417380305884171666130730400675883804321115553794406054977217059428215148861656727712409033877277456290971101348851843741186956554497457368452.18

06698291104505800429988795389902780438359628240942186055628778842880212755388480372864001944161425749990427200959520465417059810498996750451193647117277222043610261407.97

50809686975176600237187748348016120310234680567112644766123747627852190241202569943534716226660893675219833111813511146503854895025120655772636145473604426859498074396.93

23312971273771573470997139522911826534851555871373366291202427143025037632695013509116129529937858646813072264860082708813335381937036825988678933212383270532976258573.82

79009782646054559855513183668888446282651337984916678394097613537662517982582496634587719501243840403591408492097337546424744881761840700235695801774101776969250778148933

86672557898564589851056891960924398841569280696983352240225634570497312245269354193837004843183357196516626721575524193401933099018319309196582920969656247667683659647.01

95957547393455143374137087615173236772042273856742791706982045499530959188724349395240944416789988463198455048523936629720797774528143994182567894577957125524268260899.40

86331737153889626288962940211210888442737656862452761213037101730078513571540453304150795944777614359743780374243664697324713841049212431413890357909241603640631403814.98

31481905251720937103964026808994832572297954640427017577229041732347960736187878899133183058430693948259613187138164234672187308451338772190869751049428437693250249816.5

66738162606159417682525099937416728839517440669325496534030101452225316189009235376486378482881344209870048096227171226407489571939002918573307460104360729190945767994.614

92929042798168772942648772995285843464777538690695014898413392454039414468026362540211861431703125111757764282991464453340892097696169990837265236176874560589470496817.01

36974909523072082682887890730190018253425805343421705928713931737993142410852647390948284596418093614138475831136130576108462366837237695913492615824516221552134879244.14

50417568480641206365201703863301295327776990231186480200675569056822950163549319923059142463962170253297475731140942201801993680350264956369558664259067626856873721103.39

15679383989576565193177883000241613539562437777840801748819373095020699900890899328088397430367736595524891300156633294077907139615464534088791510300651321934486673248.2

7590794680787981942501958262232039513125201410996053126069655540424867054998678692302174698900954785072567297879476988883109348746442640071818316033165551153427615562240

5474473378049246214952133258527698847336269182649174338987824789278468918828054669982303689939783413747587025805716349413568433929396068192061773331791738208562436433635

3598634944968907810640196740744365836670715869245211829978938040771375012908586465789057714268335827689785547176871844277261205092664861020515356428406323684818072879407

1712796682006072755955590404023317874944734645476062818954151213916291844429765106694796935401686601005519607768733539651161493093757096855455938151378956903925101495326

5628147011998326992200066392875374713135236421589265126204072887716578358405219646054105435443642166562244565042999010256586927279142752931172082793937751326106052881235

3734510683729398935808712438693859343891757133763007203197608166044646839377258069092372975234867029169104263692620901996052041210240776481903160140858635584276095370865

5816427399534934654631450404019952853725200495780525465625115410925243799132626271360909940290226206283675213230506518393405745011209934146491843332364656937172591448932

4159006242020612885732926133596808726500045628284557574596592120530341310111827501306961509835515632004310784601906565493806542525229161991819959602752327702249855738824

8998827074659363557685825605180689642853768507720122203479209993617926820659014216561592530673794456894907085326356819683186177226824991147261573203580764629811162440133

1673789278868922903259334986179702199498192573961767307583441709855922170171825712777534491508205278430904619460835217402005838672849709411023266953921445461066 21500641

0674740207009189911951376466904481267253691537162290791385403937560077835153374167747942100384002308951850994548779093461222086506016050035177626483161115332558 7705073

5412792499098593734737870811942530551214369797499149518605359204038302357163527276308746932196221900642608861836761033460022554774778136410126919065696864950126883762969

0723396127628722304114181361006026404403000359969889199458273976241146137448040596970625767647237660655416185746905272292382282751867991569833907476711461030227766060 2006

1246876477288190967916133540198814027579921741676787992316039635694928515136336472195406111717673873725557285229400543617851765023075446938693078734991103521825329297 26

0445532107978877114498988709115112372506042387537348412570860640690520584521227545338480082053024504565176695185769132000428167580549248117805198326460324457928297301291

0531838563682120621553128866856495651261389226136706409395333457052698695969235035309422454386527867767302754040270224638448355323991475136344104405009233036127149608135

5490531539021002299595756583705381261965683144286057956696622154721695620870013727768536960840704833325132793112232507148630206951245395003735723346807094656483089209801

5348787056334910923660575540508641115214414814346304372732710450277686619531078583233348578402971609252153260925589326556006721243594642550659967717703884453961816328796

1446081778927217183690888012677820743010642252463480745430047649288555340906218515365435547412547615276977266776977277705831580141218568801170502836527554321480348800444

2979998062157904564161957212784508928489806426497427090579129069217807298769477975112447305991406050629946894280931034216416629935614828130998870745292716048433630818404

1264696379258430941854422163590845761460785585624738149314270782662151855416038702068769804617474000808324343665382345551094949843109349475994467267366535251766270 67721

9418319197719637801570216993367508376005716345464367177672338758864340564487156696432104128259564534984138841289042068204700761559691684303899934836679354254921032811336

3184722592305554383058206941675629992013373175489122037230349072681068534454035993561823576312837767640631013125335212141994611869350833176587852047112364331226765129964

19

1713252175135532618676819423387903654689080018271352835848884441117612341011799187092365071848578562210211040097769944531217950224795780695065329659403839873699072407976

790408267940076187295478359634927939045769736616434053597922192858705749574816969406233427261973351813662606373598257555249650980726012366828360592834185584802695841377

2558970883789942910549800331113884603401939166122186696058491571485733568286149500019097591125218800396419762163559375743718011480559442298730418196808085647265713547612

8316292004498803154021055305970766663627493283089168809323592900817874119857383171926167288349184024297212904349655269427264025596414635259143484006758676903503823205729

3413298159353330444464968294413673234421583807619483121933311981906109614295220153617029857510559432646146850545268497576480780800922133581137819774927176854507553832876

8874474591593731162470601091244609829424841287520224462594477638749491997840446829257360968534549843266536862844489365704111817793806441616531223600214918768769467398407

5171763075168498563592014868929431059402024579696229245666448819675762943495353263821716133957577907663707645695702597388004384158058943361371065518599876007549241872117

1488929522173772114608115434498266547987258005667472405112200738345927157572771521858994694811794064446639943237004429114074721818022482583773601734668530074498556471542

003612359339731291445859152288740871950870863221883728826282288463184371726190330577714765156414382230679184738603914768310814135827575853643597721650028277803713422869

688787349795096031108899196143386664068450697420787700280509367203387232629637856038653216432348815575570184690890746478791224363755566686780676105449550172607911429308

3128576125448194444947324481909379536900820638463167822506480953181040657025432760438570350592281891987806586541218429921727372095510324225107971807783304260908679427342

8955735559252723805511440438001239041687716445180226491681641927401106451622431101700056691121733189423400547959684669804298017362570406733282129962153684881404102194463

4246462207455756439604529853130714090846084996537678037932018991408658146621753193376659701143306086250098295669176388460567629729314649114937046244693519840395344491351

4119366793330193661766365255514917498230798707228086085962611266050428929696653565251668888557211227680277274370891738963977225756489053340103885593112567999151658902501

6486961427207005916056166159702451989051832969278935550303934681219761582183980483960562523091462638447386296039848924386187298507775928792722068554807210497817653286210

1874767668972488411395603494803767270363169210073508340738652616845074824964485974281349364803724261167042668708319250409976153190768557703274217850100064419841242073964

0013960360158381056592841368457411910273642027416372348821452410134771652960312840865841978795111651152982781462037913985500639996032659124852530849369031313010079997719

1362230866011099929142871249388541612038020411340188887219693477904497527454288072803509305828754420755134816660927879353566521255620139988249628478726214432362853676502

5914504683776352825876521391564809721419296755493843755826002531685363567313792624758780494459441834291727569883762262618463654527434976624111384513054814498363117897844

8973207671950878415861887969295581973325069995140260151167552975057543781024223895792578656212843273120220071673057406928686936393018676595825132649914595026091706934751

9408975357464016830811798846452473618956056479426358070562563281189269663026479535951097127659136233180866921535788607812759910537171402204506186075374866306350591483916

4676567232057145168861707909846959323267249467375830996070425892204815507991327520885837811176852142693347869218952406226579210436203488529262679840139532164587911515790

5046057971083898337186403802441751134722647254701079479399695355466961972676325522991465493349966323418595145036098034409221220671256769872342794070885707047429317332918

20

852389672197135392449242617864118863779096281448691786946817759171715066911148002075943201206196963779510322708902956608556222545260261046073613136886900928172106819861 8

5537809820184711541636303262656992834241550236009780464171085255376127289053350455061356841437758544296779770146602943876872251153638011917581540281208182556064854107879

3359892106442724489861896162941341800129513068363860929410008313667337215300835269623573717533073865333820484219030818644918409372394403340524490955455801640646076158101

0301767488475017661908692946098769201691202181688291040870709560951470416921147027413390052533408348128703530310239196999785974139085936054335996970756044601342424 53682

4960987725831311024732798562072126572499003468293886872304895562253204463602639854225258416464324271611419817802482595563544907219226583863662663750835944314877635156 1457

1074552801615967704844271419443518327569840755267792641126176525061596523545718795667317091331935876162825592078308018520689015150471334038610031005591481785211038475454

2933389188444120517943969970194112695119526564919594189975418393234647424290702718875223534393673633663200307232747037407123982562024662651974090199762452056198557625760

0087081730832883443818310700545144935458854226785785519153722923795554943334101744201696000906964156127322977702212179518683763590822551288164700219923488640439591530184

6400471432118636062252701154112228380277853891109849020134274101412155976996543887719748537643115822983853312307175113296190455900793806427669581901484262799122179294798

7348901868471676503827328552059082984529806259250352128451925927986593506132961946796252373972565584157853744567558998032405492186962888490332560851455344391660226257775

5129162007727968526293879375304541810807292858919897153817973434961872329276147478501926114504132748732429705834084711123337462746172746265824153242710593225062553023147

3875925172478732288149145591560503633457542423377916037495250249302235148196138116256391141561032684495807250827343176594405409826976526934457986347970974312449827193311

3863873159636361218623497261409556079920628316999420072054811525353393946076850019909886553861433495781650089961649079678142901148387645682174914075623767618453775144031

4754112067601607264605568592577993220703373333989163695043466906948284366299800374145276277165476238255461708831898108688068478537055364804693509588180253605297407935386

7651119507937328208314626896007107517552061443378411454995013643244632819334638905093654571450690086448344018042836339051357815727397333453728426337217406577577107983051

7555721036795976901889958494130195999573017901240193908681356585539661941371794487632079688800371607303220547423572266896801882123424391885984168972277652194032493227314

7936692340048489760590379580946960417542796137825537812239476461478329269765451622902817011004378460387565441517394339600489153188175766505009516974024156447712936566142

5394936888423051740012992055685428985389794266995677027089146513736892206104415481662156804219838476730871787590279209175900695273456682026513373111518000181434 12096260

1658629821076663523361774007837783423709152644063054071807843358061072961105550020415131696373046849213356837265400307509829089364612047891114753037049893952833457824082

8173864413227100029683119402033234564208264732762338302946393789983758365545599193408662350909679611340048670271231765266637107787251118603540375544874186935197336566217

7235922939677646325156202348757011379571209623772343137021203100496515211197601317641940820343734851285260291333491512508311980285017785571072537314913921570910513096505

9885999931560863655477403551898166733535880048214665099741433761182777723351910741217572841592580872591315074606025634903772633739144613770380213183474473011 13032670296

9173350477016321066162278300272692833655840117914194478087482533607144032962522857750098085996090940936312635621328162071453406104224112083010008587264252112262480142 6475

21

1942618432585338675387405474349107271004975428115946601713612259044015899160022982780179603519408004651353475269877760952783998436808690898919783969353217998013913544255

2717910225397010810632143048511378291498511381969143043497500189980681644412123273328307192824362406733196554692677851193152775113446468905504248113361434984604849051258

3456832664415284897139723760403282126602535166939140820499473204860216277597917712347510975024030789357599377150950217516935558270725339118923340702238320775858021371747

7837877839101523413209848942345961369234049799827930414446316270721479611745697571968123929191374098292580556195520743424329598289898052923336641541925636738068949420147

1241340525072204061794355252555225008748790086568314542835167750542294803274783044056438581591952666758282929705226127628711040134801787224801789684052407924360582742467

4430767216452703134513541676496689012747868010102951338626986497482121186290403376915685762406992963724930972016287072001898354236903641492702369619385473724803298550451

1208919287982987446786412915941753167560253343531062674525450711418148323988060729714023472552071349079839898235526872395090936566787899238371257897624875599044322889538

8377317348941122757071410959790047919301046740750411435381782464630795989555638991884773781341347070246747362112048986226991888517456251732519341352038115863350123913054

4419100736284475675141610504109735058527620444891909789019843154852805339857778443139338839943104444656692445508594631408175122033139068159659251054685801313383815215764

1821043342978882611963044311138879625874609022613090084997543039577124323061690626291940392143974027089477663702488155499322458825979020631257436910946393252806241 64247

6868495455324938017639371615636847859823715902385421265840615367228607131702674740131145261063765383390315921943469817605358380310612887852051546933639241088467632009567

0897183674905781630851581381619668822204757043759061433804072585386208356517699842677452319582418268369827016023741493836349662935157685406139734274647089968561817016 05

5110488097155485911861718966802597354170542398513556001872033507906094642127114399319604652742405088222535977348151913543857125325854049394601086579379805862014336607882

5219717809025817370870916460452727977153509910340736425020386386718220522879694458387652947951048660717390229327455426785669776865939923416834122274663015062155320502655

3414609952493560508549217565491348309589065361756938176374736441833789742297007035452066631709296075919896277324230902523974438610142630986877339138825186843165010279649

1149773758288891345034114886594867021549210108432808078342808941729800898329753694064496990312539986391958160146899522088066228540841486427478628197554662927881462160717

1381880180840572084715868906836919393381864278454537956719272397972364651667592011057995663962598535512763558768140213409829016296873429850792471846056874828331381259161

9624761569028759010727331032991406238646083333786382579263023915900035576090324772813388873391780969666014696150317542267511259933155296742133363002229649064809345820081

8106180210022766458040027821333675857301901137175467276305904435313131903609248909724642792845554991349000518029570708291905255678188991389962513866231938005361134622429

4610248954072404857123256628889317221164329478161905548680549434410340906807160880282279596869501336438142682521704728708630101373011552368614169083756757476372397 63185

7570381094433905645644685241830281481079983769185121272019350440418046047216269394457883770901059746932197205581140787759897720720096893822493032368305158626572811146379

9698313751793762321511125234973430524062210524423435373290565516340666950616589287821870775679417608071297378133518711793165003315552382248773065344417945341153952024 2444

9703410120874072188109388268167512042299404948179449472732894770111574139441228455521828424922240658752689172272780607116754046973008037039618787796694882555614674384392

57011582954666135867867189766129731126720007297155361302750355616781776544228744211472988161480270524380681765357327557860250584708401320883793281600876908130049249147 36

82517035382219619039014999523495387105997351143478292339499187936608692301375596368532373806703591144243268561512109404259582639301678017128669239283231057658851714020 21

11969570647998140315056330451415644146231637638099044028162569175764891425697141635984393174332702378123369380430128926263753826677950341693343236075002481757418087503 88

475094939454896209740485442635637164995949920980884294790363666297526003243856352945844728944547166209297495496616877414120882130477022816116456044007236351581149729 7392

189667373826472047226422212420165601502849713063327958143025160136948255670147809357908896571349261581613469018069650895563101212184918058479227206918716963163300448 5802

01028606578585912699746376617414639341595695395542033146280265189511679380745733157598460861737026878676029436777805002446733913324316698803540732323882818475010516 41331

18953703648842269027047805274249060349208295475505400345716018407257453693814553117535421072655783561549987444748042732345788006187314934156604635297977945507535930 47956

872093167245365472083816858556060438019770307642460834898761013457093948770029461757920619525492557571090385251714885252656710453498134198033906415298763436954202560 8027

76144219143189213939088345431317696851018401038444723489488695209819435319065065553546173358140455448378847525262539496658699920584176527801253410338964698186424300 34146

79138061902805960785488801078970551694621522877309010446746249797999262712095168477956848258334140226647721084336243759374161053673404195473896419789542533503630186 14009

51534766961476255651873823292468547356935802896011536791787303553159378363082248615177770541577576561759385512016692943111138863582159667618830326104164651714846979 38542

26216871614001223782137797741312689772667129920259220174087700769562834739322010881593562862819285635718933849588506038531581797606794798408783609759601497334205727 04603

521790605647603285569276273495182203236144112584182462477120120357763888959743182328278713146080535335744942976217967890345681698895535185044783256163807094769516990 862

47100019748809205009521943632378719764870339223811540363475488626845956159755193765410115014067001226927474393888589943859730245414801061235908036274585288493563251 58538

438324249325266608758890831870070910023737710657698505643392885433765834259675065371500533351448990829388773735205145933304962653141514138612443793588507094468804548 6975

358170212908490787347806814366323322819415827345671356443171537967818058195852464840084032909981943781718177302317003989733050495387356116261023999433259780126893432 6055

847102787649010709234438846340117355568659035852449193701810416262080504299258697435817098133894045934471937493877624232409852832762266604942385129709453245586252103 60082

928664972417491914198896612955807677097959479530601311915901177394310420904907942444886851308684449370590902600612064942574471035535476578592427081304106185462198818 30090

634588187038755856274911587375421064667951346487586771543838018521348281915812462599335160198935595167968932852205824799421034512715877163345222995418839680448835529 7533

612868372259353900792016669413390911687588039888288692160023732573615882071635162713328105181876021048521806755266486739089009071951380586267351243122156916379022773 2870

54108420378415256832887180469879525130732663402785190594173389203585403956770356113293544825856282876106106982297214209619935093313121711878910787668720445488760894 10174

798647137882462153955933333275562009439580434537919782280590395959927436913793778664940964048777841748336432684026282932406260081908081804390914556351936856063045089 1422

896452199877988493474777291327972660276584016678901364905087411421268619698620441269652829810870454798615595453380212011556469799767857389201862435993267776894540605 0821

88382279098336271671244900026761178498264377033002081844590009717235204331994708242098771514449751017055643029542821819670009202515615844174205933658148134902693111517093

87226002645863056132560579256092733226557934628080568344392137368840565043430739657406101777937014142461549307074136080544210029560009566358897789926763051771878194370 67

61498217564186590116160865408635391513039201316805769034172596453692350806417446562351523929050409479953184074862151210561833854566176652606393713658802521666223576132 20

19417013726649660732520107719479312652827633024138051649071745659648537483546691945235803153019691604809946068149040378198297323609300871357607986214254220964190043679 05

47904993007837242158195453541837112936865843055384271762803527912882112930835157565659994741788438381565148434229858704245592434693295232821803508333726837918302165918

36181554217157448465778420134329982594566884558266171979012180849480332448787258183774805522268151011371745368417870280274452442905474518234674919564188551244421337783 52

14238659799259882032870851093383868299065719946149062902574276860388505110326385445404191849588665385450405713236296810691468148478696591668618427567984600418687622980 55

56296304595322792305161672159196867584952363529893578850774608153732145464298479231051176357494946229525694976603594739624309953433104049942096778838270027144784940690

37073249106444151696053256560586778757417472110827435774315194060757983563629143326397812218946287447798119807225646714664054850131009656786314880090303749338875364183 16

51349825466946733161181233648543976493250261795493572043054021829748712511074040116114058999110930624923128131163405492625713567218186289327861388337180285350565035919 52

74140086951092616754147679266803210923746708721360627833292238641361959412133927803611827632410600474097111104814000362334271451448333464167546635469973149475664342365 94

93496845884551524150756376605086632827424794136062876041290644913828519456402643153225858624043141838669590633245063000392213192647625962691510904457695301444054618037 85

75030366862124622786397527466678701210033929848733750144756003221006223580293437749550320370127384681630610265703008722754629667968808905871276763610662257223522297392 06

44309352432722810085997309513252863060110549791564479184500461804676240892892568091293059296064235702106152464620502324896659398732493396737695202399176089847457184353 19

36646529125848064480196520162838795189499336759241485626136995945307287254532463291529110128763770605570609531377527751867923292134955245133089867969165129073841302167 57

32386375758200803635757280027544903279530799007994425411087256931880146679355958346764328688769666100973957499678365933978463469599489506104903836474095046952260638580 46

75807306991229047408987916687211714752764471160440195271816950828973353714853092893704638442089329977112585684084660833993404568902678751600877546126798801546585652206 12

10953490796707365539702576199431376639960606011064069593308281718764260435734253617569437848484952501082664883951597004905983808121052211110919433239511360514464598342 1

07990580820937164645231277040231600721385437234612672609978703856570919985075956346132484601884098501942876879022687345565005191215465440638292538512763176639220509383 45

20430077301702994036261543400132276391091298832786392041230044555168405488980908077917463609243933491264116424009388074635660726233669584276458369826873481588196105857 18

35767462009650526065929263548291499045768307210893245857073701660717398194485028842603963660746031184786225831056580870870305567595861341700745402965687634774176431051 75

10367328692455585820823720386017817394051751304379948688223200443780431031709210342616749980000730160948145863744887785222730763304953839443453827706087607635420984450 08

30624763025357278103278346176697054428715531534001649707665719598504174819908720149087568603778359199471934335277294728553792578768483230110185936580071729118696761765 50

24

5377503029303383307064489128114120255061508964110076238245744886551825810581403453201247547232690875475070785776597325428444593530449920700145387489482265564422369636554

4194225441338212225477497535494624827680533336983284156138692363443358553868471111430498248398991803165458638289353799130535222833430137953372954016257623228081138499491

87614414132293376710656349252881452823950620902235787668465011666009738275366040544694165342223905210831458584703552935221992827276057482126606529138553034554974455147O3

4493948686342945965843102419078592368022456076393678416627051855517870290407355730462063969245330779578224594971042018804300018388142900817303945050734278701312446686009

2778581811040911511729374873627887874907465285565434748886831064110051023020875107768918781525622735251550379532444857787277617001964853703555167655209119339343762866284

6198440262952521836785223674751088097815070989784130862458815226609635514018744958369269177990471207264949057372642860052114035812310760066995185361248627467563758962252

9911649606687650826173417848478933729505673900787861792535144062104536625064046372881569823231750059626108092195521115085930295565496753886261297233991462835847604862762

7027309739202001432248707582337354915246085608210328882974183906478869923273691360048837436615223517058437705545210815513361262142911815615301758882573594892507108879262

1286413924433093837973338678061317952373152667738208580247014335270092438032669517421195076708843263464427491275589077468635821621660427413151702124585860562336314931646

4691394656249747174195835421860774871105733845843368993964591374060338215935224359475162623918868530782282176398323730618020424656047752794310479618972429953302979249748

1684052893791044947004590864991872727345413508101983881864673609392571930511968645601855782450218231065889437986522432050677379966196955472440585922417953006820451795370

0434724517628935667705084902131077366257516973355274623029430312035962609534235743972496592110106578178261087453188748031874308235736991915156340957162700992444929749105 4

8985151965866474014822510633536794973714251022934188258511737199449911509758374613010550506419772153192935487537119163026203032858865852480193509225875775597425276584O1

1721342323648084027143356367542046375182552524944329657043861387865901965738802868401894087672816714137033661732650120578653915780703088714261519075001492576112927675193

0967284539711602136063030905422439663206743235827978893323244057791992784846333397773765590187057480682867834796562414610289950848739969297075043275302997287229732793 44

4298864641272534816060377970729829917302929630869580199631241330493935049332541235507105446118259114111645453471032988104784406778013807713146540009386306481266614330 85

8206811395838319695455582594268957698414288937434670841079463189325391069639557807060212459748982935646135607889834724199794785643620420961341238761319886535235831299 6

8622689486084084566556068769545012744866314050547353517468730098063227804689122468214608067276277084024022661554850240089528916571176174390203375848778429112896232470591

9187469104200584832614067733375102719565399469716251724831223063391932870798380074848572651612343493327335666447335855643023528083924348278760886164943289399166399210 48

8307847777048045728491456303353265070029588906265915498509407927675671297950100982294762289618915914415200322838787734851309790810191292672271037788980539641563623641 69

1549857684083984688616843754070651210390625061281076637990479088796747780697384731704752534421563903872012388063236880370179493089549007763315230635483742568166533616066

4198003018828712376748189833024683637148830925928337590227894258806008728603885916884973069394802051122176635913825152427867009440694235512020156837777885182467002565170

85092496237477268136942843500629388144299879053010562173754591826799732177350293689280652100253962688074980926434580116557158867004435039765053234782873273688408635400 02

74067678382196352222653929093980736739136408289872201777674716811819585613372158311905468293608323697611345028175783020293484598292500089568263027126329586629214765314 22

3335179309338795135709534637718368409244442209631933129562030557551734006797374061416210792363342380564685009203716715264255637185388957141641977238742261059666739699717

316816941543509528319355641770566862221521799115135563970714331289365755384464832620120642433801695586269856102460646069330793847858814367407000599769703649019273328826

13532936311240365069865216063898725026723808740339674439783025829689425689674186433613497947524552629142652284241924308338810358005378702399954217211368655027534136221 16

9314069466951318692810257479598560514500502171591331775160995786555198188619321128211070944228724044248115340605589595835581523201218460582056359269930347885113206862662

7588771446035996656108430725696500563064489187599466596772847171539573612108180841547273142661748933134174632662354222072600146012701206934639520564445543291662986660783

0890681187900908152950636267820756143888157813511346953663038784120923469428687308393204323338727754968052103028215443247233888452153437272501285897476914608083144041258

68181540049187772287869801853454537006526655649170915429522756709222217474112062720656622989806032891672068743654948246108697367225547404812889242471854323605753411672 85

07575520571311566979545848873987422281358879858407831350605482905514827852948911219053831956242287194847594078593980479010941940706717644390327307121358873850499936388 38

20550168340277749607027684488028191222063688863681104356952930065219552826152699127163727738841899328713056346468822739828876319864570983630891778648708667618548568004 76

725526754147428510281458074031529921978145577568436811101853174981670164266478840902626828244482580275320945499151045185177165463118049045679857132575281179136562781 5811

128881656228587603087597496384943527567661216895926148503078536204527450775295063101248034180458405943292607985443562009370809182152392037179067812199228049606973823 8743

31262673030679594396095495718957721791559730058869364684557667609245090608820221223571925453671519183487258742391941089044411595993276004450655620646116465566548759424 73

69252336955993030355095817626176231849561906494839673002037763874369343999829430209147073618947932692762445186560239559053705128978163455423320114975994896278424327483 78

8032701418676952621180975006405149755889650293004867605208010491537885413909424531691719987628941277221129464568294860281493181560249677887949813777216229359437811004 448

060797672429276249510784153446429150842764520002042769470698041775832209097020291657347251582904630910359037842977572651720877244740952267166306005469716387943171196873 4

846887381866567512792985750163634113146275304990191356468238043299706957701507893377286580357127909137674208056554936246464126002437968454377733902647251281941632007684 8

736251764065967540693621758879307855916478777247392720029103429495624476613082007292507345291707642266210476730378631699542374551174565220227833240968035246676631908610

112067458562873174135111622920788651329412448154716281820798771683463413223622341177882310276598251093588923591620551087632980879931651725893800123781743489683215159 056

2493347370206832232100118637395770567473867102173212375224325241626358043476253606806696163571594551527817803921774322823436633772811186390511893075901666650742952758 384

00854463541931719053136365972490515840910658220181473479902235906713814690511605192230126948231611341743994471483304086248426913950233671341242512386402665725813094396 76

219396554073865242298978797821986379182997095579247473203032391164104459069079778623155183459593035305923789817515891457650408025109479123421758482841881950138546165680 30

175503558005494489488487135160537559340234574897951660244233832140603009593710558845705251570426628460035440282367876855098267816176552037579565548167789603892749835560 8

791541177749423573400764161093294003899982199267257086957326068774974224802023307525187650255968420760693229988587579898896460744381788170081548895226516722834045277219 1

069914157646394852311267947308658031950764551976756289574288179681209002638714525785831527761510908863174024369568056787301523542780479341426649522383370711751126537550

39423720987846680491394734465307140796225972871305030772587148755705025825734668666138023514260561161974055434365486980054448792959702875903522584097826835986644658604 5

694241390729095266249932902973440568160683805726626052727088407073471496060064561454070734432782514087474275506722304845357006092214390002992981608211717047917614505191 0

0813267037521493074056785331110605835291278100739174994919784511291591368110739405517520801963053935074024850955377250036705466516233043042508744232426240463211507899733

69299854070416562610419767002024150948924118560924096376044296120023645907064497706272079190192359648070489236369798601982830872842285647523531628827913242955248144475 05

52190967204608068954518171220493032185374062724742151974030576904360268636078079200477623242955182947352202724437633902772139208776706571624163975178585925442692342853 52

743288563368507896519620725194165560618703705502184628454342578503830000953745182929584404649188386857934839611512971605816657450967036774958366666931218817636796449436 1

713041603724305065848513174926405585519401800518090847521186822461697614924323831948643441590855801107307031120150224341607315792952875293683582039700338911211417068521 9

36658978945950315438958901530382714300192958907414994359289408309707707836287591448403704503861896697581120185231923186865996803858381237032915620757883594878094168820 55

31605128190152647592807574958154564221341459378167056992868299895611982353837157880480478704584175394665497690173220310890070303362911767308448450372145669644401469545 17

38574341578101586187838392785526093991305702555755590609470514980934877332007279757303824598946680968082222134848587382299928179409082566520958165547247524456674369759 4

47468637633242890426977610679193391098330042231029372829879890320939109268283630617361017387812367989864514931170243712828588263048629888449220741564060714705913740552 46

65756971870217355287245439427714809179364437650637861861324348635797411258520863459927803688792498354363298457687650165065115345008695721239507544785683173631557153527 04

65242352597375134088254616096614407466755142268360319580107215246355106917187133573168548563128085783443562367095965094994696882066118511808603420282133180124941099150 2

60143545001743273079362511307029825049941799428445114647932915459955590958780762163666859179106543596606525352532027365072598912125568684280207724648772201099663182955 95

529033933122843648644759735608598407609472983895424339326231532399189818522641808312963335463568748288634656185048106322880055967378445620009414656034992808794051153100 5

75871295525719641115068503407737106043803712595755969859493620584775120263549473475347481892622541903526716144292848998575367406921652716300860606543737368235565886264 86

343689153218095572204456777137368310458075584529612832832606319629728527966674362974800821318627921869044284342630735760703996694307895081472697302538173756949227517953

54326156912040594832860949992366412287881226419148504856328072066418557059520375030322916894489427578306090910852410601400683274205583969773823150734996108758763704255 56

496408685507194225634496673243065625925047458176273328181601701969816654242637876360145303594653845032547667499973734083566513818602516520283637389171016545414882674448

00910570418616262683797112088614135727961109908829297022969212818097879895139150427093678644498319642013456683390877594300644248562301212461451169792193963440950808322 92

8129427043659914648274998437594211302041829730841717881309037955854560324717081919530277146579455547554475428443440813938890860977601785738930751866190650501807716500184

07443258540241843605011182429907023234172436745253653495947990633345407543718126993998337192184854187359798453489345922685150681826624900780293350126588249742262418853 52

52663670282766249934982948874833106176420842901692305289960897860413006510902817980504058710767117904113021748279668235300196022025318557678984331758680637835996879160 15

38922220236575765581586611409199394861599209159917553341783033347643131635012705390697079326567812415906434284721360235218236741214733124499944334155915274315931687477 88

25331550927703362029012225977948098553922000645271622808553982789065842334475528212765176505726632676911410750348458718969964348757751384791481836351006214668185850963 48

88708145697672202016799119946241777668890791713686594596072646853881077878300216136827669702622345941873747673353799888440342704680304255169412715873932039844437460454 78

16113056625176412759821181939661101850562880555942566060032321216180994622129301002470913347150682268430458680300904242861682025562140946087900065191099495570815816505 82

89833407394660844575657806366902728434620185873282529247965052866814085035385198375236374519256227954902905579070302839501048548359298345428144873043580470533150815105 03

00152142811717539364913316617262123540552786330800208317705563029496359420165433309409417719632623411938710516157010179805355167937086029136675698609712412036858381295 76

95307798141365700174761356966986146068491439699573837631695824602513342108072621713601943018087209888551415024163818325975259593165531865833117126857941527206612218422 66

14118251546574848783126103478345467492583087299854474212064450952332450508774314961665552517971680209917200264093749219075699369633028139164720896358177173555584859270 6

52450486251641954055080134351032338981337830249770182275490638149996472333407961304146973947637265086927334710841568560843092131624043462986392084166005590459850649124 35

05264766067600344441618186403670083774114101094320588955598658670077863671896944089622321374034113597199133135946553685446692367652589012108413777432482191812747847892 28

72648929700323718734561579815998348391004126010507469645994303319788106349139238124905030614334079183280040639070986725961970983112659601474737253305268537177421465540 05

87392462372761736490519871336806772395257078136068668326139501432950947485159472466752720168431658660880751276858475554118438116901162200555211348448896066825922743131 90

07963011587084670117654935393046563356225311244727796669005831190616101972663073970542531439818457379449486780134618217875939076999602029083965677287846905736401564015 04

76964489939475414746083399186968892711569423454926512466455077925540281050376220359675305586018564920560628790907694533392088088494778288948511221547432301913832455629 93

88102061449026687601020775321091568497783074085964985796715261701003947549453991769879132354655010640735581699940975624814996744327842920276264418979391815839456270817 33

01582160225519659898769376164019861207466755048861110855726764507052622446130222335852072273620485057289238815884938754535229186399714380884061757286220950122506515863 10

42588841343554319737298562177530720226294755524830444453404348887858117034134534252235431940787797284676018158322709774518092934219318981581248283265895004070485520609 9

89378390034191416304463916388054965878650137504634169565515661829887863070584230696766025405302481147100789978421183048901046405689653970288559553092555863605215895737 51

14089564905844156774937105859648014315874614491250549253191164653821585197370093280194530320572628452658046046337816631429933076646646530760590548962888724189716060225 88

26175775399220551315093772006248630855628204935757527249955670892216342339836025653287310291940070411769192208500151167356701019589710017970195781208929109694177543699 04

36820256302405482262540190569650771058157424072149633956036527028333440730575007367456226058464988611510168961218111905847171446106871976101745658737379674069713742323 87

5383903031720020020720592848878512391174647167374373792328388196620168762219134623389376259952702567213862211245898021213050140728890430032253550409586681872413936993819

3069148744717186646183111942603161664070377316487001864799600243044003242241809402278533309011509880870678268835317200767522553138008818780431690190072804831799287414125

476123089606833095828377667688287578688683092976001011974533898331952588619630132917094385816615374171794496319177154312506959853481285684619377669894277459170918802520 0

1274990555940728969659479333167224362156789677966670803522903901848573080627567086765862710476940920356559302535274341896592700222704923318682999156093641375700498853730

45963961527346293969749517480626964517930187199867885375814159757993148066085572326583743052827641756700502880404894298995809481035348339341449278859252621924155472319 97

1433850866373209266327282435149336407045896838523456247443611752567669877675972234392063575074715529181027626140129924804228839902978799254185174991296302839907296355885

7989059331779590876907390564602562353356722155225946883829845288292296627513716242217295467867071584092418408414755758253938524096330205134970474069539956789798172786092

0462286839735779815111868152659884606949758965481314651150392626377749513761557248195116119877250344564710738513435927355538712462375598193813214238441581929070046389771

6838872079163617414324970791096581627464297170728717251427458983568970955346268201690835610894489840710058192030217694512077177458879551951047338418473998079630676788 58

4516757572990430697154264238349800987086993367091210839445350624592243231234827854966037465718801489293794514787054060792457590060121962212392872001721558866634573497140

9533721151655985757941724419889026167016101611557834315025460328781198424027484608510722406676778760855247617773833089502610064388350550205456324346167859451941795669874

9685152448838475136181806671083161655642093692705206118985172926171417144346555087063060063551012949400309759167799158426049197120954322702678432654296572403272088714321 9

9964531320258710967716512854966996255269860731176371820749882739977060199136209308323073683820645573256376598291257813149222420427971241441629951265945639792759380383 80

4782623160424325399132851123032247037561942321733047854078576244013291717992979240783390715757981426816864655382946847399205888631655934919867896962840447344968024077092

8313764081033522552427174041076735654244410044833474401017264410529547872963458986405012036080244511903509947449397361718157527709378020923666813584163626831926340671 41

8279742134254622070541560005095967404561684045177174795279035325493258912048338574659009678173041600052108893461076875400424197780308288518120017336955912713771419501136

1304409753279190504891583246399143483531648681548579178632935123925552510211182788573696060276931301469661433449642302114382483705633532793858895267672076688971274435 815

6320881066501495681435587965769098577659027687074536592763649755534496173080781609871032480137951361703677634575949756862080139963745517624251477806287222659714554829067

6929571364357215267446898788941882075129222575650914355282887461419509786242752788157156640076372103780319404309584427254926998716923433189002214150311399876526068876156

67402101972017196023908610829749276395695411530322754601738707956259935797853024434767163995914623179312399899869284379757024923695515872976838540052276514956144471059 71

962889888157109415171701518114743513643854005116246202131174800791983749700100471363432523281578911355450453371905275068229156185003328469567926226208190442473340362503 8

927920715859600393631533688427243753667996986479347411331983286194414606539227840999031438403545650470567895520248271760118743356436902435030856313095590552503904927316 1

331173492258464460902453507919018441129932169977045183285358648042855682220873721361649058630325636891308410376021567992702000532235543980465311933977545904404507856802 1

39846500969342954731026924994758646605809166998416068464608729394380827430828581747969417287299031101319267557389798409136425347969494348037770336463495847686298259010 34

70727861218623001986607987782684245933835638919570206853521603211635230064988744600200170413056985365154668752023859375183280372851143274811699683692849220447380570633 49

66187112409478359158696268586435891413598542535776887749327436345147544886408688180303696524317556883002058607732569597160864854158344684324899630770113713446751569302 44

88548207712413355773230694945806726784523594363150787272815790157307003317879685443627952571902362327461426286873273800949774112285623766321490465329407202619753907174 04

22259539242888164559796570030957141389106936845036268231053986743753240052701534745893325679514941854537808827063457295962169085383535370381418115573816378209032561519 86

97453576464121254980760051561417072980469948135934831505681166427932193352798227147157673401860887215187996693502527007575560997198828630642854481282751392806947027501 48

16328972731434734852852950460488327167397898156367880478044360210900732072736974934463049973144257156043313369038761810094887312071348271081588985748326585420751007795 31

18326861708037070935927614936782530858340482351003632166378957426202550350116861543407379504516482896755698358935522020173679548075781909502697981271148703431190363112 24

61282953038205128704309294719745946908210256347889954317715243796962112812245034260663992688521330791963702777804488579205730469908009234401866381132520971230964760599 89

94792575985100817303960682221997532730160658262852758257669507854726034938298133582528178670608512656002268871781125359782933734779141273628418865617592083287944741096 97

03879854736984025458063294835022359393543587480223989760916296250110473931169449100666907230634693130169711820632535269244043840093724284428209709364856909468920087371 75

32525570305435398287278123011398080938670154748858034456318713196026785487938933162050076752641120443902375833427242986996547863685341028488573702547255023656634186809 19

03838867078790720840361940216467012153483797815183282647257862881520710108149955898033811896156944175676134071704653851217090212377788433364965187211990540758187739439 75

28364143953044245913903178813004188791887114553148267469987055587931040240388884083850687341625071657274185134952084963670955542450439483948040597915622828248378793415 272

03622633695618055563710768148888936192757426599358235594315308879330527675587475123650658439694756042971920023198680243517199378681003611023125683642560795974105741536 28

29718004649774857371837863903703901539737491165468549971645394161121641761071714540176519056505252066227788312904571969320599024137539598386198260320549583950167555250 96

44137118222561496014003023035407899209698677507867200038074267970530307167932296015648622808518403352350170608589512912223246117830253163628943946073652771336511631646 44

61990990212249224123151689927678558637363155260025034884878132330019101893996167027314169996265119457426367619650024347371727290284622097983948710659822700099549188776 96

18850543265321180221944428222842515255614118743401804194614139451471287252759239125596443735683397289633126767823491035633296129471910151571431157954909339032614119186 54

75237624721531102079369115848742205822747343201735580771224379698579654915806279502740977168861148076163151618553068566924571717692204436684331273989337941116297224516 9

99854685622157024175947117699529165502116855001089857619346394559088262707753114657752238846343519376539734984802454976076024403080844890106838786972612370978357824516 68

01171485983679405529046198262165669172027426285482393396001825459940925430816969103297841123402288560019054934275022318529471282960969397681373419770427812130014732867 76

05719405969979275512461718434956985641712872481183465420642318714551824152867630567513116267717735061751124546338799426529127010578995671805721436557918350691777930704 07

5732904397494995822410623810514917650238504182730096620171750940590805408957283755406355152219965820757351315707592361539863945921115586400098809755261053838256899272158

4785041746065161511337883360976012114848700556016581249247068256844272045472896309420306650445298646223594226008554991589149953606498428034579492757009497959450602378775

0194706246323949549578230822830668408188025210766390742309737209162853371768062164469354323179178553058331714208479886303408465726426939557002685760575393478885870946005

8272323051910811751423491268733658596079989173292891589600181509181633740080603547520005151175102901229924870961545928026206076169827218102916731554892942374085196743307

9166078499055782101935713662435990883613859808516156417476946054785540081953530670803089697630452946868233210532878237438944115685176271711636309401479909649456354592950

1307390036268210073263700823561506912696431833517162543903046989893142615442635951136346605737865495124457475262167895470362890483048499680403772251343193737344123661858

6944588064018584073147633792940386340435919419872355263015654608051868676068043160845128459160424413269879125385602991599672787661951950531764883134693257366894644382558

1391084862096637426745798313012223438725831244220330945714575414704792938758582389977385152135237238955966431223564326262860114748908681715928106687270840080820337718 69215

3523526926347226809082598988984002620815217828261122931311820866007099686036540981832680755824776706950410997586143624355216194535302920025466736799648504337313349520821

0751199258926638995647569858707901856123791578864374469037871509500112550210038845311923652965599461900474846620642347942329670060529003709175578188708193522146871427235

2776325598980869487211138459800141238421638278244127365424467488333816797162011288619141540193671290947899026466443156098372961501968624228250672306166720943 54657142514

9308642488778598682759588749065077260250951829536765181182368616944724360783764294762469226319498921964644068316928766161506050813846319415116202577907863071801231159458

6038965625265542233462344545073947886902681594975131168851436945210216883190446168629763325229863851818500492869357276476682385556463655449640063176482855757858 66610228

5515648599088209586894443625469867952382268611596991005636608292679153375381606611224786953132615853187176388598937792918890299879387981000369730784895927062541048485931

5854323395683104239029907026344379787569185543408976440760130844481978626507947644083013494243583428188591525929347143631753374958970107287350127078898048163504567666769

3207553051840432446100740321676471836083708470651269307076608498252990003178503058536821395127350386382460564251033777558098646433980171862081426630741725922260 00511091

3426810746701290143016541010649332122837908275150010035300156545975083237729654396973820477416265710657408216499606262274961879533479070659889748717795643340648417456457

4790692517014949981009535341354890875483632757952240720698629102467170357925144176670388660990698572626058124082533622521899200041897574576531512300006444571593170177168

8635483333051921582055946117357716321132233931965320386199005116178171334001070576652689919708169202219464704323795356411866063920558609034457064151797782145054722278852

9872101978588460700474200284688737958442289499743336562718779917211379161644925413297156528795295326397595385359209501386333805075613695308995475848830242619627589859415

1378051580502576754040178579585244883117210508927708922727343197382388468730716823024878868585510108073522781405371406520758107270848167263977098731455162646911 42328610

3036932984330300323676162714264067587806731883971515002798163374779078775038307986759404591073921034587404219617034925808189907205961291586420202885734009114955 238865107

9113714953346397639881839488045300750747403722809368205354304949519483328334700751619790086872854399629815756058916376247230691628711111376760864803237524966493 04117539

31

4613646433780467116505550467067183622128579504806716563042762671142999911348769844705037063790018109688862972175795173243380278061747049630204249291661917188624335559928

2093243919445711886321556320161654247055375938696624656334121541014032286990930159132885808831241242882876373872742838038590710292748633351503090445328052597795658920554

5624342979827941348917563824007716121733247364285401606100443376414572207859217155914010378320201321338330963807789040957238105588293927963743816606868351950592770195153

6160172215890428785678482068291944169871819286273082704441630396254713053284388337913374768735826122116258360272896162455904189677024745382758396652299371235163048983301

2421417455788591594256059792427721819908556279848605617453684478923796907975594555154646853163024462325674034895845462256744858202042457391994253094264224504202689038150

1526836024125598075975236481628093048912746151196231546114008220563967806585354076868882275426503812259991620760170895567474465242344520176616503259456659129667863246213

7991922296145867142248249288064768032108647799410041006003390679275237362546027742960073478803835668752200348245769490845686269605771570191917489226063520812973879744383

5483286136939562450392976805783223402171676555917766840375723484409461762931288492689936871389838822271060279037990019045583360079739277410926655739233147025909233890654

3884223513241153880185592349561399302239196450504503693529270115663051533519186418648234424999192720272953459599063048723608041595760029668121116831723660381105428035914

4572024825645610571405546242082134352094810841715828957244507206354681600230512014084805435874252617101768185388355755871741542477544977222141926131552526910917556333193

2322243218525422182729149159810583689702503522813002141192486014248068079753699647771939490680468355280834732761030604940973309169031678309793463661183278453186871646268

0738833656704566010423768505801395074436479639222841126979451347730049249878649656367949099291327125289776519181754279628060849323755208153611132403397131655043918879601

9838213858500077324246177884918758145964264233788979333081948816004011312652563569324465939840063689031525472292399141447437706963389357619260391892479361780083310261141

9548543605157787160049557886565797066588551042882466363057207778902266777042512681571979533225107638903681976284402861025880539233932947467202408854127649238644760216116

2620824212991660362299184923782236300983478119522913821847326342285759120979805478285250591837983368017874112426447460022562414980691400740979721023278539575615128345806

1654111179267104279905793944971349463289504565128688478418717580205045832838748531373691135102550620102775345809439105001021833973245650472889476879298925945019875076712

2363791875864720121496606115128048709648863056228440839369443872169212084920085515838125107074195518720809374694245973117281172105192890389637039423577686212766821093182

7636649840421249381440979598631142254364839654999834790843070217643855543512574368282815303222238083476795111355701480631820045322072379489186357214910624252699399467 10

1536684623410515333814268477062758520352409920797208699145373010955164150331762820019691641154602682072366925527514184299699205398534330730680573723805041671972211273740

5078927266340638850686734458560773266648384578027718911475801323105519878413365218519071460681389868867103147598264611293795439526672867275994833590259744587868768496462

6834844344141359177145877660880778435718393293719373932364083563375766884682111799350554102085561884901020160050563954168745108220603555410817666460524124966224 4228045

4524321603203601946413560979200195902404979292367329892455399010198011214029086869992057589177718807414612220502472858571536753074781438973057178726836636015761361007722

8631963885264623512553807731945956356796538236249992655180433079635962110674552852142902629498265675533527310046878865731047246649332656792733134512295505918623293739332

608607745135077530901574443829487339779605322849358301361837958626480321297368474817516476913662110360369509106666505171711508278200932788358722598394046306837631811 8089

0442362621998812368268078579526219721668720174551747262781803268305854880397097704793483103543985590784355277667603313988460527150313885633246768892710459585193289513916

78238577357726581004798256393551935200552040800287059678249739374788605283564935914978380377964960005212445834779001756042465866651998077028839438516380955043049219 60324

43609034008517466042962743097683715194598264473594023424821104475729111777958773134155360952759570898612586771456252399450075938020609355024892008476733229308574222 2550

20645569023912654366357852427242905605320575403082101451238209021746697579765347517250146583747884808053773515042222404295760361375432486199655891939220504699982106 29316

09675651790751322960777857555331026585842576086686764535520927748275567545177169950878941180593630524994496701237598006553499873966639539441701705969810151271933311 840767

92327185395398097640485278467438723164329100290654953086128333026640075801296184992070220025559721569575883761687843643467927558635739722535648841330601192895746428 09357

85808113233143311528748217976603971257952890036407198923328131611640416937736628013259738222374268189176489596422703380390592959649696482133114473166765041976781108 4909

66469425717069457007871264014486522428469488976172567465352205061621073001019262483146821203551699501522007316384004132030333242312167082685468931758436630430784350 78592

81044784926639526523987186441733800856816923213474297545832694021612533328379009606486277854941266795136740458774169455961407626566250299006922672627876036587137932 796041

84883939339346926354341548095183623323317522937035210291464133127520371171667548720634738923293785107290295144629274154676194794274716691603049782928896147458702649 97970

79206387240825023006425544995904011974108535167844409018806462937483543439614400353523310304041178457228902958180581032123743825898702747370401068377719252126453570 65083

00921479258349892475127453622006105854575997369313529708143742841340551954446721489415057452839171603715453082525558343202512542416624457524562964457910769717152147 0951

85055003550543906316882581057850746356562047914667680556984384552027709969719889807233714869563567031776877637897432734928293439051455670607446079704769316462781214 17138

18274378561462197088087021064211057377851471358837377388240765280451914271374881105597447183100939375197659802100241012511230813682603384744910877161322857660263938 84928

49598982365657272042635720263748256494949126291419171306462805956698254936032613201925280434617043902892602799314043613702658201213128514881585731117821041310335728 88718

17295262711200081475064026830464189887697478791731737038139991888242416994212152776045185956711909418073734793310997092831554681656395271010461137625406644958618385 46389

82208996778329550114314995936803982223037136329574232173574464734210974149174364199473195884005263872695923183642325491845595504534377846709470450959420120211422086 4191

27904935994521373924871107432314951138042937936554363721726348190757113531270930795272952211247953149896990808946657476955651243605611420086639905609900038030250612 42360

77503293413472890501316772809713162683495963409292243031195084878867103533520023712730202916592975252657039210421496349523857085605723434621576956985134068304548331 54590

75364711469968242091023214311717692277385347704177940764410013010485960927072113205231853822744487024332710398781147912754608083611568779215131131045008366363100751 7511

02590028086427715020962713662397401075288445468331618211502789264307297635576105511246203324800531059951115054314848295534329598305742724517378865271930007323217362 37587

32731489091094553740270481185557199051683938745352067970859211896407854895041094056996598871598863362077955045219321563361246853031747054439402941829263552401554523 16098

68255313897018801539704596250169179664812501555593231148267300563383579726032860177847414960045697257834956205873287301245145557634523029864814954410090788352980120701265

41095251846066620176742045257367994690771908453787482060802904825167017661982073061833123921935356900407052154989390344659388090475077241695436518580750664904594431888862

978723571603022481352204601090635214508280639749275512847694354996203399164488791974379020957188863200247502079102379073072963746326336674594275563784535691367345524014 8

97125909480368566282321005003940073106632075257283147115192633289285206967239347175098295260212549476433019535743835092582831113391153906337661737307723630279889869985 79

94501659237690675488379889294006051628261400481504694828140330839164342486509396354589091328059511163345503656348245191505831794980831827281347950507727173359496633718 82

149192837871164639035669257799424573943554730449355593968480327902086141968150826064810924688543383329866390745478052636291615627988031878282707451630327863907566533621 9

7506322424864576945975359667320060389826293000076125149479800895671245256955982758548576901246368659494224227727177151849641751071598416357207241224371968067203927064789

42789421712842641334271183184794413346064724314115015509855117124146682433123520628406572269260690474791964472975283227495698196327787281625954012020538073295825004974 45

93080978240952991296542331849878800771681631986086512088315867256506594414061844683749631892913745993421603484822883158289730942161473689258516992715531155888876007217

034102445874402084434282730046730979555566681150130033889583802314643138290026007632285034758307808788951803139810207627889851743534782251208467594974300244378958428956

80752663203627696299460180834941994912706559130840005862656399639110406851041282007153246256426371456355757694528492711263557719632506589654553648212545926335525729259 52

81499341587877651569223119151023373440716991656476398200089698462984399775938539811213321810328198969945792617649358297483733877523528594640351382382306269453634581003 19

36725020698280738433341175283157314342639896416347127053034775699155800311815918091137880268838547576972923398882860323029977043066628869553012102727057633959897689410 24

99684794981684201199256134807564404065594623837087236888125489491487948734808614168105521140018455170084444842948475507327366428272220633658240174549880829130188391401 56

80905000084954657373000327477972099175074617859515799532022372852359204007425152256386166756203188398117618611960221628474319079702503674592828046781785366473935600354 03

82782818457669478233745711382212193261672950104270694095202650280522898590935002394490874562620534522173119409577830195360518503854961406218253061820365182733706211198 93

90244889753863581809944918157848783365288654365422483020278924170496896511041727594750178122678581439174864942435730090917126487716059592097445811462955422310022008512 05

22589764778114827039426776664278274625939511743807198618722655865040300284691469278646800318360346381726405702707422620342971875558099386871240465622333891464658305543 0

13155095285109726300508051882652726853353729373385691826937171677303161186474948104242151279159101460656979533313377409593674932644146370242752453933503013099283364854 07

06984034399121245249275580299798824092066446404258596620088874191649877302754037292042158109378147131362262886666945474212449552849091492193371936234029433712557556998 86

52966236450353519202677637942482082860568936231521523178850145213132149146986854835944706865850109813142058926764161151621094053567807368100897342458729327052108535726 7

63805642288409296658844777952795467107351932954747130150792208403282322044289446782183965471109021173407251397247573570085553127432199967512595825680632358808838843662 03

26226619141493474043649800024739833209241183866742960926946070141838817811071428243965779638843986478231371542498947258304114514952687242361899676305881682084632743744 12

103905527652187107355645257133601145580455856845586504328599176765196193271143498665407774514500473072711714795712227572018128864464407751746032824231733853376529898104

423224046772463204795179809715760258008576897513405948054826877288477629384645496040270370508539419092769937066804551719416040376351180185513657545109524703460226002074

174282384948178225490636599208474903758320574467795910675566064077500934712981700581876940802799269046059498721176341519148822518670439557310017937100046657292180372848 7

979715692278888397041982545657064289089858279586256599013759687500785698534209443995971523667673559911557090061413018853956006933050826115788315979018829128777653969 6406

753920808485822904755619051863754905941764720809084852392996636537774687098568014236137076370467423618029218679592476977765292629290417983927505343294338447653339850122 8

283627985150263745427966717714841975733906572871543054321575235449320534653754238204844850884634590853386677292538520444984413136863751894117684862613603681937363513393 2

540806852269214743073291344676252932264084533084493864715156181394136343503648177947550976339255988278690369632386330342579445292292377520328744890200405326681393547528 5

501746453171721459950814556136469252665022711533738181759785579504198807548581133628915490090390806077541575736137375598801875730753624873700129122382611343810392343723 1

353689889153374949378632498494176428141704528408296939917243232867725641504837657731144933521553852300178110827616363037090205259503779092534110470570046565251977925679 3

314108866326405926231788931260315285758716424211903337987257758742901290375936269727234314893572572418837941862786845667758686920276014398050163871435204776738809005789 2

836338177973884573441001499664332358222579253517110594856078918240152199828522694650958763149247127952016446764740270468954543510306982617999140223407285489154680684209 5

743207506621154487626644657986364438802325863608869187594422715214296506641613849638150279721730712659205782660027847181400342092656930703090445702459646757649018527813

931481315092036410498459690602253144748229457070252704363040611144551422276693665012542523720743940182775250894143291521517059974545931259468212143510622763303318504339 4

889512767206372915124936819357031910469357290527628876878250048505480059732307532652227792552419913159617911522069419685479187341566997810967025629939932081645071741734 90

564339865219986639055709352119852439067986150214486239284387398201876022854712303949459661572587509650320071247665759381372124801134153550616754720369579105597461067112 5

417117453695430147191419937319722797169021161357262524311647228936664414262124385498136236949635712821160368544160710823177510780129830425381419089224920859536461082139 5

648113205316073707720760559934981503424064077512331512158999246297497845474385785595227089267102479199196450430401660056217629623401492821816115205046438140512010176 32

797902693271222701259270816304579408695938850308858577776769880577120277461583728185859970177211160371098273932414719793766386484316000841579272530611640850151500165203

002001427433763904187886226352747022589848494690776947476132763910525994056603823823716369435554706581748273071824741827263627240462399440284444736424586444751046902997 6

526749734435698570853905781915995859960967506128309101947488656507512613971363292764158349130420830095085110041407455744378492789857607261057697418196336967907551883832 2

017344376439805368296268732851893953081597213840998753657746635493253113936255978954300091191426740753859254969015797341918371040169991790094567835962857322447147907320 4

569647197863154908628412333251748127848288098487610221009742783475164627905539385196688956965108760628729574590889201702386720740106024538941519547393281424662231268923 6

265027205640264302177690318955552061127114631467170389157733900654528692327208081115787573749910353244466936165351752212468866080593973805468948675560258870687103081189 8

92202421749529345821953530099156135536073159095673469906992487426800195382175246210534986270106132159075726024080430082786835629319838427105219835472751176423302799 58926

87273053118355805687527612409197424447633568095687484441045467028352365141527656270080436309747745376780982087349803849825992488106702977549495352282995165465598506 87428

31762852085719613937978285057790149962321392204623415241682380388944662426737300189654337647650363412518285095120888648562947143987795665592807491648962562185926715 41469

21767683960545008216421626056106423144435798230691965780470574714846007296818237228797756049608915817868672936323790241579204728364697021031397518009784159855000705 53649

38753212574961674875872583259925957615074339186228437988301346044540880817809685491194541193470268965059919860410997653211196581062966550051161836517062029288077609 1498

46167316442686419708923064846305675457388720247601652577608529377210933584453871074027292591915246267623538179786930642153401316337011357356351110981418211296622107 36726

26961567267483077524887444841676657370240048508393702558385910122669483580683915454791660164569148630523935977932446725588671741604855038711490317607553732194472830 58221

91558078807524536969327446017473605242058646968697577061218677619720587491045165142715495423853920232526975123495465463090613294600566507283098728033873735155375223 56318

35702537006494092638080317374634854036114660004846876242310894723791650074517970524862846727663375517303687368385644037049806617909200831710788210498183315526148505 37354

07503510822393924744563010969204227884473716968895091118573692689033665971852253777032962201670810655181267580094085251506847757921913893213809286961195312209050380 18107

65874883683178827814252786261879667606821977030909326006729615127557125278643706989835444409613917379035454851804039733137480523587910955583040481534804539187854038 24323

69073043102740626417777626573010347033840211296690848180461624964873947345844121553025815222149945822499419419547256410317502114422808652302802213424093193932727678 1959

90608112598623967339458989619071679777780259511631477576264028588262514815821643994413506196081175890461951158539082613354960388032371352224516968118059751218959002 85917

97390866524495280407827130270045377437267855325048503974637573946460984085658930184822341614986583150346608218622360580194811455490351547426626606129502687840975477 9814

07268239569314724876098280345081189383404096153431486301124867646531547875845494652222753187735608908350438370811208824417599385864663093970481172530040203058134090 44745

05115637705410350141668619124852526949334829785101811147239870453961275402222190958440508723066232688884970422345670001194975185979694940991489713853622794588740760 9904

32854228127730581830402494510870633698694686740089481097539710090849476830410711529550638887652490545659994260773886347394552511448972036104793757254472396602354774 81274

94160698351013147640236419491461059805563757044651556671236525682827015744528476022078175397233716409698626492055766876156445774644664925477346729725557053882590789 231

75970676863982496629455560193873152710362720124293120176425224644803181954468333763994613138361445704160888342225371558783580701611560271775414247233315278135669400 98980

04445823899842006407489589238923892752289147329455312404247755208380523795101239384358587754549990012720682866599985790984293038460073296238426290797218233372747669 46401

52692048814304227394388383869880723650340088095245127260013615257041577497895464274592866962164154275190720789657656762047087629102592988877128340580613171820688795 09627

35523080228036658853093027046194006144644918627856642449420816210203832761116962244213863973115713011899185316991515816502583428128487414927536050735501492751649655 68949

86881445782807241540090116176936589862811374592790322578489093397688160867085700299534572157942098099722053214575142715411220939886987456280116533207925455196985191 03842

8157268351201092367995242906867995456830838859301366721852113536417244228370492060364815444971779988618739061970126506684370640425124459951909006226082179845415139874086

1561892465930844027470147101672547160166860173976919976620111199893015535406281778132823867987398831854809365141752690405027399232695322939310360456984252059471087760223

2101677467927935625307683377220692980995213327549341076406829369625653809798299221502007619065671332333071917531109537696743144582704745219185656561730561853216604259464

5538561688375993453276738278878122231537281113417355451707355320827604407745254423078545374811259665463557459604327036854215738696224447960925936750083098914000685383635

8817787486427106882578787407992834182519771408422304894979155179876782746847540849289938647634983917539244593293129138080738765005052200666662727343844540498968011834325

5349997625011921767875580980672332416782617825708911630179808819558379107540118050962160109308042257018054929764678411538769143070882475312172313794037236592877104345544

6962665999926233933298641137100126804081160276969402287136507298106445252016551733860468650406212924578927147227426763861426823676408516411947662651437101393855680642700

7782965968048607751794922121562917386716354649889853835751532497431583541399132213650515513841090309027554332364412022530077042821114714191814757096183313782294342072543

4103155582818669328386683660726916383696779320102142029046813370491534380592465471149708354012272410065039497421641886692274473689950625289450277718989469132963467585879

264235211633546474686426054856131577840361143149026954427505648038478887943295655604844339184060202704514682784231514065070221048519592072312004933717673835237093088565

2643448419467734538241329688543063024778255435028195957175433268735831728279337741010263471725258000551089980879204274477838536427497206543092247960572140033066159793981

5697061366098396405520287669991722547240206396060964299454270591546000735367315498807739083001581335160357301111141092801541228066667058785550927033385009831156762851616

4924255092928303908770988934946072349028658560205422067037156804635003826052763710823986597931848309367641656360790706605233434111377931216120205880951461437739476835388

3950472129452834986548086483788501946767694562326701998713318455453483736084512767180056787542358871951058956527978045378344846504681469516775381369518451030832390374965

71621433079638601544816144955239351112121894430238269540578601164673736647956520658725081592753057313134383569920048999618043254950205219555020617927799305642458366587216

7535192817503344992391833256236162650208149035578612440518344040381599135827173843373404529744999640599186566641535612424308001626179337509214296580882832219570578431716

9794628455133096838246000369899618059298795066037607124327255975365088203863609588090400380017604750786697443325877232154383259983998643950114495415077009728226536958394

3808509128411041629096637012742498817616344101667423400506836167648232710388942239482025308696722292524340750602651298857635878137500851005688687432827471873232428984773

3542581504162589550238544890684967676489282970728115843511676077617260489135585109814789508429849836055936593710532020599790443697353401662876453206371886938218978015732

1907629981036125683876483872698536012944816073176186580668059683733894119826500873262426696002409088320762261178399915744021058427898450630360141993392836245540276835099

8972042185962090201621015651922358421194882020912378392755718560554165620545534719697866123505834896282128608208403497311998810772590454586337661085050958238503075128425

9642859749471596754259240349558609796434019664667217572372370707851846466383706717029954169832988691247281876802738125496293898760722340846570950989432016548760479339467

9468513437326303922309331790687303169941800740480006872513659785795859947801994965234272868898871781351617155057783915871386404057895659182321370814005871380883652304716

712718220060186088112572603398624035420675212769089210815522603293004441018906372365919571195303028824858684782564883005251812608103542135181224715840046275105924487058

37095408353189752152361034204084507641376742347300588220343231604746330435062814232108294872409025947644118910322337404979474085782776220482618219514282179811243726766 25

846895195106998673740227323002602615059706421527460232699949700615823592828222978328684019972903653781681600288411730673324496628384032435365041397536205509105219749 0957

998605957269413840242675559674863774293085831406648031844531532908153215494345828804429373556800527667018000947887335886091364949458385268927913655943428817418645559 4102

961792995812608097064547465090234261840345010812403353900061073469412097838671627721613708361451511050077201170421405751029551149137025545335020681411652447691784586 9435

403411879135071947286833389662476101183017004972618956118398981605390920089117272452827329958680838010737813140018760672501269264546450976733747002367678201352356732 426

247888048234362900999633010976573057107450862132187796828074343989648355242714487573058303218024945210923199120417862983211064561898234504950543971618030395685126538 0149

225169487847955472418638278627582327821299397820742867554710924982182446861479580814083550046687559626157906171759021927186972378454724112985575731793747953518295584 2991

336928140588480421571538074685311302335494627214184400563239744587537727518071466016570650353750000780005476100367863699111323985862132218224624643435010363223985967 0172

899284252341131543432629303907359534291441393387428218721484186131279071626858266847205954664035651133279272928367042153333781564897878724347231657710811890588115922 0534

134477675212977463550655110980181145470892170124410634923949242422673834943940786546583638685970026019915416838558615578967012722002322003168619541970289247574216667 6680

152480824022111156190982909528829342278406490395339672008649956965447075211846134340977857777364263165869169876274954188683133247514531590023354409517149140813592731 9114

619200677579215856331076125470709339611644150880072729394563684925327185891551688147209601141540566400389210281186485459504119005580079283947161996760030187700072991 6613

487810389918979927793308260333333834057919338601259926635435064710091260634625238574346352684749297906578001728766596825621946854107798742184455047104825113899365427 9944

593202443898985134425672669327861329504851702042670416810423988787766282835019312545495101087037669638120603127617996218893187778305204501948120474270520457321254873 3903

930286680853928985514539518307016773725339156792769039073362485903433514761178705177976647101075024507681616557253954820094809110586317329891753118416036402195034635 7321

959475586008320829267512378849551667250649220720609741203129313574353745218554549830258041565179862278016468937481723971338112369536373581105739391053691797392934319 7751

880325243525860808275537409997210154008004697992794342234544768970758031314906549976457271996996280332692090891558381760321398926448802376910082742090668008043739925 0454

122368497194097746704673167378878520494165644737071325437283139540962318133764738488941218277568760582754721153484064111928660919806142282295524907588525871140721341 4016

352381199891274778913139757468280934247282311021898430070244399964290644450844788027668653946357835978633014357430738552248011805785516300305948035170230529176193766 8044

897455190062298141740225468793859809142285837449414294668405678447862996873037366863397510139100798455883197189398404205851783126255609907516425666609144857660683679 3744

806529724037099333962928343483326610413687134472596294417153661683256929874607519349004367548712450125173882289594264322061718377059516656649038896234159034283659246 7623

892154316210947396500986925708950750411415781971894579948516829239976768526059094084769255556032094730179889261822947383468868848787742147478211246290050487616242097 57229

5178607339598869641860539956912742611053799648648272882147298654479372705114310364153995043024924890389871904738048121737057256637134651471541312220563195699529710744845

4232578540931960703748062432887305740374143132382158355626714275687557555136182019176330108628379725855115674172305047190608736162770832629644295804827975636308237643616

5455540616980045819644670667810243347845988069248477274895298262045169437003711201912953531129197138017595577974532179706899810786979967116140647258355731385280378144794

6186458216347452039855897512317136407974683851455920414500521772122914466992786476520100365397889970941956779542290004143845487143488528556517630802992516764442476821864

9062151219172342568685160060585978089662366883201283965312270307465481821199948225388143004016811445036211672024446204828296777616016563789757634979554872551080910578133

9420347277448474876989841921828085630416492602991762303626322504418296296521543856287607037421868140047386309450159109132542103032561351107575582873478656260809325645074

3463372334224085585816338537153069458782692020523950672724753690013980114964316594582971648686322048417952196424498327948806313464620108913932870531345561503788769211459

2726850514677135599589063223865076477828269016803601306170856982886336353398216641166133554804037038210034458380815055830340179712082249390950385660958557139537463476283

2404217519342656686392559177433783255482070386105633012623762876981734728224250946153189070215082050421810397748940765721499083247852854595100246795739308411062722554 1

5696493892368273581434607727598033462643125982788894418184917380268704496038867071864770831564787589117803543082013186565820343540734229283474557696514986839150397614126

1336078948099755916482490625516855367948247405098464960856818891720369987375796439800116529527027723722601935755572023263101476869284762636285189304849269092640985472493

6481814128316893832831257956621359883554452066740895840923148625755911051962200050308020425737002899660124136355648802803399956946560958857632199260300046853975598028765

5583171070639975066604761486777635632261161271522426710967361840252910825524461538857766602779608089830283706877813984923812545171789875779067691651324603108755181479600

1216762016855436138875351111464644596594898628685003842938167759796191272999045913439604283622782145743849108066267372039815968331145831327755737193964762139470369487 13

4483796533672088650760949443106748938628101668608093548762040629531426836790162232434421625009619198865282501847807500930929896168789351440852784485210194972931491229 33

6642838361095835911792669732105032865863719619130649857332086615243198917751756133072533690606289440140362467357916861241907679730721538960992609147780039218290966056780

5157424539481270515827865608617662808876754852826435345792975109103743243148049050997201340093871209967992266732745697219975739749835295566344453243455703262602782931368

9388962967691490051117916415739641516223459624143879984997239721062591045242665562829601459679012861764153524786433047855814962571113956032515036318374506194258790732974

7990654033781293234354964770959941597021691810368147338333306415138771322151733984093817465683332375212452120426351494801795737064857482558812962411141464692661774781738

6015615569677680806354280813392622268057358604395739162738771435084847701866265316974888647386824309419601892875891202138727709615384880950656532073442058984978568214 48

1099344327143794129234072975479326476182962040361443641127465240436917542835856614059594332610091323144864164204976494795520171710865170698122416084821707217101649482479

8077491801666631807604571639525183860958271832720865705298255892664923127405067312348772034977998295609410636030516581681903848011147030423901820457583727316520859225399

4751093890012112219426665445908677926913711549507896665766765460962882777751995705545072979236662085235078168943400320475437404007621790091881351094993669431342798599 21

5806292704213826756214353405924672023502064258541096859551282959888016794748534882762322608988214260279669494883399735380911531026157275260615166467574723112673113045630

2101644275628278219148792466989753209783265292168258433047908547833654269758433077955719520001012078724019881349498443843676382704117421003695116901118016832699946612010

0860532094157901928897613978403516511159934642044414827682054550634184830616179994602704896489524389702584341717731903153309321479805420208961951250759293649016278147407

7322477257322019135045680559997856927754305465787984285946840858678413411453824124072065675598264826257619030338341742518485385403847037100690876508085350864021762101015

6728291435673677110351164397836344042830234780735456691438177047450894587211787839154166530924726979519526863928233003716850678762078775481783910819732182904787993291396

0788741768330818653181999406597926782213227134596324714095294630761973967499846349363609758067253661551807859814534953582160148026023317625201506366399391351428775115353

2124112251505706572311520853765028432210158406189825700470439171864907241208917145612024917300437993499942065866379857873460604806192281194643315629256867108796971234962

3640619373881121802073791598180109759080113272257843002501113788034957920439189928830051624292176003376410793371968133192067582991826078485247571177524201683493481941400

5391646393521827371048915003658047925976158343651355349438431915092146293081995018359167094253026540329803249676158439634711435324714370392214861784382826113866885521598

4613445058033026369143941743559917537871666881400452968934351987652723008458465501565659895211301104852881693941568670635178319221855955305000029864832544774777199550165

0826588967139640889880567958066916065806094048513928010222769761561382608319076033245484652866146494294839667733008070732006751042625141429624471453687509706878506600593

9402651877861032765470280632572990619689759188738667230511012379493292597649574826255195927394471764009255618521185772443088894589313045709752725867071455651423603418198

9031519545721886211491710345305965784508261868074364977358317577008647587996432274454895007809667119616215136769508530892336123866628348110293980460743553427272442810490

3280756767003372271120949128434487450813568822156033050438835175410814830753443420841220816836058132623457677542793161986045430504448510555800411679433767132055581470587

2720882536047310649679318479637352788447885205873182866006563349325602359088983537772507970200505414402105594610720764924409136337227897399466397512341178836631250900061

4162322765702854104850679744981271814676430841410300237525653730495276727548454599978716332533105061902402151814681001465126285103975983941288236986211318315247764967957

7744191332394798552871653023198698020983984731981788171333103443398908379580000513196534523383390109097044471434794265026285740315181520354650728231183851986580029362135

2243797543193801983432914312502757766754316869888602865677013500372589696445868683417647387839066544421819235857731078700231917445428714160030268283724049464360347876903

5733261881143101081321885527985897303450534403303722769151404531823618783217199889890550829089662419768559805783414287373064809852907821459411264949921965113612567773307

6994605802064072391806669002017569564175955272113593375897911604759823155872535644568257143746585668898203737054970452907158469737635558706092801201769780532935796783807

9502279220010520167689887325410838931925090717428881081070862320755101848004176969682629039239983938116236638478713081932018555926786589807098502295373949421754246962535

4704395473241339247648521037611777311231385001618713047106477839324875850063619911967787753268071392468984403882659360510854652369222619272403499121038316226297241144083

5568680499807460483713592520753901701446937391640928648639190537573932945565367754356329489530854791973561811689434694434436430308714442549106098294828815811595635629933

77947392209785110406721664480320531067091303759488434578734398473707653747404793080903438244339705830532695856299847938304808177975089019323978819644747281348548648546399

736790769039302521285919509594533031379751852981866262011761260953213926339182718256327583059118937210691577643838872278422852900912261251408052315081202726247737066716161

537297962365171711830918171522805265375933737558128234864296932266784713386959887691580950811504993633735690590084289200705482525461768954164710778011758607143286624044848

305523642593775798552448696080726730590765024885140814761891799989629290795406069165098627507033091008866119931836534781106895005532321232310409943156697571284321105892727

29075626652983068346126881743502763445734873130812787853966825948045024450899453850626222815657206656259080710600907194741580643428961731315157060558113998960765684277233

954812062465492792246644108673930170526784065224750410536043235086881525438218840578152295198789560649956069827453289227327038537584520927092429466734689593377789658067676

95128590449057399130794876253979899894685344867084276328476440980465348855120943606428893738371053515595879507510368199958600924794052205154880777749983061313790264128277

37157571061281736249783647450207227756195212674327358168549611969888258311261669505222402188114669306257495384708699586574599887892786847387198643837904804637462228161266

87127634511309478316617599707595085332574602849374001043645034556580449442950345318338129078508883338583786977108498206651020627957076698334451779345271803769114102075577

47743154293290326295321149788262035159874125464228843952777954992895647543471058985851590055084900569690369399463805412744078272079588120610950182666750528291004286440111

5969091560260245872117456045510940768469797368274814597904045521904841801154566347833534380881534140372398178819077576306472338368480766171878875254440731865830501186475

632030171398339007898754244110262777492594557872631516087487025048062038162606284156754299711008457236079436838831775697116071774760197736299860847092256124190334430386868

060751607783650278916662836093176759695530149368127979354666523938986549220821261327763789820294679958162439870593623917051175070504939244293712287520721004790036952035353

0541747026881003131427531174456246407354520013033515441611612845306363822062031827141203471057333057060956104199977441294378972333619529368071162946497417467460616194428

4195771506421244911540670731221384206414126967154566438915947177796949351958346843367832214130374310734291743734376354415907377380776833554545220416075583245001412712899

74101494705488647243415899129609228298624074551605891496310210005958713471920979723983683128011017526431868611183517017358675406492657915137405821629724201883751029772022

76922807801732353658524866103735524636051974175871823490873819774519960413516046880860827255906104828222575767463581916629034390705475970348090440043043333317413461423454

12745677987258923240909151087305920242790014967370151347721514257148023878189728909931923211885180400304976289387311988687633977056903190741451762975055829507905515712828

97726034354672225187519472277750347806298881580276408830585887321140899356256544526325626293042854399332825503295028936990777054909294707962200839029322144411265738208956

854344785225355843731269337547937659943069910056990821560314508198864943894886795977365202377638052694955587145427065851747444596468235269410568519337370049144862376065959

7957437429493137628495423747696298423620404069903223286254828223354201652282912884434215751270602021538317845218564841150669394364364463390329462869215001200331737223159

459937024404665464401070954637786273667690456425997758603414233762759258536312643708973075795526996850313206909183067913265420306400314824559862392657597573177591286253030

89465412516622840716337014979026738432530161901013729788646954034256945572630522038762942326480649962381630855003126516805447885568199731089679575544268392204851309190226

8824033771201778639860463980025603720606929534601536735130093516649047599690415348442284064946435783962739597969701199959968970550071398026714315391239146116135818340680

8760534667255305042239792809656622109111184778965033519003128193081404706478740367155552114034070303989072232339159423512652971711214491591287469696454455709228043473384

1013858874280507251493201836765498654426190687675030397956939024213437475259202844493707032198240950852874392941278159586475430366953365464650438122955385696018708146303

6000681022319353567758842217066271778752289539374973944984606881958992605790426632428181883297682570878308901643540546417536779752140149169816134993044910420427417299073

1837969851312455958606399199659668996108005049400729639709859591757463495011315239540543638424771576730579689978093511231012700060683156013470561688420818621059065843854

6853522653099408095506864518196431045500569852864036972272644964072209107280506565175900536331942571882619016852091109444623049387276226013009665098018150216116189314991

7555448664845101939640892424535185862966853588072370252086290396375135442408416767961062540774535439718720220389829258815048817462632144019324591263846776453878214900321 8

7360528840161581467693409724342496695967974551295215247541300383824175967755422515486890349846758461066315941988121179713345250927530701408561426350301527147370879790229

6635683001787990288084193893922491884489117670080380387588878016977011134533491153480210658508757002551736325688200975955274871225357182551697875315095569086898546483794

8430351870614923135734029631365279127615262306104314092395653522974932610180235741449400201075752924889589293245803518893483362322662110704722279517896431135332221513311

1281302699657056542366660712427360675833767835191012510994437030462907634661496449559967303212585228400681288632060138439153523932091157906047341393629732349275918089423

3656526093948313364810290643586311830825965878597847150234490787476787995667424852051040103999757103940220630691734742020896991756005288898736759396629367401720982195418

3371282339328624773171964386256653145512699222367766777081986434967998415260451946404590163957896049279139291043490275638172684047052298140890671314915262504417545 27101

1235786801298936828493391396383366078142291794554344916809706649131884537812025796215523212883988829483096025415451583015564562313284503174185769799799107895565679960825

2916553758612233838070069219576394198374261176767691005073575014710412739177835963479441159241607409649189238641623144315098433799999574123860845568790650179660465 90401

1900931064914597645550874416917093615978054671746589930170413753904682544419849306977396303361433004032263704413884245648531960009102403591486043419567188919858561 5655465

7750890144317771286164568621900128459460742160742957104583140046201246390110210193236887462318746390639051846090824746612222586831716989063606402540489350875060019353832

3584777977777184156692711224004455167705419310730388369438797889046242175900466660910401636227006450671672563298135619169577585613333903982775996522509904038709327898622

1595494379970630648070964941770800581227055933091621184400463589378563243586419106154006820478790162140445787717398102952607173000991217971137542433348822666186718059345

3500359794062610169455894879852873823946192592738300586578623690127192963826592639378195968777634491927813839152734685103171283501167754128969634017633688033476132425006

5479448355160024231256646080107867025860376099390800451756260090655554130984042735743005006687433135282061707299033893705322254670042058896404652393614283079185404 16966

6783240709559587709423200941561095553433443491385438840861082486242898596197412565717042400678767123685935372271091567040606219434726020409941395472013175524491583459442

7491919291350235583440418720697434588605383370185897657206225466863899147406171384409111405424448924181252805873784443759903370271443232078520464193147559475831429194169

7190629769790449882130801925875904858757010280498900927646674318174193127387987919086670564601741410451836547363921120183127264213529907507531674186041139085079174044172

6580092889664003508561829937247213684341314956929570401981300160608754127957466419031797332593924021074167670242353517422118285715161832976814222607309029636948713308807

7855566632397283342252706565073072518903090395022975145545081413444428165414364410492175062270643628610175171712048366581497058246357800755045626453744628052593284156788

5798506901058045279756262857220830478354366813133031723323813526470752577952330152891663952865431899957317457801678267281460222640381899566937994842421098248974200882331

1140013410440951609308313090546550315955515473977480221462406761105271613757998628354396966578355245670079360497518767958500434785944448348747345525999632392588201044528

7895767233391108520813799484234715318952612818751089051213546549692460665576734518571774051139809005074932280070940569206554428799289768091385328823923127426496379071979

978524909030460958502032813011881918979875386127705098311267986672117810060942886033416074080204485324414144579454721054698929166499819415997508117083997585525312534 7930

7072377194823738336760655418502113337357535711604984088630696264989015115562982779223043334984493678391519856268504325200844798554629691262997883130293630646337045033155

2637520404739224157072857799880296353289006984007678189697563540217661942944247537256498457226550687359093405723897937819146080319827113924804979411004922981431 7594991

9931032808979574725337681460617454331326448924803701346264266926317342443574270517747565067556341333600591783137637375920389020426517169186542244841360946598636267754333

2217290972806772121812294501776642327167331091927523377242450590805892756556434411548443888953213270285605406006435240340117743942638314926942036676773124923344 60461527

9222871174737320682073795040775380702487513973697487220807941836272457926715860587568437382569660152884501593636057997648795546668979631727641445826714098393026058444373

1183219527357824992380530460979253217595294376466967178599565547975260104811160900192559877031603692890535464219693617900985472078551257259753257687803418428239419 30116

7647127802001324490541706871940608748642509924900412437779902567236238752134487546868018057002677165904571741675093585374663278466147170222912775381648935814037542041316

3179662046260168300583584027808425084706102856321467634921644155965653995124411152722520988557631807880862837444395373638662891394399045918693562979931691320743108 49123

8782696708173798380276007328352371305703835388120192178074557053921312504821976693406394280234533509698054521919641649696642051922322293325224980990680943829860823933868

6773954452673632194440159865904066528670206510049409786713024508960506363114749789754318632737637932022795300108771768440261821800385909060770293459786409329695112338532

6149456558596771175442461946208674178988143477802702378918386547507367250701231645954210423036825301549927292304970650068874455290889366455148193848056374554470 31971718

6422720965870630049834366571830309458525285629892546132607901723746233601382089476404721767102107836353063283918528942630970835241842688066639724543519498745949527236882

3511064759383705313614945233262990060613544202079008007844359185142381095406246359288007491747232601428291506647356469491490753130404119487461275842517254229933765619132

2834413615968845123050979229080934778061000120243079336754606956718867847589029167162815104791098481996879505375476103438392892981834755913533728342888492285393965 95016

4594229684902163576984603656667706884979061493789586623897851395030195525207115947916243038057133910441235127977174258949971813208993972409457630504381765420237749372929

236664085826356304701889428471366217962807794758141064720396869005733588378323839385156436769291095321263095302372341887763775951325585719886841563511434654449213461836

258200177391119635659736209174802895107119119312161615049356614001989154067719147406045020084890078521044898407155872491318142412374531473909585928549192619551275281 5404

555548948605304395183055163865296351143585542678957884332247030322398462969403700363866705975518962282166849472155167994010237260527619206175045604966371707626384595 3005

334443878994432854366301464142075150267652998714841483859242046843515052858926483541295999641906383622550061620298517908079995517161428927433221628069635121629029650 503

454559800242920380661131224998757487778145433349578136558008300458790545565523759643089947282941565846898062943112725975546930218879127310353002168642276336610318905 1108

633596398607097473741955293417850780136533786877679115147383325251300237191023587588679803938552970499832183038998533373533151034580443402573042275868260972398342231 5017

640803327317622631967565989772971839422971652277619673408573444137475914779317933398924359940581396032281346592478558756550575159428611613176739552834815080851852071 5479

514392667287510574413867697189020877611959245935929086383969620057686509962930381814549128073418097204032036336676649944391993264145227145073503705907609385752440000 94

748229713623772159263083602213915885590946140740697630129708656969066762442186183635520472790353317530936077977879847080239021859587448789607452374905628374741893102 6806

280481743381300198221277706092184700815252327145986723437854310497839043649305860764535756025983828162540984963998318112971881438639542654408280086193057291799568887 9881

825724409230860777086263513136094630976774099702847276682966685429084540520229190030343224718982049938248060751638727945668940849736665162281336948828583339531350502 17053

611181750210101696093737288217468389261806871887211712203206733649831281254572692763105648258790106857520808063392875136481758375810969559699338484199106269224113656 1171

129547967778123862393493414008547053784583782851512327987308840937357211004586036545449453018681073293761108679782828034366133339780538614863594371635008771193119559 8480

218289926777976940913930849268923971610875162598136464279519116303391796129190017609532216557349110374712094579004186028992215511759156830362494716086505186322797429 5613

651983035189714287602237507160599904685227028482420257009629938279147818015406641691792596965023061674967724784194741420092429772971388832511665522273033389644473052 70490

241477275647154092376806641652822611221660551903510495321695382999570174114651029984898251643905699093945986820498552583128764456838984342109365894310511407555838492 7893

699740950138919882124799162098883924355873655545237536238906410726631365733868871501437667657439282983207346131403949526170953084489658005655846309523296318712451836 6166

304277498527362023490678591557693622053472426111253263891430252893766737392628606460991664258399987464743414163952677732246933313408989228213523671609562825134923485 9268

040735518153607196567395027069135357634343767443117249084553776703266698814491148088848013249302584381770118785666357293539878114064658836941728387365708433757510447 991

235973659724344557427183847336205164098603931021959212112257203436510013963890649452967142056089086176982883163882838252297076581896118954572982581073394540172774978 3454

068776411077800440742929663980795802668931294689089150061861841892185304164332216949278213392118277190216752020805967262749463002805388777945962185683074384298925646 3024

089063366760647389704968736267734714319346437826952783760286146583892789336232361603686965888599440717090143857650085623703570747288123004277647474703779463200055437 2747

365847240261839025081850203994131095039708124812210776128324595639956407303784228294941790417991353653370609295358041284490195677174363265587334302844014814990754651 0328

181387821090721433983745415095721808723321639314118488540488247613315649935403031131319714388566683380217666836082950323604059513677592715516567968029585973380361344 0693

0781375730116130026579702426559178634319436264662301868725879630575563660782899695363498145223886600730147718791986416066143908057772551924487070829109767355499112006123

1753618478131765439557295038545292365366941334856217878926154740456150452308853118438978335074806994480208158304780292913950274218678869198017655546818144544307419110229

2721931864440749790317259939771136218099711176146890001377240700923484996330832030429732196758278940008466523507115511183481032014483529474488518863241330396067639585766 2

3927274353864765533259261139160105897206949121604194384362769225020408360183481182715855343925553745823628255237262533143596996463666782559338210091759874444027185225129

0642515745359479613052718599488847824353172256275413109599850442774753526388871189264977071705502201056823253071154389564758303122551162876396883514326272862981524075 58

8799596209804394659688932951904490105741914199814985879000053348961206916311175468253485829007683953766264145205273936078651380572241506897185582776538952151358565897640

7301488111337386924588890982276022973395124413507510038725894820664785445392905511604925465568301792236352637755462684090494780037268471610264950882620693574846316439 68

9789627008337630374301756194573890788481042430928552363102983551745174544660652976708199474320599516529159600871565211546129673541395732775167844348486459833913758485625

0554601750792209883587719338601394896751483376380213903415290428362645376374580878281537904708544575967614210377236123296196402292022895146968811447058289309803570015041

0369484170547286922751970446946972920349519697662176641436203210987497172990814400037157392818915284083197422943733677477825870921871722529840693055569352258367862538776

9903710832987797950516916962995760702662487725611153210245228717970309373800563354059293108901874004957513305645755468645881925344372271704841107604583450525724291768442

356100139456682456604288000474072926191657544095336500054458333133334866581972749276001242323822517184689306940857232461554239281388742202766169379793566344104503711559 8

2367577143124912796231411501645288044094239005173464563780362939532778780163044104275918642025167711823591081249478448954880725855125745496079956918012971317205403842 4

9096087363621351310975286209862242624187424357837677291264179188013767520032056332618924101861651303481024400685680860611048112942740900937527485785858684092297315779366

886086199644290736157490533034510746792139213935734146937826784053313199235443303460719515520570066100116930106114645564891680500914500556137849404543693134859106184193 3

0189545486385219860080820040286332226857792865947499006936075105780234742015035317373008364949891672893509093542120975207472725043666188006133495909306114071104645992 44

7597175424019965407306540841697352392504156355003902644002922751551910862941367236000269412071278113087636531615224433516226691901231017136295013316980770097670312259233

4099542352764898442089109243902756481617004072173927256602429583715571067168541418713003571310230544725937706454206437302838147841910218688218466567138326128263780697530

9886082906633039669846839624674768851145064913151861552462947924811159873110907977115298058809209285816227706527567771953931120357319433459343473372951899415721472276190

3678004308795904799926642428196201630098837148845043980122446245560266040861633199723284978761593171268140040450561896686909847082394137085518132619963768897021241523 1

3781873313001560112199565703541410653535638452439655642672721743450531708970862034765475867412814640719792280574469540684927959945904582090187317658661625178733129693572

6248763018274805306560029462418101514313186902408749384112421582150837307412833731003222668395769073506988176827548481773049953913103184653278383865626717476001806278875

8049254008878403927985646496451552789273450200154100313190509627108096288952376898272651391408194511671959166631818853731279822794780963841863308123249343682732708847 1

6848408230651068049840198996614184868219292371243226293284424830236101798391004269904077447991908376102111123906672507192997931365177067316450479862322551979702992566531

5101596604596690150887068882982527286405989514250476556464386139713909302025719458558252713927198113277588869554462920605202687675221367966827468775874528760778634491382

6995634800825441441318253472049480142126543298296784668605437790613389102060765389746783799090419822642829171356980043472466696993015751149537152043740319181079548689432

4062290945862326245220966757449528566016465787368842465402656045769732900129583787420171151005742659749253328682586625702458378115218220127757607872278753625441767851681

8491979949497649100036934909550819450205563811224964796766249650785802327123689446228669796319715390249901099117672053296592010243274158366462851035194005414571907148638

8246944690382245688388500782441039250163597153748490569452456053125403791760331016534775001998649580583553661420716997411733105525404205521107731858104589461046273558070

9471466835287835222442439519510959648019339972822544123729119753352333978820050032094830778066283364606324667100180087066628897715761318039445308517785997967916175623642

4579913187479952951873675602067243360786278316446550471333425577456220329705837065208461481461803279556572311289137915061078782367241706315742790860275826804832820482530

5959448653553053355736089436683787788779088357733165815665640463336311789655775538674513596547437928824432776177665299775378844321226267587896126638330684384900580057761

3730946043245733141597876165553722630161642335345100237463536829894247824255806480766433618052377415631403789337126999008115460840814240586928446408742389124577519366466

9946373591584411931779500858480652805204513861789723299109646117709762971698805474148640403588839279500405680966882682526783325875358351600505794585314848377702967618326

3606491366056471185080491635911181680573568625676575748362796259542314440842686944417808465459001098300832470127327673251862965281011987566742512371854719174196446109963 8

1436922527648768652429643328488026710488044888015591064476982918336443256383798347892249924247347347492558557293151861103453413733567227462578276718755128522961571935018

6325172175999942277944125127694916659641176453311307678394358755701511268339788077823089327672921967390656501679098849598999718362018377246697916468158884004015083264133

9017024402863907008831066490683497676288008809713157726433416470525153647177306613927224056325710013972998990955937477305596363485600615984961253518310745042828059910113

5615276461371873230740548644387095103762391293174413926799644747323618213633118585804069936583777606558414953328326602877854696894300229268531019343019873705871735821809

8006693891250766257084746595062899184683469499119620505628810062352434005024075121256597621835683455225766840491652515707584146144132895209700930687290227163705638590610

5921696945735131229699292583567531883445210953757017356326181664424591830719173259280537351848183098722945626217254044418986403975038451360611100621071808868929053885565

3803212319776645007978808922913907197183215533766071468815888614665937080218118486409491244157801586964737239095958580311735493963934232398121883858322690622730436915 47

9647732903620310231584622821186608285896081690940900061896442134617344682521433863060864107649130309630386061561226947756727056616419832661282955940541852670099389441 81

4526699815121965396719051384313536550321313806824245173488947592503124192484175255740381823511390261635537093686468847101525986682006296660433267158847028467252827367513

6369158934985721514957696957393793129333387868587015586438472121908813119471337087338232750005623992374477172103479216899958700150469805895623651885426829398566671272305

8331747394679893879179844757263966997256515093350494496239329894118380951152202738593619916208931559373521319380127029848188296824569246640158910245224083340735294723767

6601871908356625734394683547048362244546199371292199455216077052265379834751066676946325551156649491168070523052817308690882682380129412541814673058459343581273433407463

4710981016973378451137000361466614777973756676762211878253942360370654923722566475192700250824888860406223409811545113334223901773684113599153372373184676634050715689668

1938103585480790073996134538888268575672435045918997400691044704111628786526792010616132471999844823715233499783637523014313513282695539529010864942058186043961590530582

5975400157347529974982723095387057721000539641886970487452897359156879270799441658104944268793902227820026173884246389592211392638749541141959433002708467142370681281377

8229848743892258019606732295576646222560726500832043734636892069742573101488777832814597005506211252970943955134820697067809204578900900556359931930300746710425701791847

4679952016450985381510153969617354552778043062677579487710979913625936622349370648370598168419144090086928384175813696077026265262637398421792751865585534001802494738842

9507635936962516581598005490110797072695324488613743499388440836614685920902013874629297290938453956893091552470325456494848342558435392750026839808919512438571472788922

8818004727979105941649371664175676509443374654097289014406328130189143863392806334434942400260228810471699972553393957076410706789505905241632902212917615707202813379630

6498598823224267102982642645482279327154804587649921241321268182767230903475957930311584824894830141731719343104663561998293422660845241977274300089475751906443642507044

0115738131779480959953392691557988400547828953652395636741657296488046346305736737117215809990209894453733255406649244555657047977207914581230450618880669347731611549213

5285980811109640356420103206503138783298144430856387206579389407056232795868744608528406980628390128319940403175369817291011930274216487446018619632159446845380755709872

2129647584261058043710144144891074881337667213835454142478713666653871820712848476170700280230771398620015232852846749805171600941770084830607816307406741291585704585798

0914354160929061349459709688925710567910055675290074750437994633821119211999009122153965563172633298735935838666500189702103768105655391258112742565036589214291019193567

7400796661271382307140818828418649325456700504780902357998346296652053903452672297367971122964757638427953370303079415632893117466348996286910518604722726888778758797953

6548113309718525774883625499507809623311682394650511685470862613640217820445276226218509468771458466676588999479371028457027858288649455781921024708840980548840494289

0275863251351203276836916550933375756877423110361610668383215808025643334645427172202495621806059358605778368398254618223644983354199190818175492396216871052804951422463

7589120113615979984380345538987436863794164300305130378895831279248454986839906586006407899335281278519409840167197297270699322133907184209551782475206802684636165397716

5123457434030443246614781771199610855372843091711263519501191538103322617009607819792294603552601878766923621248636248851290354428397379232513895550640142391307665467

8114524402470683765280641424872089134513796385999449351608677107460143274772385102847494666363346194172301607736297628897251283002580846877265301516820292508730013462

9231565387199041060550741930363390184442397874423384998260696760570205353684564642727270348943923664845900245979494739486041667113357170281209226805278156883353132643317

5902994653857485218471097204771824805672156192313199662763782820670627977864382558087274035538875576372258299905067359154147149474372649839787057663311505334211612174

4089654152155497746247888629118303526040368732822025070893530843523458081507195695889241260528757183964963055076628600911167261753007281738884588123735985372692992626426

6600217297690409322916645780080286157310501383405996052151802023374674932941095769139999676638521753746488507214642227683648609183197332363921592490390006967888121011297

63583734052586878544570222146208736858727966414530176263354155887940590732122253946707378265467560810746496041804339579538721133064646799286122948571393385632976161785 08

91155827661197902337999866357704749637968223993509579545082055051118930347793570244303528304428347024105904612246808113753997074287434351207241798271000829331913714192 88

77140989863705462711361421706031603887158734107562660346260346932057574636326530612059614741096786436632812848924621727699060440035648313727017184326110762869070629628 7

67824833725218167849520870187388883526681880688561553821029179384688125975922387175756873776636521727918293598889124812904849996547644596555459515319230198673421439626 90

52453374634498603717927205427968168892955587945755341314658812833102455747928050200866695716939577801534143906770746884443719972294731420962430984645053185396521906026 71

10060566217145056523961676291582141003930733389291862567033371447241709240794482208195793496981155249255732540880883164819519948492518859797181791650718864975353694319 57

63660262426172292425480056059572174815355934092538283243334477794234508946594682954015616400884023550373234965498786621710766801062510274472340547773872282337063244223 4

65713099833535636179045129664535920772793879392700954660146105091802732697555135713654909405170986914333834373538622395662531670508132212346736878144276185478830585005 81

07851555678807693973242122087306618262009083050415060798786720780086387483147104679622180439475755630990862442443828090707516360392136097396719349408198200518930846341 84

18513775869421385970025192357210352359781475656283700649893580619427747837673671656860440124253539425460837473462224960829827247240218753734151044388427140893032903966 31

70598527274357572249198043964068936908330704606903363403761135669272008017206016525870166209246565318321783590348318466849633623177354463039337934892379583823380148352 46

62070768884177564682572717136191483552944036115796246825347099957785414816484667357356113380319206582213549678296294583794899259090657150858589924036877247095602252060

30410594547223573430761992020038703424402223490949671809511947981181231766216132812657418887926717804023857800559852923256156882467651635904834058800448384582302419984 17

62420397502821442033232378136469561291816090880705226927447850235794371561428549610330999701394772146061745007882475417006791781881337307355387867960101242192434173987 3289

76322809867622937453437289981172593008222324624375985400183726608738326471207255449130643364499510019478254452554256119854446896338619233410886119023663625200616717734 07

26844487670870786339928518785748868906955952057560806553597236255486657680659973002696144997913863949137643343951178186561697245750119555271398766331024199364961593673

42333676593518995108210508545586590240452439501495657097516880177298008199222597252891615283264328713301912072026225055993021055200593642720672067436580819591983894686 2

41507538027516566228260425584487236963423152737049647360124729374705823518946377728760858627139523599906922325870359910709275360771787312754815094035127013817079487040 02

79463643368842771692401282640444753830021680605559739911153275674304250791689664936534610664903033926454798262450752752970355117029389549392605026116732805080636191135 04

15038722553548052495030725922083212991676993938578960521919040226593296893201528053855848832676736575685837994286855431488484598780431993710784840893374197790800338636 96

65963270004800753410733130286958286013592876613508856941307268952706221194446570901350002850781700817329693606994478080116508997746983832753354462231178900414244561256 59

23619067137782218830990126205038713863746110754713824333342606611191124996043119748730035578467538558093194053656414387240871593070280022336202403420920266924841036541 3924

60325281513910602580690086792469478464151377425304908113337485925659032521084378705836901803059338532970010969600087425044814184589259856965345569808272371276255400483 79

2707641017020870006758524443257752645790361826803605262387899668752686845758711349482617027872642077405327791783966059302468053761252878362242163181476420476433345 65869

2429156194617414792930326727453319879627590510582556439064127960599605162941055835770035365632428567139724330935998617848544097181817255447779140932095918405016499 843861

2807378871881675478875650566319631976730470586489462404594926976645328510919874433735121566448881451325097829979985682830183029271812658757997491525942142606638449 347618

2366943613010007783474504544383940594638253164174696216793754039491521160083553404587300703416744768853863537241759119191257302962875769986698306028405512542557781 3049

1965735370810975388980514498281958517209632887924975966878585576268728363857714282335234665679589269485489195448742419522285402758101327257258848460465495185162225 272148

5896972726328951526610074191959717832883659455976857057726284785595448837240757916288363148490647791455348726575585011194226486996243910900959484219505011828545702 189874

1034571838981790486364649082967773150836776997335515074170081220258053885245195363983453187617781231829233845161492018946787202174802452815919012422558651698747259 021550

7624974912267375945633076166021194348403239799144070248817343433029272571092865738989424064956181090797654985118424771133900728809299016301869411421261170343722496 6747

6682598881833777489153001358002314602426047205527579931989940964316114429852831611486989731748642308262649341631684527801622286945217652487890699556099151096015879 416910

3884595635966852936125991245729283769357494996000637454051029331252353719421156501331547537362809071428157731851185927633100144847811577305157277411636321762995559 710643

7427871640408298307104630541915599371481539161255478112643743890397452120735757677775074211505082981008573752383518383575399332029759891578248805041207059046440732 276884

8308743534512264547062694097544451369757257089150573023242573567272108516847067390111372101828058046032247916600738326914541593198292432543374648604963396534224817 293832

5475111403759384377805581002679023593384798958654860787419414168840730434173496904240691742895829113388157394322771015619624776355190240217127468627824721979967626 629009

1917695564381053856926401859147616695431940776934896555906031303415790944551975602966248754548790911175326993709371263806722567514630607402334459831482057807785525 381693

4364805356807945204538688872145805202281371652698201162506165729737974807500297233921909750122049474941700659392967296028738767195222550630864385036022864184376624 00917

4719032833908399536747468613101093275450853701032488164563575489558603679918936112978761908356737312274823781827018531064263096134724871436054931890377026133291202 074118

5702039654923368590863272900347872223765989418631895233975752644826232284678147416783941758787792841341122301988055483719196620859931129697830338865167585446227913 405384

4760839423555344903316330001247579965276168526319453293095962731314672619266181983219456480048912740421657304136383304222664897851556264565532197114472973260582132 1486

1528010097276801510402989652020638606860045981614285237499912085209347292307977350301533905977647834289074748778151734815736268828728847309608641886645303294760075 330935

8538027704926007328843294119520864829711317593752534438895881425554838517329528360113779153290113357598117475908082955904750657658456860498951986993050662506017097 078329

8760630468160095210972977875136018632054589557818005958757171911723235091578713751539912551505260695947605933157635090917977332083361368071945515640749533035718842 256369

3171183439732516057365037747321453500563595638262474936382247058368475214260727919955251074551304244338336405493970033337134880029978594657544942765183034121119702 102998

7336199117764793004764932649152190991777236255805271272577992841862212722025947295783642415695183890426186293194985023980882790182277086708707193539801835763828047 217627

0261490249184634025236129564512600179754496131220872848373883319385740201890828176785029850526215633527508339394101455547212563763685464079094647653816550805001796737374

3140990894748416914300650811821039900941917142905544287434869177808284127163283349333738760189805192382363730219765007029919984095395364081929393544844337867252570777295

9596138710071579214721837075804150054413498600492907499698937903488102082792506930057423601746771263982504447987947755138368875388820775721206353195865003008391065447149

0754927971155721846090155394573325186389812858246877695980082741765554992556526378718474988706329149439080307417260367589680254871837399996196829326612241217067713397137

8809252017026230147178208006391621359820529738550555820959403332647089156195552235663806264125747913463825374925991288014312614436201171810061047225858418500286341562115

6884418566202827276600655362434165318617270547046018295233295364896057333074530647300773945817405562218010296586954554296232136268085193584500258735730586665195617446371

8111344775636102931642292999771284847399247914997524697574616765240133398871189935029199507259403541776788784275863173386202123143312232454999521022646419170590206372156

3648704410426983363333138216958848319812093693683591904114931623478727636627592154568410702417405297925694298191498174063952714478690511842343371955926123191937579062117

8580909320588479483630579561215601056518207521648952936475049978364259687880476092599970186536111131360484481043431372673271493251764067795912827041809284099302148095745

7866349377922712137553712549498364613219105479011950805481637782317553188054048344745682348295528213063830359546479755313386037131657640783340885935945737671967408625251

8097818178803698601166138834712999153771123834154528740489956469028260300688542763451896235735546181582221407196786684310268265573811519493713161823492530436547791887227

7303957729166760359890682984979275326445793020562500419821581783367975832458201670334401637519436130793606687706059615507458187300740588554185707771293765395461112355201

7737452675365027712360102626371140850249375451997238811849720048529540760537575504863349851760393434036589529608607344805531229553335688214567118057604758841942058349638

4542165377020226288732032814262719241911346980715350623640604880106101761396430655066646671459774792751275013133465967646396069944057031186056087812280632896781657653727

5062967283576263974828384647295018917985604824919850079916092323997667196476783301263846508088042831110985025546612968618556503500123610685297435664465619849209211012663

7583119546240112661948930083828438659999992833337948765982135588393330975965394351687477025420380520337338231789387828254304773685927377235747888656685873609256869105637

7446851131559478673651648492178603892046920573921373965976293426617993875988610557138473901586953800144003377394259635248692636896090870539526251096127290887376798222410

7476784882990262592141720651354432719916459983330333820509702367037918912977711390002179646455681701380894182584629598768936359243937980370126043700019545375320947585656

6862618691377693236555385533736640181142604011712634532053725124468808392504553806425476628093450603910861511948831424647739359453611346263253979030531061551540475704318

3580698889116850882783575406260470081334894277564198811046150339108196697436038560732670871560877665885891060896072087471582697016905626687199268158483351741024109506049

7613332210304683200931629481966641786641089259633540386294130152476404151953276182470723527897727695774543149145720405417995318578837498108505057157671110581585521670552

2011002403121471715798464585433248907341098761099295649676156544718806244284933719422274740449837985964755841334894107426083336152120775019298015129465672084211550763881

6459889661764643697603289243245105302982505122426807037312128083935122022554084151320294799980575137686493365567616784994713349269562577390784837182482831556792981972877

8686290363105606858009907222402153876614713644801965614811245388627165423334288756197979204855730192999750041758818622035508435269374224183477542357560534672554956154189886

8381785602926769085061313659528803917355560245687971772310603217457604975950232256294196379063093795581044909587621359167786582953930306576530923070439867570625760671427

0638526055475959525321304780063261071076808321621001457946409776926800691390719372725319228526274289573895041376854774592960335922726252666683521707031894962827245652845

8241425460630372804077747979885412946353979992464746913355243372318304535384890808089315251813576848527285891732859174645036561200688294705032047169041537686780019293050

6366957785508855054236989012229808779126706105235629735806022201829431580735552190937586577473626473699288881297829333934998697735232413759931554636311929820706537274 78

6072589973120693062721040157239438426087560393263870639290221903085890987772201985593853726881479322882922369825904643093397881652299859711143887919168112556374983131611

0931906115632552892612058651598514939761270556240876767140605906275936789728632046558940753192715912951170184437557585352369782060346030811140856162220429042890528709348

7193875366819942119678716034475116563217044041605351341390173136689463887385553138636824336997598597061645706267041304591264372849891483568904556090934810115809231807301

8459984087990904615749310986143133159197840606356831884195057075962103268508407539511046071367743150631865568117504568429109859360948634686959367227758077306072883798814

2468100342685874419533203422259222591131568718551298843839977181848177575276528687274786799756095598144332697980232246925174800848043735402673868444648250945683719869661

9833088985878352579323281004784980000165924072903146602815056472411034520315765276577171450510804603051297596390336904878227083901331040053851493735374972951613489722639

7902119889634448662018819029576929504346472305784526520058067990645390049554274873960333111513343423239392815392857552418925427533689936707673603270769534071539778317693

2998580029024738091222702470030149732148309934933241880821118256958623294651857563689754163574689598660266517287106373178211544073283084095822937176862803685645159152 57

0329027569036857129883127811874734596074173100978847315628386494861931043501661812266303769593726764588538380943049453023030268014210975025038907214842460093398754399 15

3838421377545972464098687379266027941662047086632843876627366087827215003598927765170744547706538396196028343102852384091338723785639795368257883705830489472663481348213

1719088833963367241231536397295203799561405420265235573318226053603015161076727016136677534720210899524060190190731071671157213153131399108734604994855887930555732907486

6756924991779147777627525721533153059191543757640208556243114944537254595680970256475764244423090474070144938720093148556612673864189942549493136310475961893303490949930

7284324090098660429647764160636212894769517265674169221041267919762026291755853059616058835981509438139888155464739539002210859787185924059647802767889239242804773232416

8011508809942907513006728614972737850416001553809727869101165381637602995600199876771052874341796486349487590228434508102484523242850619456464928288338024674531436007 66

5393932531690693471534111025909155950980999607771081924043400817401909049952241694593670841551263350446837423540829126465380354941695384687191594786448216907197188279045

3741758978656539635436417964211383323912726608538295677462642204374861375086965603814411544678174631824157801254897625802405672218165190255256466551041784031399315527 34

9701282746407837967734310395750011676435012323921872173693956157256120962946586125817922599712293601560483252932466059000746753828911358876966050230432754644157727204135

5353431069230209904095882802842492545660922550473678663353597767011475477937895122163950391748837006069208321431310565114032165914971605450331526087562443039751201627044

47566549744508291084491427532865125788432014337191619507424345854267127681102600799697732731091087404071388839859302056854770568128370032410609948808912037233751569 16771

294476770105736285175269226738673324904110576188363433437399317405736193536907770580699187001103875506825865123396341929847330966787573203290483700569033536216837 2869158

682248493164586413099556128076135431583947976503645798442252939980325213460972862269536267247076289971779632763346141120704154148305304401967545816235986063466572 733057

403342467568253998855700384203956509771995410026837628297511970715692877805882319026171014758008973737834649921004305707615859532250733610872957027150743122979203 1372110

315120578694581824201741832056515117533812845799817329613000972285911308282090905331476011967818503836753034704700057847836099759090912963034418276550511984294261 1742125

017453108837615277210320916233088335710208577216259509925298643641820689439656908564775124318290301835339510911467513718534246305851770744343216131691304544562072 9557791

498890485478502949425186992301564204823672996782085432770817139937297136472855162369102809439490498095711147987353263361086154490921362107195786271826589846454595 8700906

924924882052343511286873862691252933569556562440153334475671624094781182657115595475669936842356249997922772333285678478624526949813038295767158836825390348461671 4968014

138599194055597917917858281975784812372478022962734271327380701712131593454022544168614641620641854955622017580271717419329604030724285575914037487524125583648684 7826530

579021129301504600930097911328939110209284222126288743972398792999872217126802442695704364082691751239472885809766317352190347740207830108250082306867481659929162 1420437

855969070083963431749157040070491113309702304687661585748313508014447599285202072786040624690986245818371056631825492066663392868941642231681397853741745589835502 3981413

476275686616221186367561134540185061230145050641464766200254793727370169115091057005880583855287751553568346135550888143137449856363777369433473077922369202328195 1260198

833485319308413912969210345115664615581718451609186530489711953801102485257498931586472339992674537252191487877997888075626737506387237805646976435268613067747611 6156403

088981072299006136202913855386468368424583544342072490652694313192636306455791910328174622465230508681145392237903469993576181922838411783111273426609317171605472 3027485

870001047866059835368762042349093563146793544370070867604441608093430388964169122938462935021661100210761640546614532826133025098992955391927596299462782632632116 5658743

195517335942787247995482872278107931497771103534255438166350502182004755984571947076429678271587726848362361118065924451595282915230181808971672271763496522837506 8073131

741445335093301055862157197336759105167204885674541572816321725939792701826776592787907269759586524444798627848766953949146101776057760360711075086660345575557129 62345406

637758448773140658050218144414570121613889442942543012726143996039751548809684175388778709977105315689605779553635967007806998565011955361699581910918533374036619 9906618

677458653659378289515861921683583853720551718196699002906225244297196477607657921208349979814831084253380066460564654628441059597587010538378376695134144117115765 8015291

972393283182374190724182705562114292481259500862193482545185655397012584064777459094161077898448667987879836035943067050826469850650965071424287984166501330336475 9597132

945835690587596970583659840237526455951428415274309347600284805973744511548230400857745381944142354918783809292297831844140223844361123221688505624335418588432511 5447206

432849620845632811941082705883189354288454365054845356330088426685693563642890202766923084866336118299142987263879880682998086123949763295104635913382691252518794 6694508

941539649332734549729944898362994739917547441647197173177987268394360240105216610149815265541625403854517795215840024958798797410495248004753558164544116079649674 3747671

52

84221183581573767370489681657618646684473995745738638952849566518957447866597778195075225888298702478900964065318520474237695238933550121847859966007408965038385951470 18

04072345776878385607580956164533921688489754259830599175376101323206354325344240488600003090822619003730634184868861438763736494178874012048260950512759863390509770242 47

25298017588263922938707936732522111670579264414090854374014853045902503716963747745860719140542569438156117014437888441888309159229271920358412987162286685053246038943 56

50023073416708375186459536802527582405209237446765733512706016011703490806822327234121408469596673325161565758066590243101303206411537511687407756787406035925878861719 7

36349367711142654304847081133303231866339855094943143974808480787647767832770534880159671410169844356697808454878051823199575640739788317702711356439242044520333007609 76

43679699900409585495562013135848058753749472569340330909172832394183692193249151868723547739392127561179466401851180013807501027772171306420425326555361143239078820350 94

53770750843488923010206936485172849761293833257931632804024023662247707358488505586196021481895075688961464986471085846445373294965523337264188383262127117827240693226 57

15707864175572896145338291644891865204955272952633002810498231098573394308160225669817111505642180307494361107813613896822048773651856670209197871094272276503470633850 85

50084211709404050825699245756282826278137513327080529455232216084540576543785400717990812768836695374975228640671461534564901126938742671140362151382047758754942856572 27

85336658487290869174951010237587497660723016951857365090579491818691542049514818950633136723233600179192443759401641677198359451069342721729348371331527082522858781476 4

49540661682660663281738590646817084809801956309540191002303038377210748322781390116820825823892779361395612062162133915786407904096277774306239458871168135932412443371 09

44830874229948965727049696689190976787295678568374918266228075947073087639094291791846467289893503816657160323834130048221490735573101147560439107642307049971417179272 24

98893625118537718445653611243536680334158347109999781275045931072949201640040438736891084890000220658968949509883554543303448063469068362642692622526048050382229656658 56

44546381725787202422393060316745016053977551655424603074325691453841406677000933481726253378578369549688018197142075830479025045449329434408065470696670920819668718095 74

51822379033311686660106588546461622513680755807281783990499382032540352221479127873573379240505817047934361116046575203509649920300943063385151557010396543615600425020

91754083680251075696272405400706130739148399782154975269620067771746125375177474080770421469498072465669210313803655901391446319337852495607651289588470395683600524056 03

77322664848897675986472222368704572600251314653302789490736683175428527930436416844913090148229779444145397767000504764545394419974425340090220649707950657786676256257 90

41678795171932282160484279042228145745555525850110505111853205128248170449340850065111058596796611348054315799010027116370414625588451469531501613765309863467935139830 64

42172125391421048484018069955555893386469844709722072927920441600174464574485789885219133254971330254820980219920946867055130885041123215989403060607764070886215302252 83963

06106149844929747045128120643925095268393316301653540689292805651871572657874119402174780917279954187411811137373534823204924028544437285424144786673531720397284099921 075

33852137685218992027547637515508803238203451410449033687861055113974555644534413352805893314950724154536504253686358765114645577638528618422250037354433860841945720257 80

83624670516135441219360521249265478557979011265815919933225542147336102522035640035827908575507305278835431594674179374264974074094794894477957316609623021732397288402 60

16215508990745102462967183685916037890598163574392667278295029918179570280686365101245445154413181429654184524519788730520200288020433895520952126242506820736251646482 96

88831505095970100022643721353487858260253357898428499264259849382698655591574552277223044783670045129262032590728447007071826463942993971057965049240272151309090201632225

78929364662069079114189091709554858581709996939845824188862304346386468537094692019086644250014237049070605479440163636224484204946141454073340772056136753779947174434641

86961441635564294715919709591245729889392338150010412294395852881242903163818939118293640475674801320054837776422413083227337901680551345611878652637873908460298324844966

77767652671446090984272409221944208729050777247422712849199862752884095453612442608122367302636241666463676956582340509347865011435452230172110431829674611812712477264774

75584183473918296468924243908358983041077861222164667413927458084410934467091407688908115480426990464476617903706913186431644872934811624753142709479512183711189543080160

61368674233086520685683926148047844566474945748323298371127834849457568184823573812967298602509445631002138707680490430110884104356065956329135513636595379057745086334658

41837937855021385507306606203236189202653437965542409138866780517648660235568680102444381998217408186830806326579344501366069588311635276590196371091221683021799431781788

11597562569334811817590163704539548800254386919502939484296333878802324540268683115920771472660964081472974256413523770713265586567292609352131356326973863345139232337949

12727416044071653328372766636069920782898851581890074068178835600338395502491054421913694943840259289757680416479873887544190710100738825026002505293715712059882179997519

05251548135128926507035031295388797395196807146312979739398855240677107478132966112514244409425462058656056386484117697376509322232005813738988859893022336308095219933426

52281506753067731168349920030749784495333173923562877249889011049829135380994323467387064792939183298473650917415993442241801360907021853768394823719725514881388163635282

50823780875617730371859331023769015518148956680264510669556676356270331637550428218469355260793128677171630081522970525013994404111099523758782168987072283241554043378594

93648816597106019417011177530819779600610206107580954184382263771744158930893440245480776358985983864600448191306329182121252200728063408905627313615628251425972291116909696

96211674082471631451891747360069596699142308087833837868659015986702232142869157014142480704589721910542004790420726183894565916757662433748165233431013197777787506264814

47896237968544918333932544522632823898399552143508647239988246182346783334120349696963465231029709800703127298113002987487588451556284431013156099089461587840584003833614

54306275028384345168367939943115519406723368803326183813019065159316862019183963643881182869704116494587694221136576981495173186043944768192239400670145512792825405655303

24642352419083789115209165207534501147751337617613160303463500158304324119830345045973111548023529147267556528539615498251732218702811891475582192510975188147499962701832

01238664665544709627032211967352066825688348737596450725120796914516873963998729508929286150574509391835248986417115156337107720704371942989785258541065122020872198851152

01196820066851549509077569921619316805761225508410799564473572362115138442605911878523611115766746246167605894908847321882511881891653729413018475636508362290409688772707

59063075951737344653812358167205699861544933744135511580828599979725070005425695844829042157032963296954183720611253277818507824353239187267379753901060421898213333568001

49176292763589739749151033610294485487554126594588308262730872974158135998785058897081564293241595652057224388601584207810475042628112904425526350548296613431983475578851

93222267186930364566727102649599400511663086637317274044545694973748748521103317754936462538061133447431080683263084662203937077310524427999513745019352661423522255141868

05510400502143876778592990110859251867499131314500087258371166936982497699408416160624284063083328979971618705057651962404924316599951518966497547503900114739890318966878

3264557847453725180452235972687766876242850753816616792488000823409032034807146522890222308061496574270447722125026619237142356260929122601825058373181197103907517533857

7137807762131772452879479158317148432273147350683717788157985202303528005999986977666937008226708804204330427176103604436021195740531832397750825376243533599258744806695

2313140950826729742008271959187161696015340654578147571012432947034049890117240314562707007085891355513065947483050109267533105047676685100687279532443236896493872434914

0188685802176697065515885025617415207031509272651458735885771669074118956676294168134057842406773388665298433582820992092796000256053731611957486517297171140435836830233

3102692447556349630182678573511105639749473357081758063298707668034213096682726128479506043615265442170363554065832901954741126321617941436862387824468108851006087982065

7196947315316887276558292548410060026288708470726414636981454676023069064848000195089152920883475200294833011835707147486046000323180366466301137834614810208010408241624 6

4398628580275352540541481178772578449824401215358088326311157679388344399416742552671812706870485790500170018827661154025989664563822695284086125700000312015134146214627

4358818811375215962355090961869348253038196808508496757130802652210017544521504388244696353913545222948382275219397816100630815713947347571643310028857201156174719192266

7719543692831282660439606992546372196029142537779739831674438120809721881883123622660338707532678942538559169182977283327312615508417484951235989157986019310463020408836

5812328283393282877527485978705364732951561411429853246103430255313019496430116703792865376695698547963744374046951440475248627476738025589674084963027253885817383209 5

7777270442659676450234624195887257359338615526808120477513640278605967148993681237120118621234905481712924548154302380410365014875356745431118006045004261307876822158851

4426730296208404822613694974262081760999935003344619768841879030415959515392641119654647748208496035361889457612204857186264614323274971918808584172165024925561228486704

4407945280918253914446987618136633194396064637822450816138177872928278397648591104634556227172221781769229741153678621460572420158898217549455474948636317672274364708 98

0215462007325013023705721216266652200530396135167883101300856801679877138600808744144960859610304104119748536983111367107082479747419717080824301691666177077131276333 13

6381545315891337525416839840847864317750667503948846636777214679211218536122363167218880380661069859370237909631869224025911914634584614974171219255019925474796004846006

33459818646080115937447037316631953518908792056481072811877724020397440246021297391101349926966489897822364655365129497329341543406894694337381826637786050347493433270 2

9083756180110549346901793394287399056637969763478106955289619876461898507220863458747577535586844687233572491790476548077510392373639618546675333495970891747050103139694

38090236340457990307072485296328514308887866880742498163585636339314194762523066152520565896307037142091574467866737683351558224442263717555290549395328823666896153326 33

1493583928128224584932540555941071950713799703563742340097316130986462139379530870947165361256508033157850445730000941413946001474525441403816920993360411596583800506303

6825456630806282500948802003418002145584175546348018765356776441151647710438436690085370611690503253031468354371335818092924007680509581888880313192299660498665119235533

3442715995130769082085266296774031025947302259177682013259107773158578447731207588645093398775618726625393836235757625158805620309231213866578072162611618127003756053446

2263494983862525666524229234436513969720823782599576261080998493754227356751224109232447930724282802917623537533863708763873518155274821112448002459124640511151114996644

6261984339005792546353949622889243623252186402524810490595955408365028689357489054200091253386743431340734226519599814488762644831855273277494122878561306225821878120 01

55

1628573521338086043652520123507908301505963245468281892247598913287169435985142267573258150924982124899051846590727823763964923211904205643849172556431873441622962006044

7190161161278608069159705072338317990240010621164747758439023757467891316957011822646217702894571191364126858718686358249327174656270672807513674315975075657747583764063

3804494482066835217833213332789677638365744674620172883957236721109815401621327006816874023136619483325010446485646460364125317413332379607567293733052122974579333352566

1685589200437596251342030638342943060971584740953801974115495300102821650559592594591948533482273271554448735213653447294239495596453047880531794558629341890107779349027

6022180849918514125716531651374508750314014667742519764762046166931133260453878964516572908438615194431140161514230702247163939901004379086641034162367907418506463768256

6038955033477348967311334313629428543148876031247313354196709800084526427401420976313695876225859100931112997379360013553352920748298536720427612698476400667669866105345

5207287218738180679105816290748701076736965216687344878743827719973271864925542480668423833027410696091855007115354892417444079433704231825456068386702420523393305803173

0647788593322929965546621687057128180663158107596988037954190286710515896821839986172264565237272159212726998561668843085968396028717153852669414793173289354584495315021

8593008668911797136649492410539530174013607858891547134085003976803645381111572086129563947096455742708238731268749887309705900533731834616896934170930000086168027800589

5674152284436630022965265070138562656843588862975858927122897312250450193975398801959929585946674448852792346410372473341353383902594807739551764067414764658014533037551

2587839152060027305459805828008341586750878202182980291241797731523538577064067711668452133686650109064399184664729143841522843559577805241786922134390262097035903035022

5270328397986765487111297164150657689153935090940421630029212623423471285210839542166491175188768489016016350794990872514594428409076951969961803771282792923306313946321

5096579366488528671853658985428232404638733828178481530209203088315697267343925583364321632066089888458071136277639996649570648133324300804430706922817962968328613163949

8341581788714262196654990514044999490513227583290203973389028542575136640742837719838951375846035685933196763654229787959796756828399831018152542366659857278588886806485

1894597071620346737035168045678974108321020687769153105056687668773293349200238935057443695445160234297945780603067189315767951908958081128270486687856517949494253179898

9854558463511016629241506701611762219757292557732222995795702695142731341258703602132593747642947677233855393949608034943296308145907993381594311461023743648260905274892

6091149978175992425233969728695252416687315009238204121285426136163532491366251378662874417287369277732668533899905091442880593169617682577285592777855488912248808866962

9022220090710531986727332035012560832761865468606900461217655114103453283127120443522951001679479031335053423535678386919223431249052133279436125690468033045406425931433

4859893529878822549531857424881037641375414844998295227489027969508981498646907616443895752343566506497982594125032426325529441165969405989586650761215339929748640528

0830988791971237287616972907302953015863380954319401820266910469313930352663628358321962934195022055821562811510082783702191422318615775289443074012512069822362570413511

6212793447479373750708585344904025189467769147420649139024731524047392237570356833125539744473636977591310167248556425227049855871329918475843821185152491532108660870938

9477465558909768150090915524531843711016797043942272006065934727864923765594695847171642902578632718343604387060615267993199251780719060018199788961891441329681532735536

5655317827878987704548492565683154048433686635893482791153784996014629433017853591892226871356021156380668887360245242861517707711106712851439717394625668407770725858919

5186572002830268782748806462486258045143333445413308616378682332572962579538006735091060533965232557596824150482795196197494590510082179623656701477056459027478980181006

3095188896213790376936533729872681282088478870106308255415850421334101495828542771806949463381388168245190344480504922435510003314142920894225768313480195104195395648342

8383168994699706893612395299336477360596737956301617803184226182619920816348676196602758664471180876032530070874535085357549089483316670801325348249711806765228158023607

0823339041428117022941352536003306330261124551686492275338976533327508837308735465914111897983419770812110908047137442356324199743619581423276740560044467491569494557871

4935547922254176429822307573665159603939567872952083076212995729056463332797905608736019668380684152160053409822871768205430304948296407143779589677891785265134420901479

6569969586033217610283983223252420909187497569528250236244494235687350103470187419905300293809698609087614945672871126806871959924240064653277115700461234695506725963015

6672290905445568896694903638197937468465866534067955971944629775631645824343862403793489804730057570983951582613921444041889422681665534895414328206155392681993338132234

1431398790872065564411761005197910307921159446412482298695403958669789629636022480766326311185609381709075532259658171492545809500486428193072375865331093474102684608835

1017655232979279258864296905772257139082911909071964170853845945443359918962961825813795766195253377709395930937558695979150585469590600816003435570792205728418485855996

1647715619063376850432936554547474297930822840340104214779400494818065457292244834261048015204893325978936823575947758489390796539861320097773887838900230664965067318652

6505682839582196258033807020970898871414621585654426237525431393842532127573407453319116295517118791369927035391723508149986623779442841884334571492927103332266309932715

9181177798427378975014789433268497205154307237560639987729616687253234709907174640540240739876530764999282725555733397102244685228197440635674154423398952240404254833976

9553714731599039115199581609459985121037453659944243964558662189512073140201773556781853195745001591386191064089978693283136483900961375710627234780052282421184264275528

3161285869760156604643183353361039723374601999153889315730285882691609204948845413009226258837771404879655160155435937451107898471808847009606077890762206936840737849633

6096342509584708257256336812670064291029822279991576193941230501066561932438529131227088307156747196820218627201948474469147750995873774866029631262112393626268432315339

1719356913789891966066712770973432280825198475061954062034493330703784267983799417188238477857304923986255856611633528615279571343531452481039163835170550778772229976239

7920840708871158662399192331933649557410994937541006679688014265020731066633219037296882469804080705418631788519380478271412256541799994252084728832820347685848972552574

7181941141110041741566799999641975328403240933119063192104713467023378515181682298661343846179559222892272724792951269711902324963913804404399574050092712081861325429437

4946808034952740287866386243934171088576574565098594766948921845006405465630078576018633790396114271309657046386091763460387568116961674247700175701209622415995297606038

5348857001481403137001128029694543163723511250880211913858542622105689948995183018091417190615926369347364953071541759066678807228201488291988205155707763583295672191122

0357704249516850618829530889889133774280092605574823119088319103131939299334559231342822908244952580052392312035468409591811803767004110412429520600416749760555822753840

2785572289944290970792203734798808673500170223540288707487241568779150621465248917332552477018448633360423791742749855343362819513765938627640328174263624814720096570576

1727339321971370162499437607223256132787424937777858926933033596401621334413649840271139133842747075776954377860117566491086194270718291744124265444598136378594344020432

28658975463864348272914836757909061246208432343903919234433434967727735561114213200143944432273203813690857297957363267447789438657748903859180992598862969779258913 74705

28577954613032054330367752203355085505264185246835194929346835243286029416899457532838210307005971426445390140901802991823336647440778847072021623062385605597582 21344837

72962995988321194341336945834461478359693702832682714104848145288290526166403281494081840243768279808314945204633401314793187522373778064144956575621060530337373 63146674

99714281990742397055859815350366620904650584483582903706278821795170109549763960329104655406069264586302126874027033337628709008663607757172312759161950765391337 763291958

22156023957434293446887129808461218026897104243417090833099109858888883525408594227691776828812075617943969011907566345241700616320081014184753329081130030931097 58677073

03631842545293345309766615291752366323656474216904228061697515605333059925079176825022364645999570337747610841475018859988302655204068322532391058724489413214920 42015076

36619728900040605927204249627607199929997651568985047882085190980357331157415446555005241314901243989950767377911479714212766615553657000299806435223585594633402 91519655

74477372577452551736846772411482287637268001963584486242603798649865758213080512548677536718044961870015910447387934243041878546178704578543664944284385030411648 19266671

84975252670736583993025400618865946300442593498642188736746677914010289921935190341984732576022585319484839338206114648070364899786708653140531734815143246518534 00564085

30192899076360160091407670768748649878661447241643842625492285981679108129205218822915194474347041036192619822496886501832878812286552614944872433559864067055348 86676421

60769960153550823282418270715618196314343109296280405256938017210064387456093585636653375409615209936844109006423455949678992586527173749829803717638644155408339 9332473

28130954900909116944267647099606051366703401744118303662250489910202822410449800530639392246517643281963200444786431071064518182924901554707466301366585027750507 96766694

47092311169507428457926919864654796897698574424712025026199362769049186893853796977482413020560763043389224736757475383147134175467829749624477066540938198182940 53395278

66772898388482829911423927736324571601437337526304802632494216545565767197675193472054649944251600989150852653750800251075605432655377272342230719696794527224661 59738660

21741689039122722547133825915332845251522669469728173031755253671085191135887654254435790412982410354317442327643432713706542099632157063640609687138452462456633 5269130

12207892078038541203760206341195539453469466949309162079581991165930757419826929877866650365908258531021071701501844136752913848473908192235647086656219503198651 98555690

37476710947140876135315487181593027818838207813940008699967045174005890292947204951246680739095172243055169301048038278147544641937702694249327243368125202460157 1534861

04406075905633203741788371475352143957277788274638618416087213432549823690048373821826384009251021559976282492414832391100246927892536253848077699875241682775157 98144534

55921809123520162309235618726203561806371374370501246256812488635116226947566896813619087391386116827810422466418488137749163775823530717510933636515920783202850 74878177

32945679572288027029253303930356290965509081112345990450064098314626001133976600372981338813161449862460738400410387389523346770156047656476774375309135303602773 0649485

48181815798555845871362783153768046482215248418050024360485920424819532883678403638789956319332163183177839752991937552421965960896506553739404460898228965083088 60509089

02496512647211910969229403866059091378266635979448407832676362544382973826316123851271358831895110725809941985723942638965905949827824178092350475995807282287983 38367066

10020415953764569087593608209053046464560549835109037784779087651974600057493782685696226892365647368966400237614213914040885302285301422924022934239184746072891 82440158

403161965703700511650374832836121305179276792029495844961075787312193123627924907087747049402772076686395128995958103791825552757370199035639855128240294793513430470149 8

5331634148827241470870511372210732637816770795704244352542402658784910323099442185047657104762926221526379911773502945540414719797361893916413646795825081052536221009567

3087070599535110232822554068808184242461329055034112463682062595644292917574019200970574675178378709497383462000351520235096582205132349518812880974170138280072774927060

7384295786765456512232806960187359279384225029823945265456615376890950037612041625651830107353700390702915020475371427789436880591732030271185778991766663425716426953716

6959331831417686469920393292873147806545499610556358587858035598938253252627842779752774869590582901784353170386419677914076504812980943838768811633599534749783496325840

4256655648835230230971526389610852632841399355173700557015792433145571339263506491269103287457433680168470883210198318057258996356417499479991411764640878309858738876012

6224392915251351274316311424091695798544231194074263914199573700819368632439542888918921590733557117772516588695449446490515695732422360494291061139881878797814569230 08

2568163508933768853608784915097614076727220176526327006304042988298532360410002404018299071505820953486678658549147523103917653043924444519665134251485586593572530618789

3317329084163403552216474104154352613758218181879127906528210805664458170081884621209532764188236193739371584541654500461376347557269167252476102557802111982212191616769

2479946814851022110835468697760470650797023269791794466405825458784123513783915987868576580174717357584005545002169915662489343277570531623439857465121125566976159579416

5500430927839580643678562017610936953432274420323727782921264107279273315388054265718719523146147608512412071162145370705234609873528525352632985551737598862152283252706 2

3317177105764683442071121848969716263292124906141666241887604178683965335208134039931999745848516368676490886859104526807873062160502149585919378227149265333228609685365

0503986140379957839323592680910778905485558610885924282242259277447736517818270019813885316073056803365796762717845777429169997919369629629072997268103049709697061750 36

1784872804915714553234024897008651825057184139097089981443210863274307629534648301060291760317398316298855807697144339567729015294792494892573053103628809298857109774203

4339038942417749608496785311587575244607210626352217999579448328249649817968808777035604906974060975581511209516205013277091078039134611475100496986771957804672823682 217

5885085551218737882384355023971353564767531284887511145584394413075616690802194047054025092561638873057995935710070954215242402389738661449843026964361569759383503580008

6525206634482325093428912815946824688131107670648072715392133808549088932174463059788581127442534488131962175507453904692292260778682863658751566809447504786267227357076

9537148972648601362808015084422632659722114711872171544581877426158697079388695592310355347744844271027727918126541939125547604844318093436796646334042828332733741850629

8654994600120905668609109495035208441838991634030696334351997137223404510183936562839490571574119917388142068644918856489681633355195066000928843332524806735584171337496

1715055093426371894023253035425993843941877187420881455435435616430348910314815205765886944478270644910995335212843251910491246905432173805106794185988054401289425123258

9909962312324053877398210144640584965597415865952320581449885251037693065497489313506032936074481814998982011182749277815201132404643038340009302231080547259597551216746

7066592294443857107582935686515980117901994804535824717234503017639891490221449489021601986841517587379191682661098385738453765280418900933755032348767588757658350816808

4898048899461346384675835827589450046648026022470795960731123470870190122939638421992508876853711199854331293724294847578836115174083358437533109066594270132580329543981

526920681054804215521024796511454543319711530574099549378383693200170656410239939685203415131733092513860829839610344837564348547094563741106045616668328026369760 5594107

860053014854032125282532232725173232493557882265939595083733400950598453008448615493760830772932369780539020694898436522867928580781581080858064953263317305646816091 7851

471254000880722579371359859196020321176985166181382057266644879714560505647641742736841891450673424567564160482903098189791759567447997044184815439560470233784356812 6761

771579873748731652445882100164106192876715295197730961257950401327995125123044607173765330443488975837750220067414677801697322800545673449942537241384582367759639957 2285

545930783851914039504744136175891007414622681929769694988612865298551788024993319663563824838294192474319235584267635073198580303015343074861824378325227933579938356 8537

811327556538647300247674306723758445557066433223967058378975019401109845845303120397416081495286336512248395115142651395213619949280477614567228484431285659615449731 3827

859533076736960149415863707036217565867010430358696114579171483445820548229597116654702113627728249354079462907060140372016903577892399326303272607254506004036460502 8380

929610076000676210935821615488096827981804590876990755827971114967485871036597981779005599204619921086218833393864367675453578236336989088161935642182109559511009398 3753

747755465860778655943330622484912789787545081355800095536186322477894557821672858215655834857416920557822343615032535519130694519600528949869404686558645288393239196 12404

399594779057554351905822581270246825732160269935312376217316256389732475711628596069997082939495981464546812429119289449321675789363587752365870831262612976895221403 7121

333371373636570097496111467954738940216254866841463524981486568437129932566103690320984324524436374578928327453254010137873546087085784915339133018487965021588810929 9037

143501149621191972437270363318901179929310091989720660589194991838526986780058093923091737819542985085168466812992334259466707617776755886208012614126146408861560637 0386

756461288814378861816884069210573731007147127556028255238461049428731994983801419274943751006947906095976275704074256052792040373513225643720532009690271261787884195 8243

923433165252466820942546627297934824209502732777029535981564982473381806163938715477491975350493217917432066843409206201758084778305188754961244239520118964907047660 1860

635733321398793734673914908088123513455137740715586822235455884575446863433377540313871302626071462240117170602401065225491198646843096415721944924460282817325253667 0353

723004242498660648053112750195435652322568738263515606179781774903631475049573203258272282087901580037039472207847114408535302162674050665051256166950057399083273250 6899

521269752616060272524728366524467699546935694759472575668558118942585377725768098397685880649644185753871729087162366584295460006457283605313875813863629441043134629 9537

418277627185306159193426121771320103100211452565766902870974555331090738581112895514039271668752247998498749917850258926890214824595782590518644548253086709605415246 3916

486748199695691963759719630398096105803354613278943598184350869745259200054859003037296168307195762268543536417311817450957993316477690774640274025905255388091862936 8604

586739521133104685547484403817107250663691014557314738282505357057566139391755260695186896492503446864749146526156085505037913942029819942221994735438231783223036847 137

330247485594298263804065129848919712773126979399424468368139797193089451531301022820717602411322963912280618157095376184520228786336178426103531073417592039782916543 7023

953430922591085010805655918772527775548002700192946141441537672275682553321417372014074713478844343631675916143872255943304949771961233842226616048964396242053267797 0414

308024116401196808910109206342903802792551569679532441619283486641083828605441536996436593196913777787005936030482022913026514592296134558029718272438419687672437084 4926

3675507056334050268371994493547326526563206627438083699582633516760708235495298561583433119524392297003987910675268314944224875870597119750135716880807708801638578432778

1815130277868311685891946114301089589218392897133594139288856488545091637259398359742076715707460714975246059863989669574623157776860487972014803576795648458982197028876

1612319470132095592421448835550576272323443484264260117532230618223035658508047011011899193252571720549962926641297735042850437026228972358528162675637896302038984743559

4806121738739066853543845303929312988193883304118342377836147805975750584066225413362933578093194781966392974235039084805932006978991767883396869131974825886474708627997

1313256137172730816533406139462568550590727545864506864656527768255534297214088338372788201028902932403132421020026106356642443696612083041768693220104899345155973211746

6300908671200835572420529225106285030294066927058050440068181922735142563465843548110959320734012749694900025447207973603791646697031950338328483551676760583103654527085

7655498002823947822313718870396521642078414038632005016875592892442489164321079620031371107462606935918955818239988365915310970042358174294600735961247432905721092909762

9241041065662092350379244313926890303062203407870584752136844349814006643996828177728832830680829674748510726842285639503119239679399702278280832904039187942701256403173

1986705480903817290109382677032761818733382332992873542517912146741696844438416099579217349254754115169550363292946067218798381779848868362782909979843021720417536252229

9672743257163080332626794270088346679931237227789280490726906343593863344827373494687180880694508882406899726165871343751874071244353589993574950576391055026023488483193

0109776287518455556142797284284876039387213049090254184884269775140116269376139550458568990047300398762225695695285227027007070022363127827564720918907236614533815064508

6601571667250304425313457307614248252993473550820094811107402642703287961354558997238769243881097597044445727972255958214831857922116838192022376660147053550332990566 38

9961139502003559003953143148531999733956110064596295558214961621580455163249615249846254913386661556613057471073066064947612592513473986724042947052713945870057114461774

3592489199997798539858915545801175707545841985707464441715735287088318155664906711613720524842124067568833334632630939467440591539281243468652741507636710833294679930796

0121322623629719228890611294395686589067468858225888839891650188355330752331981579035553586855155782065468218332159074291034746956756633924854152236453715003886217890263 4

3137853026622744881799998738533234152500505075994452916010384924296473792314485199676400312042619311018390010745597693245743996519682211157017225000078018520076909279952

7481957223522490092455102100832943506047090382176234012352784838737727314319812533312167350741624784195463253446152082891223780469229085093862807526773733648916752751088

6718690748573151179871911275897371721222006979026862701539770333762353916857302353277808051500852598175329555080787788667281565096669161583911272169869938875911268864848

5453452898384500172007531788096127347744030045241675032393038367061707101305504380587173067566833533745378303685599937759086951306218465528579235933917419171205417969987

2561324532665773975697093217056219380046148285749989375231643513474707365882098106057788654165147324898178700946301385079255922607297152262038819437484391431059409599 25

8433446565768173968932611045987010037275435251163774416122729999410186195660514215969412063551314485971954528608097486825487452445903626047313806483937973446818662497007

2155471060019350023864838934375622763501279258494173264366237202327855359419493045001115249370114763463415754264095574739430694456354236208121224117637357697086777 6359301

9356383644402889363050783332280366747439432486570798950852508727418326835271995157792652719876374997907620843894634721262036078308173814280478755497828978622747244177 00

30163255013397053724176828153235161769069219970255699962054642437265357754725102403129943553864594831470194940156026684943031837836936554661866566254708258607489483 97282

51558916038553495064513847442211882756298620633135692134350535417532546229427385701851422160479791812391358185702336381354453571127711719432166046661431015474198215 549290

47562109018957208060623490880290406785456674637241772486811900742065578482219295010659668353520867908758553449009271325107353781311232860041052918835504828256821243 93180

97857966641441641974384650359754316704183852145907794335773149648457421486085488674529131457458931518483420505854272116027520701053028812182044257185040797177351938 26444

15143034000038965083547606952112614351514494096991515178332585172479894740524206100459840736384351138298293353702855164153281846898780435921758197601110371882601157 15212

19899280357546083887409473752204063912336289828066187319532355292040142200095154808807061007453865639725897080303279855124057096752994877525034838119148447639606902 39980

08588751011612900600807691194381030260949480659847619690480593217852139982865901636139729473334245297578429975902328892122887617453643431583753143784957460887473734 25879

58758219901935389814229423917941515613153979302514641379860895988767541369432304048702855541978092295804469899019290455890684659783833799449251271604941337790706486 57858

94967575994050617557632934756808289220291115491864882015921461776544992118272549886765689622517063614832195140603084486842947490817141227669895297665284671870107291 9337

79292824435324313828520635706158076925928260322211940276877904292408365323232151023540753423210947605321017167804788960416851071973969399186187946346189679713546786 72244

02905164443964782932669463584918661504011655032137958238846403545337067500146824508939635084079633883393164400215576294876554514962298494573570455639845828653901031 20311

99558632978985996427424165456402215526931176182193404052804977001395818569950450626908321844220985806560360396650520405092652944916311224741224398545523345939736021 58488

95956457603560112394722600290911023235818327077603181928957893191200422829722719276801057856446673340203180657975989767363004515534412122274649211784192104299302330 475

45934081486957338858531187892572435996247019581049408342713006597163643751657463710249870536291692900609979797982081471471311289950851849200398640461465302509949141 4340

35836955688421615182000667253985853203567078753447410182134499703959178739753496211472367771510750644341932067097848101390611946814299656594694998030150150504394991 65819

36434061754712006023253305100568566199539885210969917968103065156627611400123939441274050406560022170985477796442468587486319694618955103513339164119590397189387610 54426

42302464412785966320148479552732274340928634620098405982453385763861519336443209833918195749629505252717041594110329416055200707970044527426550329106801682918215054 88657

29790830657332005667110403931664289460739742761327206991377358887764684076726016450379690906737624912318156946132584214651124301808862838501872732082930493205348834 90802

25799962089315382043458746206252362968122407755716676147083325307435180280156466152412335772666545968950153512096407409879933525112423683932555080407795389065135984 86931

48572687489949014085300106254036984402433985738212677629454919277281927072710075950541945450379091805163615808362888700153867243945007027498984321855676474032471439 23664

84341116062093101960182503178068353985725839133571334493036144917086659797233388145309217403181174775203258167433894582649675275252036112627367210976454313402380658 72011

25134514611723801638359472687522817638335665589618861321672998939401494125103564658336368877609065883769674182919192031194564697809424983860904160330963764529279423 419300

23017400543432258546250947435455170968354369756035650199238514737184926705972332775797911738152474353163342411729845894129107504555042885877627373406633041603918082 6874

1726065961598933607786330701992223184666488893045271524055117461202230160366192193936615793786736961581625973005821281128255764679594940281466745760455774706739022001976

9831825970029381954149275908113373323605885877787161006725835962326049600160589914889342204736161327100754523004843943109899916372218862326257224723071197918230494443514

0335744766397083610698607144570069276396639734920292183462976443818601893766880534512777038481569085406140328036150280386094903353489323035792511739353041584113326547129

0567398884435930822830330322211659298541919655979718488542388715808914036930161717257001557061483690681274229550279346352264500686934307748207466636874761476200227750181

5517969782667374145950438705887238733896329121395730399466305434028913277468168754669502161412465503700912659179830290388734841761397234339455693566008380160943556137855

3746892071454423377646719631384646526315701017132358397487466544236302779285419045015666457881859979478971251481140502377690261728979301308065657163121212079142907054215

0888983795453659164355123341745987948092769417511490311746055224557854581355867021530900770319556558995997468057416133383616416911400992334155643868362258664428079403362

6701052266693619246747237136409054289852051883510036926818799746564705254506826839362640699442231179129973336410667817359159716289832741772887230205260980424875777100698

8196240372912716284558358478403404924364878183372432037161878814931836632132424242420147187986601290829544902098739959542872139066776982756308916794217401688235876539750

4203024489864188963690963162701205576819699291549927751425437881294676650832503512671684664484445472404101245280642178327322277604369161028807835887184371005180840179580

1410835281635163603805346307638919476150186986736706050147556545191255634854744061620273938350356278561529588946817016999401433231109528721244827047206054602585006670407

5791114136827906978686587117792043561148929968719888032590349546258685078645156073721715399533910705457420844700489981128289942160212220926244947274540561035820926425126

7803981905452659443737519428132171370336125910575516989928472946953424298072325629025888362684267844702983631329496605441625386147488283479816732288109784876941323436718

8334829751327755520981118356612998485687021734497159455814205167601363168104474987091636494315666700163412473152631466469447022286028071181399281588875163721426683212141

5092317205731891117328825980525201560900415547775952404089135009403651970848700749678332743233588694631268790098502313172066141132108607604861957063566245230487204929701

6794578145820090361928067821394589374337776931269876868117124816408491052538842393336908946354092358023108172557634997996943645975444894856647732804498867623578917302150

2698796454984277123360252396013678890263912763167334869009946588810286310223749553599501716187794059542720325680750991723040604059244759347558781923115070860386436400166

9769358844410773687028457703794092834941402822129586407527063935399304447238508439688527577798355208317581070948268654551492346771164511885672238076006299878184487827005

2720312938847998209719432022757136352039888007560979354968507222173819096427575684664407843849762359541643789860716673486049953642921576909269615170952825421086026688812

8762132282887012394111213560849984856022616743503488305211519952221309472231188245473926080854412153442104345431104283533610723224461095047549030823849762337877239798576

4670714850727011550335079176889428553256555785689141110539376812300764172733233555555695819795161678765161127158802381737058125843376445496393209033630888423831346474132

5415758340853287016214784675273660353298142198999001039965166398578162708358962481375811285205027468314346218654210028735798453064197217331119032520073461929812872295178

9824511177032832347598640395627061908545507358079165897100777640229035197705516514631569542884142437557975709689422328873125015446591323568223485632308818621486917525444

20425031155171125209326672093524453853285759307857205196311267715965633535956460663812156991761342710503579689346925609775922911355755004495468093959419880769193752888 65

02489711246859161951191180573662336507492183673283957490669396689948638812546588558883303308642792235459716140863913280169686706096747793497025136967094921185182680 6837

10393297681827934909048809926852979497855733371545681229119082889999649617367275829677225427182642232866400132724327309242950923056622134697756027497131137749640216 04518

69335895994334517013147431671669925355352625191822960685511025521066176939130589930470440130555394785866316843769182864723435324859388779733700023743444052230578338 50423

36974867005016002866371635480721425727423634716598259200059952735028634294139066792669723798730437353937957758746704387095073567124554496603097896118194554170245592 19300

96405938055229427692173509881950338542439019622355656650959811895084955834758326794413719433477706441743068760728732386031909376467452918921839273405652449125058695 65176

11562069812500393153884581844064908193055138220680810239336308565359538328508151852860249073808881939719741926645634161448426511541316956283525195911244283826288101 03084

75545489736902540358823648314244059506043336372172311369797376625377898329148146768547541189710236465877932945245536608462987170976671452153935936295650841686793887 47451

77684647019705601206291119659392716942878200104738422691208420374736338838627479266343817074600861816517701247380026891028324861454672894643770339434204648424196702 56187

91648972518386746222304165160001856431299654117598208500563323952416320466755350013353729684917467646319413499223744247224633022021859547406463788211882345939408996 89958

66776637011429528531270793556632378325619667821366570922060283102565391354011906621429239381641120806961721604380993879930032791351939616054590672596572424388667309 8839

49480405001995869954087761066913890684279935646950245990878656104815262619488029162203772854404310761915233096761345657898664927602310346708078390992627547645000231 11598

81515250166756373957401905770341261342041635904480839076537485827775259666285431629883314207477826120950407760433388366358024308924484034883541028706147339632834646 57835

79669745925874011346345762321608104239762225168959734768174285127377213488843126429868916709631623873420014694898521423020835510710710502557188462778564403760534154 8737

13405703530471604771067752932007990830085796335058913648697747150937612256284483514279338752636745707222002567691274834223794366061319862676094406210515237198485974 73792

97406177243330777353802543022189439576676950956672798124850084862425884845796771935614664646260149649514634714900618867260130216748107466054111926689184063788353110 5630

44170808357801999223295643474303295979135889437938009727044258215069279799888314467253297689672092433109678778704372547040496937826853533277996781762915186712775413 65726

68369349108729256606562281591526335044669497567922949764583960403124782609680807632457291796313570638055301817950615589346192005525020421276892047265235195908441637 05976

22758052733539905727377292458984311346620894693568462807708795934236143426183573972841216652601954384817745024429687378704478184580845669859181675745936300712509929 94559

02157979712679792868141836179452938114743834591130494949062545777573965744825041893661050156722391140633790442693276717835728234780242922904037674700397134683438554 6406

42707102611753009130847612735756388934449578014367197801389826534243776067204873056592069333281697377077205067321400057367553449808955405386888786715911240760240287 649361

09146485632435139228296169205384742208960466080590382309965918933425890790062237040800669979202979819440927717350270127336846820867383102707947935530220822775215446 0927

35620715171955387489668190846802860662680526626173073955928932432766560820558926492281145720789325877823680827930505003074177435351425876432091818543266940690676007 91908

21342039636895309452563340221307302098645862976896554724865262428461104736657509041771732052323741407565848993239270868216794264326875694735191217476911157754079991999

26682888507939039340610310421329646825040770647705217690955724326859664717698638291411537797697600025819272394469201049660042850854700148091808108172665045679671866688064

62058478809300711671419078497133939149939952552454520949465078434971981036142877818403322057069463951547694697276774770642486460793923519565436635083070252079824653742374

2569969457756462611987386294345328054150827620990662277435844486270376709248843139673126563568059785342858198446090082502281505106367269141887603978831977318626572931421

807329055093538562444488808705158512055619441373703285405475722146340713736932655210869322270942390754299408994425444590686757411432252426167234352191278258543884559516716

978299328323642737457452544546052939896806263513733587214850808820205518659958034081488329701253781235056793050818818568505731232325755554241960542735831944797643249922886

22660435558523349606680905502905216337784746351934749713022329493965510415987839740167516618593605179338950392466205245511268837311120785257244245799623294450168341713596

514025209517926468115682982031361882739642662332167644152469548755816408435821258504424767069969380375857300390579011051541477955717916931272909599822136411598159520145866

613678920666663532183944579112942949372746424648239215477975615733670895761840575322098504748583570891766352727809495354274482525113739382912378335184147182784881809377666

25946725543342069023837559765846744498857297100153365702593838609837888370559661656612261881245643780740364377558292593401364517385844624554076490296162202292447917789086

14243272492456246105728329944276796783144819346705517570835029425673263352649065141412102378610932967188631037171704617628931161672590290677122398588365964149245530812076

28570841006607616854351666353034138280113381967791228997412665524495134833893463618128225649905341150317914116709383076768774232569803429140799802919107611396530776180466

0762219445151940260406347035679935388327437858815201108040649088517527008205623802051286421842482300263243205599799834692623266564470195635730067953905724415039816423908

2136235132717714586191210328112357269933087662553440894151205179902731473868182626644475280406727464085723801550389418912958937399265016877527437697415337481724220377076

12866449077116260315441711941410834860689952950744477220336274426681847119656361571377242461545607047965087831290013343491113629297558360906017594945379686150681790850766

07566212738100117918293076118629911635574502602021275654360951138569094815424476722607340061037334261273608044855312147578890237559057711317455009411859748652962705885636

917389715951598898701417586964865418532486377943378050698934555388050523312494984188757304644473314445985055247398653997073462338193980085773043569547616982826589381003066

60241121866568598020725337165613533509921885956010788152195599298483073711416176483995033003789882479034541053250205495569358801615459891893688657212474896361367186281886

5464478617924358171011255185131787177450430735364502976150729230110833080255153495186929484971690099173039476973337895650229561487787804836658283482754023019230369038581

978853430382558273006721561304247679650997367389863963084595330994467366005278953510077510623540518095062072959121477879266263385428792589775958630580646504484526239133

538342627050430867009467036220406339767252991365187842306583966702262580562122210733541161850293635641616655779237766395860494693244550805903617986442755741294983021046966

69861644931370103702775084860153961665864512853545304815598296382985981545562592486591863288176301101499737206920153869877418621655782087885028970856782970192695827695236

940825795893466666883918358815549069436830703532763207934945109365399450972042836730670351441963155288753214822189325967173707812714051334747386080963694563512019018439

1605573384080516638291488624793513794037131979668758562594829420746324161481962682888498009688756413177902657691055508025432280312585899845828720832573588947631349260624

9627183220073181354243953643770564819295399570014455438391087844914419368047106516347403117037448245850518578818068662884417079356604269800316323634912030291975370099601

0666193896217318762267071826314852284417279433406818103101838417534997349697901352604608389864938417085293469279158345594247787414758186260672246624811772249856862298974

4043843921840245603609191236989597824880644631955555593083281673460231204066700724877475998063268452732025570156216876628405832688949305051939005049504958701540034854277

6024624858846666734238597445456716114198430383570639742666670385556096452390357020107365232835276920677221366635857460807615994825758902615564428664967372569208046851174

6267024678766860322879651197857616442650025536622079972039998656146915511996591892609987569195721982755095064759786156264742355786450113897041993509976406676557120850295

8421155914947290752355349927410085129491938559625940326382025249882249214444755882700290036795187052357627644235584183330712046012462993991548419581355125514677093447144

3309247637321501186127983818560255716314174426442103923184124861561304709814802473388125696051967726943832149010465240998150118339414506008422291319416099500996449619633

0766171680279966145964908485717408237805713129439661036877272697904349031896749323216657233190372154146103647188424635680197125709771242045599277189401630807555791531803

8863852263293491228689445871244071873985131098072996000054029691390863266714179236497562971925021288399097084846804390717631982983862589760312738181027549342610128244583

5103972461726002712472644102839306036777543984038462374655711776604274794044711025322752607088191525962388103594491210025921567550999035984902873663946533362227856019878

5244807812000092267255630431187021878325473868804409188331048255150339506237035345911575694871584408122253546614612133683291417713871207911325632996961058630638814550382

9307065076425004095978377200913542843287311066940704199932530568316953318544062180960834613197799338171659170654879552114439934636910391325853497773805380142494093450362

7616581368950030951261057084123445629601328070394871467589010166411517039393214698903026672660584673505964752748056178078679539355103268491298676656542631273298852919270

0824708774002213743115658696907606589908547798087756486559413089027045689772974195596550109221935693238497816225875176465524209255740925717695468860519010003160801289728

9870528610854229739093968150775009659717371460086115220926226085270829883643736243877981277451170822368080610770774136633479557435335472506634409792898991840821815020062

6290058136781545284857759527335953597484087245005388274103999870195212623316986282803438849726914169586295036202722974886898490039741471616745751141334602734497423550587

8072186655258735064125308324573880356085157662659100847907204770453688975071997435665063066316758761134751644189050994953044117199851499167397662294269445166214080877491

3553673453065182999775820146575308157940816750357256313082689752768694913175166031419627412271620957829974512595073689499764786513098304455391676187931636640409697787311

7158004122655528863709140625817884692390364398767943899441959633227733151062417111111175895820421382268247158558623159366153128943219165489282119597622766581435967431904 6

9318970709546254984802349550186923112936640292909966700863878400428904420862483661779064302063305933920322434365160794325702465868466897715343280772170987980118148551579

2816444921354300152529961377236010772921085951314599524616594227164157476323657025718806117063487629262732360083125256996543432189374507796744529154278947127228947044648

1314744124221166590081005721723304438700873736053316468302928700555720019069943199870645446550624282172711712459206812429481055050404705924105288357400656484547245607487

5624763472596201955416308086991308656786967875539700812791176866919496838135150988085209582767929487854818158433903895764802898509257246086253006148886286503065719865793

6561579559825729918943289477161896205693546728054418563501846263442674857155608884433767767751811195879631684185363912337497661237712587055753677142553545280102361912882

4660846856736084934133311957993354042333577358896378053183909344428049227035216223087149443606730042311797968286390517195157505209765590273099670998902005130022633264738

1845202399769112952460615572933669965418267875614644743693887290887894259922714756326206666732908094698629295343111076243281643273608630864133864864668368334034117417243

3613790860478056800459754328933272140608034447503284344114611719096701762539842822686846838817061002536499007431738470008614817616431964214609199373818877654827069979 39

8415393897490946103080608952105623723373395529906485456547771113235115058351872397486970763522933435497256100301121589126783284926464529265711611514653003449614413040707

8693714179233116662476964087635487399017477537102018211428142144824621320489013665231442441340428775298118356673485565936917962558531536751079806714527966374589942 10311

8815474548075246518531702182499670582009281703471433056490611030296600778862186439586203091262195374593191550111691331559547339411720861353588405204585927360463219827022

4071542061403310961486299075908103313306591475749539436538701843065303834279040143059829881096866287939606842134010586681368770086255041066965536224307609748692066674406

8427555947082594037595454393281264615197860109409221003946623938100024885780821530539641236263035680445023302479433432134418808431469281618392368201869189398393330782579

3915187659886158526588303130548206474199238611662169190459756562533363184467689507529925867773089781132205545268932341196377415807042979172961849337651669375621514648813

8417266217153236227108027841813774596097865572452165349267876088009188070751445179559189320746484076199051735585848891330328063507879705231316767693157737318795949072123

7263799259715349422416504918609591639298051537541530560108354141244266350884117088954264409770274228232128737818584813773935509749335551140624466436894204535523793022955

6990256888924724764828569879277717770439584369247240062220941325554943292326806265100656067112487799788039988221458634529591716662481653228741155327526411248966562365 3627

1790517082015310026735395882470223528163997240153464122032025797780827313551205019368428155208185549975149151101699141271160844540090762083005141646188552963462036087 37

1679019058251894683895404682662971748668008393029512607766929924069435225777438631812759679506943700156062505627855914341512413394030327712953253107118617480257722349489

2925219809743089521223146195766620759235563597266907679866612332910595279610183431069077020322287162520836119564948117529971327497305978355285212857854784428618168525710

7399791644850737946301947948601093793836405400303508924994891380108931322703064366040921361522517503364759125529933623450874620625221161521345333464059073152732407955 9395

6003274879097386942606636143145093479579364252820760576736682245561277978857985090507465575999523325768019785164732223573444661249477999064293351032029241706181476957105

0772801917271665422720280245480655682926562444571074844343809247355832405957279281370093179495842802006781667030234830107405474219268605401978802767061773311698549010053

2526580700391933221832551762219500495602329543188072487098922493307375904553488785189577342825125096765197185679965291017199510174647814302781333571695642231934075713767

8346086967122438121730798969383121704204911241451586221205738198926028132533616506332709612681127354457645034386271837391993894379695856116712668383393759855826461542797

8133179120578291237899892276277256159512584275400014463204457910654686673241405336586191842804226258216882737103215382122900160538955574580481497079514288287427566570758

14826054824202210612037688341073437044616953135658473158464995233288974086138926037436545571031357309787030515797674186488333083334683061776199649533323434059168678388864

52470412755314395794027884216137596849182823286006692891160507611815980980572296761164235609054782755309990228360118255687572387881258582934211212064353513623423335454 80

00376373539228441337466447546489972715324870623432473939494074367849041727254265742675895182796020334362260618434065482929109694732775810631500580250568949213398370571 06

19553810369925100616045006231958956852776338414533870921568787580321274603112849248871469759238966121654100784516651875999266072990204583456279634204397156524456500393 32

69758416152741868905281033962277286801457027003189627877077513728951374928538916013451181479091212455428351145074766206145020787405527198310649131950843193937940513935 60

86244871206328233097256310656806715935871203992140966633225119104450832165354362199377758584328122723097176497270028253352023603346945160822872847272528184687773750722 98

83391318768369026382449348885643460614702141015933553708337592611935438143753258368050686926516021319638590042494502607779328982929749310257474851915475821236842755637 39

77810151502777188467396734397418256427158653009213673388007912311266160418917228468906382678717224697734147003033770948629424678362291721791257397858895723049380035859 12

36399689631216138583104648370796376626799297615668211984659341559339167444688620035568965184061896502099578794947503421345106468294189135762409949557718833764748449361 48

90337338736408448766512857990600569180355802175743282237209829640541398491768614254113578019232843223566220125337569971038210371450536113521580875443258875177314981234 15

97900774841548524687471869828437164277567966121882258983635864612337270873161639587829938155273415802880622896032274479197315134195894883841952929056752913582847028899 7

29046824217811588125445002757734897566106936993830600284424883040855689756491161569382878286204590171592066183555970557350218309269119605068711363792198916388264700303 23

98559982585297372067596850212232594796092137013315436900473475800526697316366280876754686843154412005445181096396331779963270733270078424261594328719836710018530522110 00

49935858980934727278261324522255474466336523469026079952018829848657929356433410586192063576580213494971238154233326330818249633020386361806074300789362848049457274765 55

96897690479630772584358960972355626885277176950957548546741563189365444345268252226873316585833671741045351860168973900370511403872160749256572866941446343428193422010 78

79944479315289080704416783720859140380787192020468714895404296577827423277263006754826839257204742895691916760052052321538211408873240679725588369972297703978174778655 44

51339369528046730979198754644054050153559842149017649397089933682368179786182637137747761899242139647546815218023565700846506246212580093382393758939853525532473703072 68

76131869326125773333729027491969501508404817998772837366525506402719361477325988089081494639422730754621137974252544785730656206162327584566336871604105456555821963228 4

44258001613092292561169521705856174292971169937298798552686573679816223076859491733218637615077351715337805336399472531737904670385755272237382781358856453237660838981 20

22949751795849901416896634521878608358384118931384728325768648734746219535389978008754241505867497801560159311365405520709508035255004812123123771815210729800323101759 18

37862540565962539948544710762023852340834150142189018389630276690864606288997315830500060541661052112618332456308874942376132111738323599102671544333398090301076751921 56

06860915099297579489847091340484776037253316486633273997745741707870588584989036478250500607565276677666730181427983462997863115472471904638130827026950271552434583777 13

28888401133228561232764247580549141453340043073513682001671030489674079132204173293655886380819902402504247589799061997394494240613938590020437450817126160362783912411 47

2682090856905268374225068910991937677220777687371267701529071296822615843757149665346296154035289806984981990238158813249007282842031664545864518677871817717727792832125

2696832297664124549673971527879680434765895761265338524573915134381378450051873859153296341405368948443972255080119607926902811622936704343711583719538657786003419467130

9665344253552356135039263743335590248778009316758556650202614245175520231051803797924160186816532721349074474187926304637935701957254656870769649025628311394908306598139

2587716575343290518298830744220931539456267189136509327785275856141886915058431128180621164533834614564998610279908878315999233208349703499009644828973619972608413030501

6138343757350335026967919910039457648503139889980405347620797995103556280094271198077141386253746894200671129290379402110509931288176786355712128822058452590232988278448

8972855767643337655132098372084536519727356629454075207868377482599376950854745853778544015186687032127032510837885755352532742246745616553017529469704928603493523766193

775815312691121571250545649366284046131575493234361611438689415519179552116040327941387040597365968277235554936953672492603352744989288822044886844365275158954689558858

9071831729289129234577442844192725052768475503870270632829799758538825993879067899636766347263679970911370005004519151507050208574470532031134283753039645068373494746515

2543161640695883965696047762481007698125762324027656324714558678116653563357384133203756328577111457947736117758910978449597487134549945405008974943123702669160022779621

5160164431446321556746579869691343020917375379329537363102934825941848515313457700649437390976420858957317742314576728829679067502992231525732869833026341233526316349020

6490429708210063264881567676324254446870392133376789489600125136265235472565170222559556998628425108866896847107872600167332422156251242927213080559326221307214093686435

4996898787430352676884922123183449241907637471574744625215974576466357242752795222891504064277678656611519193319181783056716465304813810106667342459156864174457688390624

1920186541022526697061538909990725499842854841956681924545197470930614227531512984453091827577151361181673035809321460322584723528118255047060621542622432455144689645726

9382316655525095989504109342537430859997971370042588583403044972671096299697632336077767437347987883567301028647138454592879163749014540664751939489935221236247436613174

7830486884631516036592243576762762344666539589796468790552923902702010757218919138214831626852490495848675432931241826413466272282093653277283976675572672897319381293419

4305723962072329200718638674667030636460133111116425468025122894305331125098538601201236070449699785210958598753293032771622679823055107676926800022074188490301650050385

3447597101830167378268194361241656963925229474103574318517658365603412327643390095651186326079173389912627207213516175222255241829612433962825182328696862544411186238123

306403453315560164069574723203836514566355749873441168599416165518249604259798392678163148318090253450716466644267026276118597649132476829527278057032238343515063672177

0663763740249030465909628596027197972553780014182019981018139812595042348662483440439211364872366629202063939628845314483748901026084036148407312006741562291596696366940

8360326433414963712098545475250177366960197146178464515555994167263739708586495987795324215832821841093916405283567907068642107803466075719798914815540054200510730096279

6234727249970112217781656798449194332226334150338567530824467734104550327428561157455387421400071928430177447314230098365760751551277796281014722053066817420350596794105

0980466563136377825172470914099255524710368126705138246752117200528494295219748862848985277878356210600487812711440634990881645924451898010442935708329047220160726966046

1984260772247831071714390934928973795075056471053802916187491886994635301357293501873206687311501731531091296794862549795815121682207571231891909138338353447102365979448

08047123388274440350534679995291335460941392728446513911908360762265798398156424638291599904416284527681893532791356747403227351506890987754721815574998488346694622 27119

43435139572756093318776721574285783033018304222517049632971612296836752748983294723150697497887414002192670067206567729213108493572940763592895618281290011097847243 28303

84651974375857368591230981783603031405282300813026631330413992533992179415764798534708178636113772001408570386394377035291834974037418351162237004017318826939326308 7505

48756645290932656650302439445362272791708003815785132529023651056809625891794240016387149612121694699254423986747262060057131153878388338830780165378838775211593119 44935

95694917939405788488606239594441849728928723084955792607211329771213723886969863602368291622254647108806210944791323990154066781602893469428215506272126054179829178 17382

48919973382953016826679061778013533650471886339784273533585623279185357977922662703802445969829686254911874868530549785796598918486218623748556393532156304899283486 55615

41540649512210466103765481806025067654913403327386294169117762638138347851183641056999660949202044895026294461684668551060662420883145374012687794781398595777699039 87707

39941703256531935561300052814359945606126160832235899032489565956752757698532457440356076128860796818576197717887655619852375274472235992720206023891671879081470886 70683

02793989768378023337576983748471679204205611884614435084238369737859482588497825952141431676898459213193894128750695961491932711470358745336608149437546971429195290 3101

93894356371835937491483074230449340295962811166523899581106000938622162265525297661065074527135894973344740728167392634862226297134715553293624445799465082319790876 9074

45852537034570900774408153416786387014809967412400383780852342739778817469108053533050911944331387304083804306750603068626953212452290166750385631858585931743776949 41574

10810572749444384001399152295292401680667458424696605569107697596987310950784518251857689800093942863712191016698078851710571144669507031273706962004730035675368235 20581

52491868239083974088092648527045816800639153401349373525471509235270441619265721100423458480853223980930819701215864173291305325898717158855168420606503405569968593 71591

56219395459555855700934771168117983599584279819556435636530938905094196464188924341766121771175457371442940272937717765918310744305815153159609482635063365572386141 39208

13075414610740512741348138890687520896517547286443489020150187201836613841728079882729582018977486126338360371109414086804414638189975514419051152014024187628978688 23386

65288749564740110724599055379921751556478198091849558767527782808038226298180439415639795617256940909295185774478836515944787206826785963694547637062382066962023966 20665

92108127818321912746808145303142177986735336848938082668189691299983519942232127263877159757642852132151588371714648542881242312246840283905615779681998978556251027 10706

28379399431907357979753622873719947845210383168668521420822019266723155810117372442375609151489386343666526579426037168289281580693159057152379480256619126870887647 50695

08501113702578802333818019030210029759755926818216359535070641885719005949744467974174202521309472461919502772132372470257029616316814647621846436446517995358777759 048091

72469556739679455373497103221936945596277893779193834069725378841550206295838748309619542046154699022684347461767711319737486600087935443607302433663280865368473350 6687

07408900184703067698214753133731542862215155131814095414979724670676343697696458309286795212019941406654043266683440819686918622917654410364920807857292423388755061 80983

65912226537972884111201306910185760304983295326942141884259428662146952768806320825719486713422469852641941902223624118633913028417184472482275572337996970748200243 7580

37179218073420208053693574061876566416960773912090981349470212072519721369964234420930547846506923744649042088873263022615635791960630923699160278236493000344974712 37794

70

559512408582397099465702753667598133047775050505366345747155165583727731007857817871530316132768489253576078462114788603518040297656960584867175676366593087480160999279

0787178913104203849478943286084797051504283332652457188642319839993285634226860788344374530927289314609254429906078711173676695984963306217751488489933778786785978526528

5705486612173792135521247023953325608190678852803832422968075544717437748950143023150146961225494895333862756944869304674198022922555065087429772758076095106879827109193

3714229096826872859632194283672724247744390906003680485278454385481995582874334418909552309926592958848289777196750543920577166893855239773609258209069343057898674235729

5312051485090384652493140068996173731735816222294455416149357871477506270376192498036384400160913611713729557661808926386467940279365670385305779912988573944783757639092

679443336505496770742285963808721870399582714758000440222404214003303590360960548004718847304678286807740989832225262453168032034084435109374319499380299081241792110895

239270965425821958485866799241157884478152195574983222583335674226798960098032009354865108549467676713405310343499864349758000215286835835721365978208435732604661260570

44092005206437488086840419995854086974773160175053902530649036204494584764408820400538605715251822177935180194147116600865329482810600219159446927834603798292688186777

8837827131481606681284808747904302342003771308964647817855994618375106206884413586284506303464419139428937623547427775867690146782289070060926832522503246399533375667289

9766025424659795196326090274261515748186527819297798368110133133965162579331841940702696498895138692396126127536959206022969008742083472084083318841582683801938335897322

4335136412244321174979404766824167809635203566415433254150964501977910541460943749815990445792838028880133564281814072611423275972894824141887025957454934254722746989976

877162316099322885042028070838100814091887352633318358420740748446573397838429805347106002374219987211762683334909209073865337959074928089283030107207550472450851183334

7630475982066178999800446274480337019655021320441396423674506953708781697379969379061637848201169796270721270358480479488580658308963231288673402963848241128765952185365

2411256969749199057478280326299861231724793050323637705845698785774531610386670675558440682408910511818429025803298514096157331538756311438547792152983663838215871358824

08201277838409736232647584435263028166475607999322148392715632124999083709893094632955985992872843352125242743334943790238249494457851649361270326423390945448086200283535

26261752981835525297880465028135399112847161281153414460389703165467739525876538384445746110351561641809273346254142217903310714720310599294953895958436885773489495225985

2103831596420623273071483716691798967444541841890372511272835300592982739374737571099277652356370360647348724784839684203742309758998743878765428415935659735883450609365

129924492587467691542804598132815825872999110300780631592481722052213206010771492336601003182710066727266488949550942336879354810557964237715449541371774079951775014666

954655741008015579341795983013187154617138382203332872631369978080937562816985753529253902365681143586553982842832417000516419900517643835120057569334304218029315236854

1424059868057389737717209032816486238395498050084360235358582554618855942442612928921434147898267094176760452225134929872997435333820627622406331004883774527381188722581

8199822194282936766600004037991487001856867455441231957351871237930599514821595486770105404782025853908333564061826222520802864866680976315610713764189090236060395412450

535703806675356765247266803675176738455646966935960226342580015557208962384036477132142966921934724207987296298616757967460829597127485706790346703915786585811127382574

2519039782995445767430575299283028634186242545022496241979163982734904359411395898404348957578332464822165249725333811193055585614050310507654858991552554265288628752889

54577367874202977037846847563664249476708485430735413284036163491347107446832985898095111020124254488464307122717488658696436722375125740766380757796859385182321580379 01

38851324567042252278538766101351956828652339460402003567338602520551347530790074689452614361638124660209433968818299857253465435528540463610413121499376921630261483151823

46942096279115494171946607206655284400443565753266414389342772209055751842369120803473798867079692283986937508881614607383824642000815393674001886257307369534997308367 25

28101494304364563497521354531951950035076482370361845384975636163397442943098863871989880818808674749583176022298467250195918371787001546471943774402458796441934330527 37

78617450245249707149907000518726929283458717863091748438785506397547781397976147102974805258930692216662225235373449011353986266028141926476293709768013187204140666876 25

54595422292493849462711775501758620213788767600298051574112378095519278181590820663636540356683324455662009516046337522568825585458292001930638153387365651794574370258 8

75626473210773227646623152269937958253816250741193599257543470320751896392792921623091299025904455121720931896617993469495415021868337701522075911300888689023857991528 26

39867824654608874627852622681424733188588572416651261959000329224404728408961960264923773072793032869835071995091733622206904266211379357378789633982192711117924375186 83

81757621347292273048411090528931273975665464440159910892056359539550622684903481778340163888074775859106047358664560729940094200063120435623081164981455149655513005854611

73552405216715566604133347587598792044559217567756328376722769011541649211224642236039540336845501134265424744898954599679203644242966524827350687996495015740462148251 11

16740163812882370549267667972800574632906194561799730944487237467063062834619376926372843710250629430239838747180411277944515182108640001558475791284640128739950977629 77

08262226345882505207818345760505308157127681646160126745615131039107176973845578732241330030005534719511669012581135208015630373046908093097927353658564913574711350904 412

75907649029919388200826217393959286123336572970664641027058783855131893465796268593304795602601154503596771014005799333688900402207538482513993086371634336600792371240 64

57617650036410612205435688688177406253057006023018982911091534071177517124423703643637158902201162317102635650130243991215404270127303916604348528921717678005443537960 26

81447698747940557159937783563996621006692741927146810896204073611634720258986246474408196120403368752089701088063354284436925218017425121196785699110583349449991683094 4

94698470780636754666776782538372304052848929117305480298931061328228524301397442127840108229799225637499186161909539509229235240387265633496244744690348057513565946504 62

50309625011185996363024036541878244570740245894880605074168390715058032424183755862679604489403118420715618426638993005968351960880991550054081911609426156177996494555 73

89362335095602169384530294074153542201700885059341080215377441689697655239000700113109469280003444356063607661310302728738927422665249899098159012376515704327731921850 28

44881119332011035710571944438712183523225548677264408667340454413536740399010464179288114132773295705233233998780091602670028929046700345506321135518225964545635655802 704

62153147060321476780387345442039887757315364197294374658678276336231119864674608317162495938051631791016021743160036372135135506555681162767164832287962390037143316348 09

58689243847116904830789651005911049650159928314383120189325251667689558973105180207091561282127947857682315030996548701378014203423508621889445113091741552012125037797 65

72630511758844557918166124319147934998793718897466767778272433292270248264548028499985675549452694687032750378394003665144268568208130902094905789962210081407736696556 62

79789587599381603739294081898326023119790605145978038449412185507347234440464136333171482978197669866965514005181845419763310556350448849713422360339130058979717346782 37

3472329230517388505004636025681998062728258112455591586015018439090409864180971710075461884773934911273571127107533095079036197946170873344664805241788806067731106455884

1428743120553686450754131237892050164182455985291702855298234917568151981749535650404537358800409736931002101619740994088572336813989068523058021522578307985844449884900

2672215492888861292502885281352717378031820762808665819870213391861211336024618736264912859838570424605478859944208240180919736271175154047465634118048628864398751105260

1860076320866403208005880981246682872769158288851453555992972145134318817716645564502666336275157142261212702829023587031467862427302335998951338331069080367912289759223

2090053533983610528084879743470505105124297994696958773290081207079728796535839232426576733921443804703617065295956729932344168693092018662571582035045922274601133491784

7686783106363023672435537093256269498230726186313109105016432061267424608679167037793094066960713544777204124017138715254147871337456602291427453682810092920558890079508

4837232678718659556212837654930431227464459773811156396674092749919903096783157044379273964166675109789264093117468241878846539287943914280719137228194506211199604942014

1675675141552265693285969399005410111647767529256494404287958357100368450907034580190874999930927342332379066474107462898117101040277833821450983160613718505842790389533

9496134598694553433217338838044229221868482471011714851583471060997578697619681601243733023068446927105578932616600129599349859749171845033446105624084001095249031129151

3102073536606699142509744167108918044279263850255766220625664347056888812091343129654781619845396751548210810244160624449318587351214286010858155871519419397655261062478

0925408142475964662701919437855071869834968769265751713501764020035993835301783027817671022024492886556546201055956741577115904728583016542256142005482685137191627689825

2726600077033683592676892717146614588644325629544170512168608373571659761027823884860670144632963682136373033174648717632014278800674249348568445726886782552555092500615

4697582885492108122247668229027751168223695025439873245618612099967380501457521453467701080259152981604212231163287602645784892088144425417823517877294636849168637871033

3559880293528797513166009650345021350087861481652756934254915758254478587897790042101592801135480971581549325386490211513898577566392705820047833081031935861720959285030

9837197795638466498733455490133656606295899331266703542551795858953425568522167057206373166820932241554656528706208202685332600866580058396609069504970302254534936941841

3479918148540317521615318893601698982971238272732961881513540418704927348526256664081364863788716802997434199218404526700361558020387500409637218865537661056462525859671

6231120914555806149237446224865590525941467834123013364881208645131781450546417941645672385775090452177054997583332360916182468663731199597425637392431936836066334687888331

6648939977087099239751769429327043157163405058351989947721259861246595675803136402007793328797865113011947679012284933455937274544677306994245626020238875493090223357391

8303966428565992346239434307543557661485518612844661731439799759776844709297927738276470935627949450937574975809402297195543701438592212160580810042397438533045434671191

14387122662709140126153844627736610886518271556640204899738718538427974087178039858785748721689263629340793705516018371405087714962816078738336233555978837136080966631521

1893228751052274037101841254829712856895416419492794385063945483861715452863298700743447464614650341446025619364938925571934232096238572840936220720551764698253040064322

8756038069773146999660101861018409083474528089280983391290914925830365117302996765473925151845027724484495376804763886401906348729677479902124856127316639984427361862308

85517318239967881715818320630969964851472957372369464794425482501448372864303542669964431539815277168679844685777773176724214993063597651813595392768068710323045802519191

5603646418455272288614825145974092997199452910599833472410418542027208513605430735748762273840792001676346615109061471910813300876924398905054283828587174596002008845764

4825190313755480860179403410944189883726523194071831370537998352344375954898132153424084287482442809898880471971054529233998476551717751441096350331443841574283608079013

4130163961579445590873662789091442759845229763054393408666782643140163757170561881345065363728887368457730018975435386415363938173762901822963330494418919406597305753851

2133986275646249847032791841511149121135250104685119008961170790218889188062488253842283641190655874808838120731232314134423335314443360965627192108247640392720608888626

2852588519928301333058905765272829571426194979164995894363177324749580959841491639960872405594058974095185184537010842391107823544795389772207975226175997379931801766025

8416783458521545313578584209699130699520991878609886124440106074119863744715309935103342861637568094850359275704744265896795661933828768847466738762703577987555965494014

6628998920998697164854072303398883936761101330378404511307837997043311605332621995442577030710396843975279691973081280251126223600777540005130859749830464540495130970480

3426138354091344540564134101462193716056552804448400880453039649492973826865022745282299484577467343378675502800997560510091528866646587902625776895712418793158394872973

8877148353842481293119168316060135430299784836863527731202903029710778302774738958134651942756160667428436070204002387686104592077696567626787819706560612033973047229654

8137344619132198858923218674391232241525774192907822570914140181569572845738336229188507948683294933053359319357209167636459558136799238696355674929865113248271394607316

2855012413231173726487739829651492342674132247288632846021041366966644267728104149594302767238763428606644807904842677191598564512608618704025727442774514307901736151561

7731515750059883996401418804973069975506691012924075303749581557846276831148373516100826421056868786356840858920119268252437039035251766690092384082646752617092602697104

0704714815310205739799768157918298128923530414649198759361563221245168274617227796815733025325573522302296833982779941603482649856938263973605905623213929485507427648532

9426710589699458926421441196000845353331145040686537313195714843485415041517234706871596658893468794776160506525205325518877942762000677929174286295148036393715562492149

2192899450678409720543460019562984744096748624653637113020873814175483338166165615185111911346847323655382485319875818181450105386941315804289410531085026258281571233111

1455512388549044534798670025707762174138029189276234523893914028052930968645560208707475029630568566687239774998591135620834859426470223854033139665551229405206772298210

7716988749068312321865667889253484373928939824303927063104601678559287530601787022213306811299142564872664971680328549439089540115982149377017032767620929876361529476102

2963864003909946657142860527160651172140132509592705592948397361299817981025671385331771066750331317827873251215013278377486502070331355062275581304810800294605171798864

1864693830142722291579435039799177801649052877130295624570910278494445900502500126475623251401612039820325502752696951967074235168421191098209014734553452473851605450220

3448865261194845794673910324946017546065949133473564878481268188018707352591838083903679072721987136512691128737953771599527413426674052986058267277760841997469706419659

0299595629560556000221763788281809646584242944311604341015403241618371124118334133690860407381821867859296600601526097920430269051432225681436574696554200716104926071055

1621362987930055591213266525433447237515483179612564078677742430707008766220291814065502136019166384385999886123275152903529870349532107528969061140401659802188280376805

3487149020830871917804775313608584141065967519604324017985891535324434233629910033903677261899140466810276148578721543032757524359320503117651703430242737608232710089896

214950638496910029025774167136585044898207551353446944194185199121456681506843573075876541271316654235668122273722233387587767393622874603410636815651866493228134230424

0400173053913958045034059680448257513405554904641633578438160588686022799179151567769918483857458197898136201297053787672660688951885072200744261029707371283569426697793

382087809127052674022819034488781068280495959107943088810046956183587209343232330440969761237719689629821199168787098339611345262017695940034586738339782341473219138249

7499140658967734474340283580315374798489967619249699851822401768193052100224551085786038469056876636412890897515543665065616506192218185586063952563520434789459161679812

3236057496837482343890555964335042927547726071932198308253807155388521773092429941314419026355802110535798866536264151014646071198559829289549075514813694409360717649426

291034038171821614396041769852817328208821036906125977046831405987658981426853170031066274257408280910233116815975958654856178161460643447651930287173430092180010058739

4834544037627464013824276340532946558088847737466836256169083430972700699748178242457419426260777097989142290035008433211923977359095594574681566469478110101036963569468

67890309337571102762086607087821065555377926088164133752939155961539106238153813148131766276032319888049792167796110491023883227506396471007696524744146098226259447672

5844810138808410145215932988735385180730698100946012616786893086024378498607208280266924451098153916959736018213287944079223067584129848490365630343690870142583141989910

5413985226930754669501999027720119934380998096571948285919876724145591715959557500060243914734649999094962280732089018531774166215707333892838863916441375939741779799619

064527740965797692825365348782886469722536557545252280316884710392694299440177441505654108392481853809722462655146890300020782127549450527915436981754966618751347831918

574125583553974077373341601561144528150171617511799963994061191100863047047957134095315827911496975506142525966187901940474528757334388929908002976987789308698733990273

4722936187651932972809463921588010581209173303560696088255232179700576000415904488179392298445358037974607129470760820065151683365645124128112940020791146082742431095203

360028505785103178129690111967320860994900367426065788332367599890317467841882737621128226150437357826128239235832602350621502538812050380876107577923411020066388359646

693181575042860662121240253081275797002578725384490578740240767751761182280220570076803317314373322497208953222317699148309229185252467490518771658719283391451272224

1076880381467647768256051991624289946664158685700335897100581727461170013131727201526453957506701723887331443852719496997537245850451851121245323760092430472399543989332

7632584741996602126125298056618497682305041205726830278895901298479037010123634777326720390810711639303289268969858980276042853098125791957324080531453599950680281647637

67861620499087220505717926326447013802103274475785095981537692794373539955990692011086845727615873747414978132199221009794636168368837698006801932672463563313936198022

46602908257497088761161261939179889891414720360559936908838930580535936933911450316665837679068253381015494633685052702160528658989694225709635345492408795324498345015

0231036833493083408235168291518964166715750476290195346765505045433189157265705149877638414907912672838031790537940390655134324257931330413249480760881046973124954534545

785626432924575397544363110660436528940344384293413102992185638619690395362293619010163993528535010572993277183944687864902771924119694776679674321691661740183719065604

3900076521196114835072075559291017853787705695420746007253475463298759180083020271502977497891528398945332554071951666575323092649513942114255404511537786456966234680050

0105576656862225655975320006948536438622303798485693682238749031954900491665783397436698609183391998723719472588452887254012846450566305472362710992642785702458292237304

2200103989251437607418119767998004961158488903136574404814727769349793351969079124128680495050177445358305674042673285789757264025168112914401728938893906007862203398066

6196507858085348249079437151059318692320640496738656353128130407910722221357665482187805198585300198832071946026351214279937006940708565595872468136554341671216007026774

8292362040145298505602122441854833782595541641910011069844160611193613415728438557376822437027368021054904985965165829729445551918241516040655118397072027202084640204393

0729863001390554348608057272087118125877938449849040437052921037497010016639981519494762949998642849373675253631752188331308710888079788392417704627889360773769147013802 05

7889504947811588756399045026857550561741605589946250346009210210935213094767593435082242287365273888374232113471060109204939561731748853780227314662884160388678815342375

3915600377407866832869398483480806707192360015857192029231113417351022174559411995983544456137561917963110704018047448038094383975482674455197750593665932950078695139834

7929873388781017794560817544738135591808299812498231500373506626543377645218316617295923566550503629887119356012041679383725200771593141903519272458016944939389396886129

0001191170558851515798078321975863643962234115591247845187082904022020705526888567677675720843301962157900852947127982339707670466783431019043137939095674184931794875599

1990551409619689392255733193871822401654043894242976165912825960645565767896269500675457661057497034947209854964172219226415181027989110590330653915466596702202149545429

2522568011997322331862993012889772600548883051801907365617848724489615573216482574475381614346708401825711363753394201684005114429600303082324276272440249343956105593 93

3073782790939544010805851085381144126655161542809528681170509607828910789971952998934216779462002016998984965144055336949093144156637478982789278074171170977983171522522

7691017629063753678298686927280589881500482306979073492161899553670790703347937543360494530720796487701633503866124771679889404617208901243370958171600709412249362154964

9575492391338905213792848132560065407087219295204117451146578362108110624285418078166194941845801094848226063578640095318038055317525908824674419440837169751263837722218

9990351660818503784010369641914891107162792024978840785035770140561634787422640000635581746899577459816171765424735821336343905574633670041826313143611418160332966762967

6001679942055334036403518166605499089721637891019331493529788162093954196820658190642836426624132370590392568046546366588202705763264918291587136360468736384505449748 88

3936255634469025847997339724378268667920048942544022238694891720046655825732880233249943539810894646293862106782615852789957371136482549194966699900465148330478673621 38

9610739799157349933726567917382050679489033578517034801568487119057331503073236481575437192770782678884988301848646539821183228847745990682339746215612585382662372283196

9860252043378627735575155507201016059787121746506973699977488505823686182397405962823899061718458168239669939404234038027152149341645806650609425516633060493107167197330

83603311809122267282616536491778154547133361395136935789707903812910081720867497056696396275052837274379791628764864557681076979043877788536898437300360143129486649262310

7727490370595621425875142932668782847880787628478186245956781686625830268203644597885098295089252594417211353558594114239951535124148861005811481337144762057489372241692

21906391313782116201787876086438886050658708324086986394651945116888379527746353597780059964225118812756016017915902247521397691067986320928338406008610218467139811 8661

2051037717967858647158891191978082065110498720927329367444664555227833315561279846519883483619760801583131790674551241002086775238220655461455939748978692163232155531192

90602588527663367002810850040367760202462557785265266977034006959215776156476425963474335647160218551276752648922167599801547259118353017790110214847022673250795852527 54

8423162615892967491280149800575414928943724074644381116109689127925533486650439477764701668970466249702173347336807947732108619344342264216085801303502463711108941668102

6503673752140328243429336919373726378916898338751375557910265452952313728785719567272503523272501149088524011122123152239535668141756360827968894201993232452749115000568

0661906710073056211312564018295049933428178361120044138708304541177799082829852391031652175532217390838633772407026085160186509754722845119753912039762759601990538382949

4982268416069423707468528511859666876879722986460275718450299124692064899469489774830505013519745922277894280494588893626617557558501668491138034540655338404509254779564

8368531413220672805295322177576799732975000077209030258510443246399340674510943312435873235986285879361322862400827815076560815539481580746263592719106228645291219548991

2738899347689840630693501530580839565139440243232506662242987659239682694103113083415119675553613783985422117921941144956191653884918597957076432673697459359381066877511

0459390561587596349661795677581631290731439520021171624160236387809632990133247037859386552914051894854020217477434941180746937803653261313994620894555749060397697094955

0080264968790833922207730630331527199447948997819818289566397872623599650538450840226160712887069191795534834127291556345078392909554962176378094144760392676202991209797

8526101261615871270753558678860701332293741480009805349019787651172350139260320282831938137530461743861854870731522878960438016260215193217278125010153660955395939482629

7850096472147696304142092856083675536971457674465825137792689300349818767309536656154094018181213814515718934852114137632397475912493597045511811135126263852182268414222

9057616723362217416385969594285558415266032581735696873478715282538546866170831715678979869207966857938520386732793631311097017509091536397157747854669238984188000859849

8493115913728360813414373935984087726813881576316452990400331700614355158713210484908604086309955458293249851269415089658986620196441030335698575787971913718747479419890

2398557644542753175912782182067919400959794488980216551994749107670219496596100713430683605884398062502933339291805043787495845783955975219184993991566568420764440157702

6373497360798043894373233710357794629495956975286424169076671625974311702801304335272139209895910162931503144061670330448713108889303292520953246042438715807137411 33334

9675218946784204352012005101715588810140727903370901390608296232573654349022515857159734383964325496828242539277124223574763651474626863042160036737390572809741151802633

2536477880629778365031696247887663049465890413981453683496483767637972403013154653988084173969361425867179381914048143126221184901047710700206513634019865582459549151893

6088602077543374425879392350378338502501167727052720609919380261179501593818710043945478642611784251461725602723804763286997966113116147174598788554203789603048511936947

1921087749508553862899585037777990447987976819174494142900098035975024431445721638798732593334748433045739408125512479377208382988604655248714452068120314432872021824917

1333324123894130322035955727280514404956058136514098616007905100746445574326539394539164730244554090449359189950093007018061723337167982022317326195486552606537659624 06939

3912796408926084161148803306464994054074645973148240396568634321593129943937077821600270955871259263938062653090637227159030214540827890353670167098152389832704238400 16

3481012749869965938331877195079369730835694367288565061102509453520470419059542468201989827961069973226213662816736312298905624737776292375576101165400655938523727391506

6906457679463920589716522864977434502284140350888083093446181683666737157251416382605626840092010884137838892076012300086402723310587403779236807772363376099963345491422

9514026340740695000357192016355410390524035290727326982389146458549806071219716833951774684572732705783001650434385226732890660418884113825083413613799619810169705273677

2471979593261177622344539819532047244073394552129375484996462128287798947592635564711248098447886354857684404007964540865343117844698666996315507153553467533182789421749

0529540847002365155937191300707653206101646242846057544591362740802494426430247420387231136814035011649167388033979628128823708626314777371109509252116696677284000 5669652

3553312357273447122505849334210006543804671533515249182317846165102580881801644604999405884890852354086151838940436071934067129236726069448064599978077724922090290386244

0744586970142059022198880684609651517094823060009646570143644076680663727968756940713515678279906490790632772090445245032485757928070822520326239683354851586069314597838

5283616945633618631495689575125205427583408433765438773575558113323322445874169130446233328853040341681853276507962953332567196803046626222693929242338176147406217694 38130

1816446836250602887868826388486228440768649870505438229258065848704040355625732567495201114319375477540770919620795371814882506166306925483188887162368525155484810807574

5353475665069108998902579006747116536104463548486258453296011160552481236658133484434023033838766947308531146822952900928520339707297247845950326397304709692877275566 74

1078792127067696914719162964677848426437554185756985108165871827151474955036156500371320738054521273860737134932832995067948381046672169613166745564983864106655385195711

4779798978040531315313046953559560125012230730135122823590308974131531872460657676506927312075405356228053969566709404554331001699200934016030151096700708703308962865869

5481117116447222429564592624702284383736316827618263814545273819011571508952201669555895255462206792074276777692308115226475110682443391341650025240406933025859894563392

7036419440753978120082182135850455473521574804387050965377446147811346871656555888351972792131850455068355394307605036900936296238783640866701294439893854444187850881588

3750876290001144690128881295585567752375216598684934324220643331265915748872953995338622517538068215345051124474409469110451163239958510575512312582940123674771515 7902

6785466343783299768437518167132466395464302078876838451413131120043836272094728966779743948889034039333934771491655609878440626123581223512954925625958583191636244 84491

1239536576587105078807396708810976691210513367202108849092498348141080851471979311983256014913407118169265377176694886325677385081130992939242797311269677458349060895901

4679711219754532868794910038496940607005645672377105349189064450652802955744757018578593533302534373141442050303581894725025659741992239008550333847929662370672334636290

3759950087543075025796397415020398849590375857682910017344100301639017484856330642201175201778857984274575912502607278147035731188031082540722337056184039814643 08667013

2641399107350024418877255635131802012518580814897540973697308148730540887173734744284091929164342440210244964263928735739824038105842003734586953107032797985022930946626

8069017927241787098062523829754926742717401093133908440060317715039883971756517156642506614086351975457345974732854603525770816379080538731580692280533069866107176170472

3189417223854132675676864108506936197728508808999120593229478701723659957911254740490230421735611643954893538440386619667832273836309911100528583707824962506145518825693

8516357639930307559070740917791768960090942166268639945930987166518752762806121677655991791029060977988760291131293858955350180182828284251271767414232794377249083446 84

2157094679010491342939738156593513336070619125183966348987890494708726441344580810241396525389714439089177752252411780220168874982430162373290165423895880298757500628 31

0445539487277012493083152494797471267648110417923690325649790862147591407233856998568489932072281268315037098991931307602227680917759601941966313606533754265145417078 97

2656421694991277672019356518712974238974200774227060081833146868926602940980858395534529813264337429483971371578266348938881758528599643215246849202217050423643862971531

70378612025782854723968550109472648686527339361327053170918496084286797306300436165421346267661010170035987579790699862232054880264185324862925109616879659807695389 76545

36145457445540016522391424814892972938142790625588597012238728348902405738552464234439119934502720657717152104991279089921169924264097040941620723180394969416889854 26561

53032807224682554245811114270095732327190155988537895755711619245963123390013892387272152786124203816814896467821416667587669182854585244394137306771464037343309404 13644

76929357832575675472246049237725453066312261405501756381159994319702788365614699745356186625199217747587896680220466677625977438338995660390403628298614827021386190 53606

63668457915145149129662414918969008081539878655838537811570342660344304822550131978660476762711151914132960639612967956751485560535966427176487338775484216680732679 34468

27374535661080150860574339919862152957878761118559244723527131690090072760229277857204073949284081028003889856654021555633375622291458982640581718488090352195923229 84559

19169463929579675300915498710901410398873834792493628931057971150462061769010546893013669125649607645519105336273179156006459648274765480572318894713984109860130286 48665

61626662957625009817839445743520393794916318616232450841043616455539817023339682807540816067678923505102764705204099569714819307832159932255622579133690177937093754 25041

78257657070596223970542412067164187424641557566178175183211009184626487177650911903345723077873178804849377654425394524714942409147933707351348787631457698510024967 49829

67257183895783784649478639854402312145464070231609321036055594619547608318410781549758552449473221438932052337349758294772936978542447331921658533831755522494658487 59397

46120313681769249128790055178403707516106150863284334456738495665891503494240520078974138322121492467179808528463428682044702757836988286573044737017987549883391821 64436

32043836027526113090044610037477990279490761245993811240516191996059651390077909634293583119034305624356715734095056163628748278205876154899881362284006319511195201 78080

96749067049765894282031932459132425596171141643166944161801552406618863311739950687964377815538097222929745986867436434377244602250082170131493699181402454209157676 83955

01316818107403428304112686525498680326457932318450295097745201389805535819414101913198483386798555481994017166158488361181485049186467563917728633058374665125190995 17627

86218221778373921604284381236585503598776831679887691766786403769600339672806404573452597520193285403903843278475185561475310226367335938463979994511977345685665446 7428

38958191103508750244754207501175472655793430406416448164000185489617235736987650020914631244006805066561821132029336859567547226664660246868542008039260740566298299 66228

27886673306450655032884162938829562588755409694680707065812198050857892405667820730191320067060364216836491652563135377625483089594842053609872295558275245931500943 34190

09078154671122572710798521222737556158326103023920531679278861411832982264597557653404542346104902957223958753311957961950228946050308356974031984824793750389739827 95389

92068971891296270970268160179994882368515441024906345037933046385059805006893688509215609249345461701869664702253261938819301682122648436877795239618136877735017839 8760

14287972084836641077328764288692535871469783972618881033845033371181517111484715753572829111377136313197782120245684609697774921963796847374969109456442146235274675 272612

85301800789573543032753550589002760724171082427797722732735900626663866690013522302560850972996315375782433792507162172074406333137963875171443926623811455393900438 86785

18424717587537903066366609326888319312103232072351409070336054165750822090603372016681138850314684644519169504365588661521259506938284344581528708712282931407275559 33699

68121099035159101642125607110757656344306351170567316743528958194754952161142519311002589041345289007507758331818122670748616937051137474051447945461979530174760061 067934

5348377394055213592998818346545867982587586043173404055460118233649355330635908322439166680612102928592930996275724503127489439096429633208730774671500777330083393431588

5963701244337695776945482607716097670861548166679423891035060904614604413039687136894886799835087804068064381621772406347917811916200629577770139937093439443217249722182

3195212537941326027533674568558608844105908511230270606553796894861190333431182933910876196185654145709689387436957061234280197733557389624076816315844335848770733607207

0640126367241684125509830095138195768861512464866019104419000405387333567120152878262611453144400194905011564171801462355300334608021767589156147995037146733274581502728

112721182646692255544413188398595093341962398594556118494767478653220714920141404358734838912071052581634490692039881387927289928539884606794699973386287843222510037432

8066659264199306084693616756774847179955538224978574655672505567489493096000388111650259900059937401738666047062621238852848170109467101387683522002537004909446671047557

9000862748699860758010055989739775274853207418346619399378999761075399302511442615689204855197230784075822784838312358647816828634723970507033770155108037216863941507175

8912025235200309364453816100089088130502039169341159108232754929969978413544833229671875424182246552003796227904310697702416765482939497616404950028309838939426022430461

6904843558047477224038717866934915392885786302298924314368417304703015701090230606750370244720033264134872856010032197236565201590949293414826212299982317332073064879601

2037972764731556363037609293837342346820918332034208803758319968924099274936352907356498472392751796483564603811318074452718462234585979449722843180527406250005784420048

300523823875108485542664866184058788046412068103591989839609872713115064108184590455579927609435421840067176453548615105282475682962685981806029377282987924425294387085

4120731025294049832789179127749003152175521488252603471416018195384541767118062521836875819415407036768156157661817204779986923314640336138033465204018426158039026418253

6185722468448606128873686999272027416268063766621120692903461969545811364344741598714018921166046622658266159054206976394359312366204552875603421650034736011943422561491

4032015794171185171542756396517256864538467095452483713059489858255974527756437837209390373760644875780538089666661399183963055434635153154818588677926291272536342628898

5256854464698144974618924149586366367198140065068588860860224267337988127687969406497029915452452721325754281953249173115066208586652077490952965100753404049227356548282

9570256906293588816904146510697177724209554461302585438178630485080605899063738090543069502613842422270537535459859099326696733216515194817253453274733360274472585272453

9478578704905484758633115718366332359132347588259340641521039072871936326796375928473313161233978154985650774595742630192501361344218177865732684594980392574196969999876

4598249567094095595490645143192997532969902929018113346849189397316733740473761021534979028013172233791279986391471010573645808824964037793669144260225224329182203596947

9652296324150462593037636643284086561602312161099027177979404824423743772421754532743690307492626172588806522332614106033816532093232026699108708475868198563990498575011

7619963690596992541043687532918190720041575982623454667270157369711333570414032093793451266060707990655868796161579984934109540903212436544310873161586375727374817450178

6655737939848669229117599204342247600648597600549782806294187391474966456601976892636165982896565574458040991426890947249706735220470116191536009452736325306664402010023

2018732278197614868663489891327347014448203242931178410091528333303769111970512525118902970829429748839813714997780527816493437056043260053526981069189865886898161088992

6992043454781553574046529382255475792396512578169849867342182185312407343115296082141120919999406160101582191273730165017695811861903668977927569046785710181059393731438

811929147494435652218962602863258663651901745365921218638768107774209158364690909165182739807531030664998062448492774756188452973294727139148997268407785897786865604 8723

3057524228571724734436667418181232708415917914786168197800328752421946480119315937935152394174090419098412578099090388940779420467270491534500024860427455307306836472220

758930821994452347214521842825620929143768143878021396936399786022126322109822077357144129412640653765264285434829706936551168067283061481553500677992734287467174083 6666

02053729220484840870252301258571791456966579523963596862706459037207268058794398134006676701411765812522334810483868677174058797368961559962178465497307393409546604 31458

601540495615776456167344721264274687461408302063879359804284966224722325504660952317681724586362611284833874076507582889567745895887361095197420723212623335561519338 2471

760218186838953006797502120769043883812835625814501250071190271565144346270648149505901969961390405607906737260724112924471994770283848353186332981761999473128331449 1596

877750429032797703476193806295512384013842291003576769296985959054439828926666087801034405967055905207668370210159521951394544773118741072352794560844354946676079288 2681

9356764666589161362140424785642792681562544850631668526832752465600274776524127942705341934828054626702815992453912487393755809401259197783465336557356259376680876997525

746027021669649252998377538196938462547085188615104752264751364891883357381891697121958326810041957653770211921182725665088966824675048666996598850420411916153320898 8567

2380926018142811762344795582942880858136988607372754974912130687437421943014866766259691695195368856911749382319697559557940217938873722515159971405447289708395510365486

662819650368486846610392745568414363590177440387565120807714563648777284438331563572496836756314810394389680921192823374145076186305657777377941453330257820788793371355

232819306677810347448788145330227671159824630146773913180646990171453231819572964017138299344586652810423902920440420503010850372288970722667320416407583835211625547 2935

42097706718652620762946720436336432735056632041125212872258941499762804689148555075973614650511764125793028369745320360465061569238119721325105618061345213438176817 32762

84141810216441341702491755843656811454797775195828066284424975037917477236204204245057363060911154840242699564336003530143666597511828032803953421054049191030567314 21075

413023579348772342390739593893004368913281485468821970534826806406133447603574659090656460980097171769957520437556354521685044224390827808348072156800312138051374475 5738

663358066128982569525355723266307395195406369794691392739195092970992913488014867607271497851682702690507107678965765304404633626268887226274293120287517384975542143 1504

998907456461215695542535701520147462658735882915448693671682109726387703669298762703332692676405112356591787736324770461161187528378640880382801363490414508513182955 8738

033657159237554043740366009431266249374473501836940883501231805702551089418469366688303806236043616899681815046281454399370843946723795281953032886060996315478054110539

798317391048191798793376309918240071636953592567835866999085256828346179992044849215828682554266066694806590558837547674778900630307639773201191626441931231373328223 6418

11934383050586585544982998689911466813117119421891601793736025759632185321048687499204734702994267871276133342376683428225657565015748972028043318032062448495723097 39005

715093145389184344943828397373451579980515625916432262701418620623694475930214105085028120360491099390536847805602166270463685525727413290604228995635102522847519070 3082

532738499555533044957803309025927531635221809889882629115980337112570172176766904545680649223051574746891557171017567240354189350611288873024043144319869586521866760 7330

385490360277460963545019525296534070301597032409851150252930588656719011250832847149680648943813007837186239244681790216217356912288723948021648464737751768189421222 0710

3565559650797948449900827193552814791434040371387172081760903198856458746108109910593691774379471287936895032477418648580648196799561464367082487089936839513150720305653

0300788685982036072066991673761641475665428771935359104452116916769281639636456297834073706478827406183405435707741216138320287378790378585893274553614956446125505055478

2668745445509088688946927185988949423449507483821850184341303236200466807001917504592628408385053643126768698034026811580709810345898604130841735509956944151799175432334

8063307326133913997978800038210913276601454961115704285802816067516233813538630524295638330950320192878164132492230047611795358249560591453000424644788062130246897865596

9262925763574028799401356921167553904002699664556025682972369504959099602717097381666168678004832729299059428102968161574696306060610106226717021253966747951389381374853

8953517578329451364260069361377774400056699301740201136677317877044694506012922607496911061576207637893345041373365119560032907835235964653265747497435286721116298075675

8510838501606998969358671522596463057009139087628741649254170816909684730954886023329828880010165296280897698772319690742102700934880934455512188188336519784318535596606

7476350722161178728735639465655434276140685594124259121417078116303080101925876219938098958943050939682518277130330334926648853295618795266446319063493896492797477963058

1823776065803540227966910818093226725142614287685085502340363959705621412515137620467224442897997855454131901372056002945670634038242991780751257244844635771525893472233

6845268000504905794076592611834046276469983532198933127117327105112938774627200813172632086712472478310369532505711668466933919993838341653013309247729429358470743882634

2400701321713209728274947961163566782615071566282502125206276389751566585134060455290109261126381662423668992710649041886270014421128735923939982500565800643250607509 45

8935850072180729322308246108258185874671334067334432680515475652761122900942154655618310341268701722865416226907574466637402578505819839033726685391283425113889776103701

5547059697842843812110416674964236193689840635613876227959545215146892881328239439636612945013878167659092925089753247166912368348275154607827280445182434537596550492568

0648532309992814514753455092755527796279414245574405255218825323955625085721179963530195675811774436762550973197511565148451372854240749048380819955880915161179392191 04

2619092848578161390514823147445531015161591784145999471223905165969441039727295741158339745939062050077580709596765898924961437343487815637744208374999146183451341178572

1356576510028821653437883389069722998862946226868519562883232057053789421789593678448999728380822512958573742956531487043040321954330164722045813979057452880207472858 8103

1140858798747211863286489844734053114604400480011189647421657199931158890372059003146641812617920911940887850066778010999185461909348926691850911899853281758015614963442

5272742021323083262625378297537478084964862057473772980595049021455501089693480645109743330059979855532033131826917519159195006558488436551656335235079497441486995545934

6826667617498419319232391985849314290703397249074104335315507161857712844527045561514872082178790107580994599078190340768484559080348525612112446388323887760077256405950

5857614506172929101763421062016322763133570869841611333843351915955219934239104565565973100823169479978456319583411430563705888413585723243945999371608451544322473 3325

7474326902932111340553605260907241688375335436141287023423782527089538235765851659641732811004236702247942658948954200246277977930989345076021362417137031065132759759613

4899242700470471725333264227742623475663202251798424952498212765595113945681412700766231548515825735963353826717922669936333918066855094802896639701633580306357779206151

2929958681816023327826828756138695348983385807840486775650291623281022321262862370364711672590911162207449944470337154671719355796034064957218399132431571758689056357820

1193562277776342162367247566768761722061015213389428844275546380391429829340970637684665116996845497749248421623454987901900590122307110248805880499208202673415213345261

9482271804045246592288442955419081544921458089045704939272983216883413429953682586746410836728468331328743219812300745130314440076835740244627809770130021421871235118844

3907814286457148671303162743814053388576038852946905077691124396402844275392116011246848558859087111204331369833551298317704475439527351180945626273650721342386754878162

3509663987589890781058433372203975442049861899214534394570010766993023334045070623558228321979391210716339044784574064959683736835999610270559810934327545718226581916273

7640832649194463392541412570472789300121814099502881776632184985453085666790070376265770538273174956120917523247601272848854422298986799882662839508678945771438815 80967

5505801148163295101341784549495324847435024616682604908605434986260707695220684576930888101993638860205690810166450518721815005743472746924004566280135244242904323796847

6832649959785995418767960988824064974266522958440697297761914626489783151028740066979236203320604044535479441981998947525319571705208553161777916410772764613921571177402

9913555501516709796619652406222291416099722702986540871469079193291110174604012065208119033479335074093396733543810667764425162109106409782740393472409226971399 86419324

0183367625787951661669211485708440347311098577186041438065703058114089652279903370706792777717324019606366905990239660061747596602743647723341713211256406957304300738718

9697608262537784076168904565036964121017260551105063180587830717710457167891720931555100513126248850749712570882776081846297351566606413813185775692321942216169819818338

6159109111212964068347645414987489259676939155208899974348348251097717173347748490424157044765736574577528870313708739190721825057729931772042125966177890946278107374893

9672753336694977597757614013390964005994959912477424058226022767434791404365977500711749068727626934563752876279153861033480986929574289849000870413755216037594678764 3984

3626959137289972377200731453366733526996836629260856570583903882359383118961476361331704366529764436074940164969717395660023248475312782613511627451693498591497284257801

1566354011257488383449578205475651348675253533930949298777856682473832247136341278156638245905023073398325360830038730248396399541840286629808766899600543606746374781759

7385932001097038400943290848252148607855800720302839262481484210735676894365084782917239431353773078338286204525607937143479890247754480001573891165778949114366003679354

3639320677626302110521521013592154694544997058887628365793340606132391102338123478911651339606482327343027611585343257078225967456692639944506549281999054027881507062990

3621350504523173250136706243941061807667613786614145637857660442074924229269770349845012651492167176963310516767267848837295490056786572397844276311347324977198906007618

7591408960730665821513844913456155557115108451321597382911001114287389599461623938505836032323442082881450433915073780818631937207838031136417583734875197650793352053542

6106483964680022831803234766267824389703438282856767409938801224558418282506165887184191773971348424955753553615428351062448328211410756609679699510482522794215870693151

8686822890659905445428976768340784284686635169507359290059446525171865184120454432711637452459124940510317749743271643304786104252035794032712103848084464363330137152901

4988427523779498345747998752815119651594344029887214917065555918493963736202353463688218131437208134936589804188526181461668564638974283915059558370394712383217345273 48

2136156961283263085165038215295650842082471083485563684152892779777775636659828621565022125074611675903211654209654147012294283854575672141738992979980052464181684733481

8021732521988211950351846705828084292711592599970153750974790079509303318805655800101398081229845654681618715385799281286103766006844083084983972797637004166030652461 4826

60428311779328935587420559345188132647605029917983382716373959860235887851346092673231951588729729246677097635098946770140282292245909364931925193174285969415236656465599

51119254685886480010377793163073879444378711516343580868385509803616734117746795525054156963255517565930011098710343833359200752031771871321169429737366650481616228139 74

36919436021979331891137147757928460485105745428328748179328722787109615896826041519103216035886779474792592988509994990492164849713854412553391673798326983742448741274 54

70962063371390474048360794157529363363904481795414111734936202604851201230536055436888793862914480787910052555989328252773354359247067399972692273992775653397256696715 24

09677307351761299221420255506531429549087426170533585033317774602589152893788654307666139718694743502046969668750567663782500133665443412014978039533409484305880408087 13

86953158989175443230557614844599297128725620764700238088330825578637544806735453859876380858476365069526736280986119900402679811302046101019445980359274335057449296242 27

37733955201691580053320669755130197311337039125611913305432979416991929482939055726795795281772696879329114257773205002147601996981522020615245179423898458185268279789 79

74405416564805621008330286054730103160503201457405188876198453502969999264657956500190448278241738609540179103702546381436244525088702922663923366404351967800356654544 36

42503625079529648921390656484140182736971450214526431646621003738099504185588748963082465740733426300253099497815591421287519705271001157101716377498833787956568653934 42

66119544077414439924874232060422661597714780065489643349623583962211353125728982013559848156570204574821091737378662802539307044655436889747490658477949409598122447874 32

21866015300973454802469617267129477879519441513440173011883236744872044780623663700352425861765683808485736885690237092290882122720834170089791292565435419407826899305 5

73151916990118070501895885964740481390462609707341954494227067754033537102964836062032755617310215914346844155390956590974654992791537633294735996020089802640794682925 36

12778983205853605856186118945140611322242712861116686725025703363488631585717397116032882123866374918745662604295318985208791769279853205968708273206077204908756249762 39

85063918737880636107372115907573362989705665746883773515985230620783485667267399450197245706160417028656143126685079557168950806738619613357613077933856652942611789955 7

61282203962786163036697657292746317600522624881300262015934469599332301304443908514144069612949095143511324043728180378030527016147459666973182339616557550910094477201 12

31301699270821104372848353646580227984353176487604395208073622595864078734581915310594558252403107537721574321457134477818430984474999189198857234881539219045201566171 92

55542998753345400238110974852027005371147123897974701015854753862680281323170286201290386585144236448649286552358171372029388017529410102252028569118718607964501087652 98

42532971631174418650752018769292746421061313176203085067906658825606161624512329833385602122519961320728581640702090623127183448482300789404092439163274524087456678136 7

35570039755142780739200343533132128228732868954206188188329141890423929582934929305948139194192101543032435674476992430684895495202395456455030650470971258953086410011 56

96873272805753462888209289499803769427671523724690745276874941173761124890701919879296423247494837186391329220403340476272852904805702005887722635976788174715798214217 64

17664349005485153222335130502004742660842602758111430811587735785368141365683667133917403160705650379085284322678216464359301519207099123433844476197489701363751895168 49

86306334712439357199345332511924340248672268509671224442280954620967064207146038923593401068440047536963864975135973583310937832600190925157443192037612293770890557454 3

62847738453351375649811641233689229522888668519991591629961786720819183717347307006822812701860650275302983390146699957479446107026716111602787170650023445485265531804 61

527980301358894310966438222475439771623046764635318028199644937566237115116051978758708342921467498000152971410916725705796936154875615717823583111771013590125353955 6871

274579972017592606546190059379608978461902722181450723587954842714991331503962030512411091650441966975582513218186178356942755406145597270535238267357110231808197208 5394

860182205482689866360286956681834866852454461440840695218263328048760444691900826696762964549075457223692332744166491319556486439945889933877587009880543318639985540 4932

306776159182859287438960578046409842089740550683296113972392222269790364697767875517303966644741574726584654780659639564895358195570035797166891226946992715128448647 7227

398117418148866331928994659470600892113189894296771965704861852768613423688150000417238002829767005277922765540848554333486168898497387186788618987323238004240096386 4067

984351716251126972592465867872110705380153194957716494850629815798946941714282042164165586659907286198493849175480269584619642294779314981223836415385570380897890076 1390

103234971796963254719656491227455826354132341424364357459474929792785696077635914847280121218205712372291254433245566053407484951814467690589598069520034923001249866 1937

621085005123644254782643573382132966096697316535354256247308090288177611337397206298364305054086194062218385024498547566872126006763397437315325783835487482440978097 3933

614873102023904533809474159776645603137681106298921409016612327003900505022947613518859124106470656031298014608889949278623547812337075637352432127180061053085517170 3405

360330737630187113669353217698428260176112186006358489653414360670914199779246409721142744954698914635554828643401104014722300847400589719394255567755784403993657012 6377

009233040177201570971402261897254902499639625668908485897750415713042927159289338014627628178042451243346115672917087218116986695871312610665581097155155569633481984 4224

937727899849000169133406501392255837452536445871131537749642845154005364042185976909329790072008362962024673223965883741317529058662626944673521044260379193152110356 06156

132717795833242389410113783086245462951409581718941653188260985813625507028147207441010322838569770121266793214647280115968243771101640588292038122298228255056488502 000

903159908409669801425015995974256102202631836172557111491392144361101853384556828607031576644860576825091946215850695082849408530180664499120714286759896688412790646 539

483571531979162968219164287626912567216947228774367226907463108534195440611336084871630083220781537154815435464583702594850361674707073027584996723132809601629361233 5740

850516758670470322859872473980864732679403949139379130849736324141390294392845766383519842186763466430180686906092143383196042210094720984397666522822544368304222054 7014

325651194268703971992934248063911831147159679288276590160658481161372294419943326376901682224343559260799088203400234350990585913929771576049472707702830957584270709 1369

770437135759020267227121355344973304305067410370544077345958430922769374840395638204885926470438636211199354225500002561144970850527050091628929604984739830597708931 4120

419837701870006385807844261773612787580995515950384666207484815725018212354082543379820275256807457417795530258946819457764606537499326993288820539515184740851474436 3568

210924250171052586349534579159028715221299536595915807697373406864938135614834686259397425293493723922991126095352789014741520946191693752339607918005888183785066885 7882

217408930223767078926490559017835733029049861047566240884046301744209164607927659240335005213697505366658033405013128071242798970006273729152590701666746437245667874 3256

001955542420944533633439391956768045908713309834479621296632581163765466480890071124740281515643434631081445327339715432334544926861930744837353290342429022706323445 0908

155379512377571610492712179818535358509941553871821944942708611496926208222831105450526017646944214984796916249383350864389979794372572585034947332112364201564544595 8323

21025867338212082720110236171813291628134769361336231630000555185416994512370308747176934753063809491488282318114601484735917650196830571470471397168054171207608541527790

61484081543527705471113866193551918255549673768753756559018915892076743526148852937632107873123875720648408237375303288464023488292875867457517411964259554732547129988 78

44133769717447828951480606297501598104109651792487372542415860607092634345112218051515768050395232079839070384455908024897675124281118683112235359449536233280515642845 09

11638253284682769037798940560973049659821677479429060522906427115409075963049075004578669480644240791416042498973639622383553426568824892490309401353227598292267618021 39

94468018996032067580342939794795005811235633986972790236701977628984073319861429708994553409037260576822344748869201029764887194618758629342131709327277769165187100249 89

96665855083811335678961748113924008069204414625665382945223391551604594824870277524026560308024160840633583104991593130853947426907323472088411918202484707573291150721 24

54468985268553115998511179393810802097734738174933999888973856536994038759525336217482394715478348005894603936659188928996217521047163046603844441237345091038293248389 45

60012983649341732042243216564275828626964629870549437086474634277118529382480439358216019860700621711965918325918075744936556603190574210330697537073223014044293913607 29

97423148218620571279724084322974222530647847022287728988917204454136416830186205926694781065001577273034683998295511059964275198344209099429823941361558326538840685298 37

50196100267432796083226766089824373992304583819359944552036471095908251899166291593196398638609984425245591944423740980765055611750578961464359199255642675963119627783 75

12874653796523866525681695700326964287850720184716605737922725209532109907611270241949139628074816965149432043193064899696752352377801336017153557941527226744043835477 4

78346154171361080719710473088602374503121389008132562431720497688680228292550243349395991782746769594115544309850361643131639442573259674614175806634244925060402322028 02

94698737518293127065541377829886195848109350266364520750931961525082015023951281069618830208378779175314247378006661366107125561380956875490976302248182547336957774425 01

36333465052729624195150307001629823433990910006155502494673712884321972353399452938067223419621701616343292479375744591466577156266828511079209801382360277908524908614 06

13238204702960336142123967891694992342321715288832979939112946652538472686405318731979322677027601787151571127131902216836416945329045195204615033553473446978714152244 7

08770708456141208314980110666716520746555367187872873746904987162762468053085758352280419199560732766591756957730028792070628730635622829071093172254124410289965621943 93

03393597931272982490188505998207530280581826873453626207698842883892896355177269977552708928071198383271264164981358505660929746681941433220376360160317023864540103804 19

33756887455935123839827605652993796946311157362367658035157564368020797909806973589289503336310275094784781357000647031676531798437489385931992844679705050146584222778 26

96067591977743142284898346229638095752245329234359223513044120345909610127448589140505827476775911036197187481126259602724586676557147275493941205551336228916904263820 83

57399520615723946451744498991297021106570959449855647567105391013640465590616591177906243645957393460718577711706111845170015454508099885004059315587509512061655456314 62

00738739443327439156540822226551667114981361350737399548933917480863741966480932781710026395502591404777519347205269361172155291928489112851231256310527709723493867708 9

30988562479735893275980846342321851624985305036273164555086002448011287948708902187528734539294131614662088148268086141620159155491220419860259848860991001008935521986 00

42743405731012142734029475943567269776285427772767597954067832154999870802605838132869028183862100061003376423791980019442370442063331998932146516974334576991282231826 11

4090708986144154081991473747433681644982432526608166696697336196953321277769129772635798430150971081585627795241014031219725399500985483700699157263817493342319841708678

4859633091293669773835628870784080023922357823103612292313343138708713756072647955306878567678761408679785388418397508504683509284144719683435669245391453969726703865703

1729624183897923454368707062910535840958625288172928169247104639591376543397510330386186905416278540679671885645231462534683464300203066362430077280418391505048399774 63

9065235270047668220337015691585237701399054126383476484663841179107634163945096265761345248340913898753793488871084408225150247944719876888399200357379260736576854930155

3430268438483889314027219668203872768490406507864149839548386239914432710035484142857146636758141086857568497496424920588569843745946865744801849342280279582356377564388

2682326241877622162260709045198459326773477350182854360693935241658960117450737611406406895988292994464538186860664747288899190982296017892978047357912479632186915368703

6559529444334995425160580908304927359408810125128045910650504766496267582224133933158027094209234354824137545730590871565675676701092095054671117837732104759766979436357

0249991724776409909961842234225938966846991544188772094653007034437183115728705732067398759578214079407363342360849638323335917902277126826303276948775320046801847577053

9434300795119666775243961591663082780839590538322951127233820755074415307889777628677166251881091187535197330098637717478618815476411602223903031955967898153733333258298

3600451889734131385796009297774588061424010459150147820679739436362919935582276023675103478275564866271218825552853178253586010351082258145026112047470924017186025646900

6846173176799057349011007272876892619452758833587582219473843208345786330528075574549382895239005984568272914136432348817148584606783059482601453596876215967081249 55323

0576378156493445652578255229382496265751170749498866876544103827053374089892104036867736560454285845169503140223236337570264179657785208175448924655709924081323665578698

5264535388801118891932862519259222135566768155760927630615575930662647392608983278347680214605557131593915751341973623814379449788885811963343728292321966320335780126113

0770108572159898202801245270141944055108211091262826167070806274271062080737562374739117901880630391519189825171866613757572759710389811462813209411824321201157828817875

5591908312841658981012599397245828288503470905302829222517897243132948789737432150734138953199236450940330089444177994978550611469561529485655311222946235263060515601 49

9246416469416535817179065975346477474751894490338863769454910133847537957012434353283214492982757320649460057035879741372847308556850024140578069949444209656454542067040

6712691770342055893545921475139946586563795324989394807296345359598977325035224253669755202625966190223911374464701515637749468173440379453288595394926777638755908647 97

2478088680068172355853131435632975431766043925383854057568997429850951712778248854066092032655982812701957135642813925406613098387251913882874323038507006601992157068887

1565313698645866717928365735525658627188144014431714367937481059691370932121680624247162372966171539343995336805414569867938971784889101598603095412935298746169109192523

8097429445514885583184994985947959024263664255486555633149346893561503414883748846312946530056598074676977360194559815286306276126267663557735976758113842161237331459708

7298604713841740124888891879713327363626251131133467653762929984089037789520000853939976358474428197985767807210489634590077801542669717456754261723840220327733647639755

1754916653384472733685352496691562976924834362650474619882335945593854238873901364177046945395798118752712159776844251179958071694546686174982003891413675742529519923536

3028640995847738080667594169715580833542623996679137191811974565095014215741445024655942823086685483454755041753280904924396975773383406920036596269823806321058421168311

5360806396000302983486986250141895196159770985309893415906918264474621242983607543464462434641837589126993823538014384957364332358908803400365035629450989571731821193875

3606040337502573311825257046693447027575665724602024343580517617980430500157661220685910711916447836787755287555765149553864629003147784335242018223228186288983604995567

1009471307362511750465682887493345110800437057008572073227508319673684109210872646030862698701217604409040280697949111839643660521843149230735163158890444548056797832255

5962857732503063907386237033313393794864907355954685179650429661825531052645982451735375632502837394355101805927205309010514290924335273127869001515130558740539553595526

4903028131275892668785026838027753980878419695424447075982652771476368013445122704646965861601405000598135542660045553041256453856489931603128055964697097277176477513623

7931869535642851430445630127610428264940367974802933019013443998609284058688100268883287786069070057522291637884595429511271236416290417249259287033518121777245720634744

4140710457987011683486538940860879346428858140532372205220333882782491606550751428498895026730473386894146657715191706700715462849334130545953261782576247455778605289070

5568120939213636020794840019045348227175019919985035172181982915553739405244748816084011428682985421981541739946151944675665399110862570266172891572116167086128078644225

9687806540552408407698092622979890897438864871881241215285862107144171314682739415124540231527184641602104587509694837594318073620219293740011902747502717656734359424736

7066180398677673056064018589905357512300230660837087116869147579133638408623253843492671860661390466843378796516790070950960770044545355631362405135816161738439970087207

8005727973747669733000195181571665144821563406218186569555695230599732096135279604360849164645440098534029608616745334380968998239436938249495509471695416706412839424517

1412601916247382708669769311806960971015572581478531946257461453503976260554651254350214524143432982492413812177132034032371867152062735928163374410315471046396311177083

4535015448259457608665977571739145766095877536672498405172457008259582124548521961643789313109882481700384223320632885004325420849215944237347069301246933339188876395624

1425406832442961396568624031641284030587825646523428183365665189585503699094325953094250608082468414255314370373906651098967653980873587499377033458953019495195170217522

6452073203602754639413113711611622899004570802847540361406381477089041361896398146793609058294820541858067653745684521520114145135847400424917141834979108526755786923 64

6084051029640568547162911479605166051298601164218923584706774478369791634369220302198889638490152417264835108736731345839534421502246261553366336148347864323356138053007

4966762084287575307448476761221295204640268425615740219898496340903610921847604312888296926638081824774162432149180517457605165276714859567236058544814854402132907532311

9337897054213766925989414624437580625161532179973469637179645547335354924900140560743663106474176679985725698630252839944434637993051824259841257983586495906278906953651

8549591816028252311296488624546624741498780234896199049347281966658307147577253201586835547077836728257562399243554639377454477973382960292239539101363924842457990895204

6563980451217891118683684611736749565952373215883192247696642959496940073685050284037768906265808500246717295897639915288552871279469207921220672013205280939645226 3227

0868221231475800660317858618410693450455257909777395661801233418741794136221578605007833903459125252454048575436327708936373481376250658388020484570153291728775365851028

4304303738094645827944631703476961344868288054938747420783607209181966956077978078659557407096044302978609309079720730749607010815586850159480975343530527293411617148317

4733966100517521700230147990108690345229770487759840572862989418102152969007455565972433483618503777150550860330111505317616416961877236613812722787109865173452038573787

3021661272225806239456342389511827106389993281939468089089171426867874887032369742369825435588832460845820284023483626235883643493266480175590460342821517219563963495304300848416621782715149495954090117944885259504947265569119457920367936540376113867493761913528838976548861213181153030471606196867043644288434988225523312968750291635458654268660889460469293705964951284489740485678003585270403993562604898282215255571996993525794545261740743270859891130014290671400225943274690218982995179553442748718111642117429343466185812595775017541353411801559140012523994839390176617521611923006320392693503074408005564853217364811681583302031705897646782329202381824767644909239971574846690066268487079269797454361050266597917187256545818722599567184718953968963563982739116945400828677220839735648520196059606726455519342925230681863759467716574716398510375801026645131490589465320117025939029809267215532618811216870595841647293227196915373233155149878813034739889489625427170463108520020308934976074748096952314381448595233628759639315703476429000525190908752531657407924492946317614112816050060433236781492170244718104090002523572907450940087541909448501323427737790282037687775988388910224289465863071885783632584011400402958201515727775055220497617418065229681281445359630774683991974361507755608490148304881526622616887549680346283304068496724884583144896108984111641853472467904954298423368792950285053562273808693036290643496106389027342398164443712989675246221349980717909832538537518245143182298170149804714744052082067717348293075742446071778472524594618574890493050965097953905422530692123607017038379108571469830225772485251738459145607910558853704706629305268611514962570167775660509871522189311993190860804095893272714652003159911804363740649554498522211631107924025308412085874327508302573604673559050242876200960582178254070725358819424274822906126115650670990690000964622866669193502613356980448499030606977087917964203449470664734358313049859323970595890765212059389769761799954609902557501292529505175646332819377848179827289216268839791503902841548928484050101832934301693930859769188207609832728889211355169823445644473332530729623985792356457676844465574078818475328003206204091248503790790336969799857569854811754811838668849282624893373134636562096236436017604756288482557468798352316689203275812083119267273807762830879194416406020746280318221576402945658339747608798691752555031704962919619171215072124527733136375472863049900375024593485960032115144992840662158257436742274475501063912242188903912068857149902812503322293010196259879383127482079514574663690869011102131053057387506102876258248047297829759703788665270217441124608373700727640915037133336149717400905160213542870186599060553712590929896988757268780006791586909108457407802739901018725834025027067523492790845564584723383879369483932121937056631027358110963094423462935733587439546101715097484176032594835362175167124900482878786934431786340777895613431476530447271015873083591865442275335060945004542709438295952345006179508154991222606770536954034708723167037735800385888201853606077407859203091001307366861533205143094832971286108366025245559269732660010329761411191743742767827897475103029546503081040604842126629274922587131958043783255838142797282061046716445453966782750663376119561547180811410663728904450860711216506060339892385555376753205387993468505034921858653621561162560773785078336813948450925056970346054311689014345656230772431780451284414990211879930904828898966619644774295261486897574573204681701313930905117056129681336246565767032752997884256367261046845139557876174554261407949927885159419323455730645855363767666277904565194687523590507070290264235937689217411258335714398554727179693347166990452457385765734636323402095801122354476444172330199688759484111588591938802652082412625415775923953557139009940619257885762438343967082535985086771745203064771259716871629271981108

7226407167316203119950574953533507855790580552280567687094003588621450841939451102129664180301025071900414351802625839184169633428710839244701121728427303277470134379841

1733012446913775974881728083780863283584806041092422086576772875220996324008042994492930498688498984582499837138589166913141159480537977042001597068934711183157338901047

464798780815652192644112417536626682168177076943238146633641986790863825847134143907867852662542025507987500598344208643353203403385407169700485895423819416463202364499

2186969351976251487589536447516344494064161989416711341044350148248437987463916000978580071488654135135723460466234792972728314241559208002510346789545427521942413257040

2630697694654016135485468798571442948680303910184410863890414481123754437128533082399373283668196231302956918569585662741137703388585366462274719316725033611040733156570

0765207124240797569995015171682190064511788702874635229298088187710072903397299225664211305601375775977190139941236326728084538940031909615421499319261336411225553601183

6627327838526740198754781876353935733949284710295825287103809975654397325671294875582247836268074527390349037453906581151941957264558587882696188599474918395265496354475

7136504122860593117832774704317170217555427338113164461220577791460736577914630762301569877794279947080006669333908086631285203725804287139455275693441864382832163075424

9357674340668984292481754076245634843585997894799507358408972112726010980185913187269858204360215449353734228209983215127996754771510867255688821989769067943231991850034

564659754689420908586518688541565005301770434777447943867270393095250807148114388066769440408803370027622892294039495464568694673656276512157444272755618547272970923160

771008330327612046440019501088255436661183840175064330798789601849572564092270264536383848782826443778467467889452186137353554365637760647667817089845054355114691231427

41416483679764597496100751751595807391647993195111269366016858482939073318797391690878819561867435883137351553906131409862751521295444877104580897721910582763398947484

9102783391699522543776814394074186682375671244233235148234658676499645194524763308705364406871414568326066397694548019309437100867957512398911906085980795618977028617046

7140202639004075521169679039679720071713355977146779184571361114940796671246292299933147763421654122778357482586275349900679211197806030788574954697328419646448724815492

3884504168748844032656277470952960677481195278592514810702849070915018652287534294183631406112370858732629402433909819838680897918601274620818939502098874883195920209202

0419914311024328861840438672146984472180588274776118855314334547759499157081121547243048811098267530850187927122360672654247225549511677834995137607049303136759892216457

6741761560889488299509311427656818795044573907260386601558121381695557713584204354934789536420233897464954927668535362013175286570550588094409771668261877850356569324883

7006191686881325769898892153770164294154770352800561194842247298518748777495354659986473128376681023168478386420803571506610437608027590099241762684102073191041168647523

4925060364563867772961804953661056181453874745364735629557600768283850026539033822042359255398293284193944942050008998872489180621128704903913002948556514749543445752448

2287158340654542307467849427493096347001514326312416129821109765779780864620896364347208805915536423226483126451502160519650265847206670613012049338696992206072120550748

4698313250445679036197937511451056109940597227061024553163431842315931353905307277364315636726458013226763772668618633479296099412432700177449539823043204002544985446412

5821814920560149218878884850042818418295383774717036791912893763087010427207210527934076159095560569028788415435937041294487067737642712638215283791146308614599688138515

4958969349477753009109089564506288798749499879189773300553955499696722311303299622373574385677800288472896421335832669747258346061528037126274322172432529339257592492447

41154505859760314253954019027327179534824534471181832675337725688313193570020831617831857934695550625098741480805073837262014483584640035336912705562119589062989355562779

1783990885764956240379714308839097111026422589746293176689672234040127892499944003014646792202038304216198807717346647351646810982056584456771849874997429162956952777441

7095631284596102690145422333953644324790898827529451163209927530749937938189831475679444124595963726257884877982459217012800557580271614757921687730287677241427839045907

3912739294267884385771923968052299403405334707357333451836725155426582639619999309836730795037244868646114937304957612941570707066203289181107238915527538336663817080430

3305567072752616774194630606087015556574452308746558396124050301480579106145816309013148988618821079382747513047641248682780160190849678442189011101839255967801588844050

8539397683873601419123096060086888484095875909397988754571250989027721154014401926221727965465649895992143756911429020204111157924873126508075595972847278699682789162682

7869491152476745851922811086526770098192794353379135535005146898579382371388273535727171789383014221648571741029006997282353232884846219281128940817079740244219054436 93

0380174929970320843401108733211145368423599993209089515669085964915227766722963969424242341883180352109911640280643484735443798320000362801909768558492561175239297741 6

5820418448109089512683647084624741173961271488101932945586738194325086126535883736855992025907815396082128790730606533354490598834747010168086963731773732582913940201499

2329209692706783167596814766966752080774826807111167849293584619841862617211092532216068525388159185598806277687216438183516049553852793642109031310360781624389147125433

1653700492954357313119180428419650321615780904250201331248560532036085914481767170549766907392764895140552152060925413291657024918171664437203666136857876733251108285903 8

1921544766528622524635921917501303428439331954912297279269763953174862232078789202684450835204124961260947859764764050513446164577995797419209394138731127672405794054104

6207559701574182718191565118320966587774081467691697748376641838608175254785968515246948077528179062939927065252312930894866450147904547107615155399799638954573549956 16

4209646047474598537724051945204244695155110576014942214459850785628980052001501442172360303944283424448708887381117569548458688101588473050820293605143013701045069902681

5819894595150453748572348798071613236899892862049934134778445964826238868934495641306886939680162402256960972590583690359144783046409691845826547581534945007765160758 19

5664124007433611728666605780640979068506843839086550136603417156612177413653524666292403852731631584292475107182073761997425700526636287342485276970912414632604391164682

5719384474976527412791288332764753400458797875672197285080251587765490565589220471496205252104012016698764453817493876153962176209981866956434937752040786704550275749542

0828343826527279906640404630855560157935415801838870887714052300577774791013360758346370816037414033581662171052377954778031082878931138989447958611693922813772482097662

2315374165551870724300378446838734446101858764966248302353552906856208893670501291450337471297708298592055240518783105621652132246122880034754732559325711750916176594166

4823949729133523705438862724084837055281700296298090586508047949546245822401357141584900673308445738818119674451429115214342793926996556225719864391960677530383746866722

2170355689084250348329476678415036783681989747562636076056173959495904537872887088011965418789247653078202554123421527555465375278485116421930021040069601544428550071743

7035400703359356505398651785707006582099804195744951235876447812486104147556590450055377874542573276631373882502017264896961612910104812793263745673134738711663877099845

4206702700296011172263762727267452101811865591052586145338819701289911963847350290174000706806314623013080539754577602872474099075069178826192831971647212849587379180 35

91

4955908545006179881321198211429825837530563509789912359405448601790248626076719483518442349058719075337294433959240918535280295720103839420962334277562087477231701175275

6872410122453028153879584509634857983910451921318634956903039125641139059641254019436292949491921353071715344840968420710642282289811864026763507160796935047021002318781

0985613567910659306600789776255319789825006631515629833544391644492580570078010616280321022019570475798656581168093324230005606648967369487624261724101199075962069434685

9404061641090571153545376809232712872353999074435919539839737211013349491656927142993028020314774560505446618457532305901707861564430381970179149576539404594356060660

5070173938931321346258503810731950386744835560194918228051677305768502688432549589089560814199507525651893554022636610084816146203865446471707121879516306983362021407461

2432738560733007417290853413714676466208233265300757784578012755078726278301618250870084281877453332078797789429648585975819284204996482962042735407338531005469939541254 6

1947347217039952702270357793858512968366440678898374656563613544478066683846697651766149996185439896235023717671102740890919399583145146872361287138497242069080950442236

7855102876286804251881635840261629851807211528332237580970905745356917840193816002439654325782504544519694034884092434085789998277191512303094738055403082315560818607161

5834339556172810637331106060152649641481568404319462356043630174317509207713049088685604727385517309538084750144865176049756773607833154472292534335963845630215217104987

3859253102039789581433366384155929925508944602780701796227452105550671191316326265279936696359892383006096981600167088134430020391177191630701637780038300371126418241460

1498704171480566260338497751756038495491915791417929536849597062863297174522149452643640004011685127587387943486668837572288996134299386640236405894482452912428201658004

7816694184780636604784838176185651558401746037289782158565908314893065117791985723171647647241893043153199908815499713774207210118331968619684944051847134805103762448875

8181727723344272157087400085249391949339810308319995228854262630851814551410496748964257681720314577419760554011665143719337637218818650652485445039993766092267777074793

9980142258086621498719124701387469895676580981634240795135703733486839946074001483818809109122785049874225639471028589036178924869455199904917116231709297408195172163625

0623714208252611907131791876380037382099894155167362903105494029053725379895773700088178658870490439209102611278111129355194227781921470074363737351900832947336873223965

4947632992827531835181262400711892034565885546809503026304251921987977374432637260928911109005219868755372281053055764147614824639858637189772797761937308767042649704 41

1511412890062126113885306037367229595816117021395030074141766112848332886016736716732035880470158024784864639807999576764707923310644566260337307361589922617875267426013

5360072952785114473129927914524506233402909639717592321979580111914669928396066090546200798237452122450360169104115632219532972695445551518240945395507867404093718532 36

6038779628051431267662740364398239831881760597852813466394695605558118458939807050113575168206806948943855291350882854473482815176097154238595221327308613220412923307765

6558692790957004784303955681994015964223359594550228105654377995470862448559080037069501560801930721464897267316393892553815995861702205913973027090063858841495346162764

2604428373891251918478198500254982630358510063410344376334073346990312035583080112435481586611646799047040995479238128789718678010981520497880605047666820366296725093 69

3960731366933747203672430031896662049975065252017547135786573464383646763792685019584615106967653639024474899543219972319061169329622877894612265668306435189561638995745

0772210945127991885324449376487789040544224233470979587265217693777637079604073278970245582953869616412632465754921044812238089336380764949967196348615540609091415949767

8087746191227708286844605325272718019232463199946676789260303597957811827900047139409217193998740760009997069581634479233605106166460077875593945457858135619117973484 72427

5643954446900185370303889947219983080588340583672047301059337164585377733373826188199786074067450906292255488990834435844707186834628390792947080116986839485118508116438

1346028123477126650093708643498081061192511699839069112411278849225010740467001002498755498035240563673715644084834835224610219675040334226809019609191836769753918448 8898

1030769313603684649075185192089980796748495206425281503988219929450163122444192907905821755002124641564387801147363534860323181769769925688623422436511614137835848603457

2918456459731014580144233910343661909121175141432240043381471464957028036817478157339925527876758136335975360559539384017686767430474620112279164213879016271782936332025

7354064982943159500554714114406637280225906499077297215318483539621059207509481870223099482546729701848378246112123226391943500731777620066954059960374037654410628719241

7866624208821464827827943811361974467588435956400543384035329212785957488140007374970832504832785755627877386652839778888430937242195945555354368165329371988636343681790

7518019612735093458041094465517258849680261210153466972738116149856071359331842125178208697644136628031319592662998008004999380179915344918391418600149176520095550574294

3903063235405129132722852895331838612800900242018592068301245576885159986036670882144036179949757693010158168404896369505707194161819229869985461811066431624215434388574

4834338880719947667423092554142522046461326333021925412332655422517900396237001187722488012189973986745402327831011155813888762532752346656497928456091175839397247695692

2958164791705775346063017007527431401713140934708262075805563651383657797778566866723143338997158540512838175395172886726792433519324279446404274845572204271370383384 2824

3358563142551037569101814745835641467895345086856290184867609167706199880065923053443639571728250888425996639289712477574013093554392473324079182045553817663723533222093

5125401324815902989026429742592787894547500654994692121246662066260821434932002080552240572239843830449363282819228454889328958740225793208207774992126360859118902964660

9838958881002923882049246229713322929703775199313610098035267052448685322637447804638432564630100081448629121994571564479009944608429368284796527446127653332448811033242

0251747386222344906972367553788033995966640307214375360233103696573323152377229874426905668961779628232189871890962510009667381607994856169911256164664511017559716379949

9487451735865867708404716844747708499097699794470710722164628013946275459233625336185844576023868690373404023429979586621889714158804575375650785042610212069743697096921

4409411346899426308743785693181269940438714594895998350024823390435296021410434798710306835342977082983320775180715284808828337014599515736692340895220746753110902000408

7766773452083041072554159390052576034609120814028417742037713070084243715867293605934806649256089060062140073524622034669434043776297371417295657738443594005475357783364

7871738588319957253806398099602645405327205628731511229781571608629573746845665423600829004305765331541271689897462285133851076706642751110612635113161609242434665697007

0932995217809475711291051481146984681636209949121191573759093465466617608592524996429649049953008495876078196360047102673940074073208436244200346144947032423951767853242

4534809937752802805259240486576284671412186729974077030839307426174014500322104972620715170167184912813438981955969766521920190813700036727820825480844407705488949773529

0074093964115215928130985243276068247595225330802848314091145503496480588336314363788086926850984022754356109530387770701212000289803251790104923250778850259083263414 3

5485168772200998294771996023052438855804487383316035861722692030067101878742606941786040706990250750049323362692993214094658099646531428589402950507599383742467138926 99

0433088969689317122152250932961528322314040152247200696225193121876948328674200291736644356806635821053328041767261515623021242888554325019347747162804383419392847933486

1222057352401287750089411533192309765082818111896355024958467106845292644845557427712134742858861286553169290497678456625964182472669277834400266072794741444083706645758

5123767092000483121940348372920856002290277977308648541474618588111570240287314904233747102205122542105996607035383950648233459260971244250184778080977172673924731940165

5039607867012304739576533280658508348412526008894658467311915729174574224177949335497989190078206213086108038655162237581246483554065072432929291154509266713185929676930

9032438050050470579802957471634654618685966633481647375637582530295286292373436789494660108835291361314666383280711983717491455000642982081727450617511533164317850686055

9958041447050514513443466135434913237773369000477575850404699354928650821231068855209618997124393434111680987953962481708529338557609002702279497412417974047571783589155

1301019483114208868324943314817802337650953324299552859443683435519727318517804946558350991943093702256819318024684488318677087309347278038059155250231204064223609470988

5301607583705436778374146441160534217600459326489012510109436883949375010807823551784950908235899540726474620437428881050839645588426654869276504840328312182091922525499

1147266406362203079345506603988385724860051085280789875092295415205368592926198779168922159220522055198532014792614710859188431868657411596378993112160998976110979633565

44490273435909832894786788397496610903105796567898848146337793780972063390584010251514800188040123088613129755420721536964927058014552750216146780869136607498122516216344

3548510063344349103534205339346945249747695508626569362762126109157268918512773037638397957159366930679492986483866338445448631276085510051818945811041464708454015362914

5822745729015341057988169354055283186736038250883523047392021460703493319106727265127717991413837226705030042889515491348029236323928824220795773086577354430078714994079

6063032576958992623920596601385554180341966989324367028510494283621790242444777381925654599726442141645880183324265738561876678959852363478423536809059927119995597854014

48359660621563295163890730042756825400192358883203001911349752328613006510978769294553113966895558347643708070331569246477056683170873581839480453887530751714783371195077

6879764077288464866497918231845023153810551758734071896253879853762143717243682071046881420524923936550870448913554165562666671315419217636666644546134197065884404803958

62840116136033685481345505093590314165725257605525869878703051193190755363634544025154523395823365871603816085282185007141035499360846627840750612368387644621458919217799

14259300649731241854636559956751295850203082278410326137151135198390612415964633398239008789738799337167288720356789486961278469180298840007702404614168435709646116237944

8458002436915344623192970276370458105466861864892006481857884470461376294282061421070348036308441116780061481502391369013966575803093965904415912512096952973581273097903

25832924749399692824383149206259029330928601033239174038383427457113513984577478320244883860219680767163316534986730347454013999215982111671115125442585376091673432769922

9040058449743475905564206307694693472835650649910777679589265079832348108401822448911704966510406187591847485769721036743268151484201956972207473448220879286996086829355

8763165389047839004874274793515152841972707861432196582841706939049681577256828050122020751092363465517693151572323810501802258551751824782234136431788165206850661394226

26534469728935978057557926078321667388980661251923880001695343252262005288995610861328799507595735547495524742430832870191803056477082736701434453937593763155017509351855

15879850506946390523195829252309751407405254595631400743663136531859885757737887841902614118689544565042160337064786252627247085434175977950069264902695112027502630954366

7405801647213159267945379436949752610184466085800078312719131602123509138297042587023364160464484612847091863431519865365466090267618224747180122535599681323522386679559

7369258068959753037126480244820458137018203432869086371659833775710858330103687434657741106218143176556339621006385831674674809188717077376535589866220294998738274442554

7924116346546794060365024746024561895259093499667359512271273205240111113918523027067614277230922947288101331929633399781855031829344326322064698010129517045570430006325

0156250431508365762832674123002049506397173678418891005142525305249688625673873847779662866216497362402883693096173543733731366775835835701794471457470812490698022977681

5688257225708949890899120660554025206080132245048251208137567437661986796426412986059431128300054088818812331334729747312126744443118675943209106681841781026557335326213

8773747375534578279041511593132185182858877744177229662046913866020934969425605364506621333173259619876825885942987942670938011410541539939189608363619248741879169306463

5784480721565839931203794011832448020155671414285986372785984697074047554361823481587960114136623524548067203456147569397801228956585216225616967129836900853900438345103

6299591188765554858501038242000728257383191539907527071272412720139581995535663735512890906976251070304892027945259155650048553046678249437412341710845111272716022109419

3145540157999752575970623571196509207796535145582849694641247127262752026385688988076835163458821474438221218629646612627082674226350571950473923863507541211664830262009

7127671004898267161324519103527836207012854444670111305232522693015087030633845197418927063406924545880913768037295753346806779350468647874587707376219253052874813619851

1385757845429291076654292834071343502250882818892915244527783987472190081314380720430207245467931758778466626656846528861507688403122321218733296072442684943391394823440

0487947318100156723452617702576708867907794884357597613518177223303246999116657596635615488804049732012975600952659882349602233294960423551178727855292408028587766780633

7022880002218103354707338193738262267571929259333563709540615727780572441483111640428020011737810453773569712267028220634931269054549816718499502811568901079688029001256

5891790555923454722921378528723282181212503451820595480967722776607131017786034388295733182064235872369934100981061723496382468683009133256534492346840680693901747353201

5044406113834755190428877540607758191617811541300399257403786443591301907846113155981228743338330985378397280207272868923202989439733948198177571392474916946466667896123

4291093703387932512337739713880254983507066465501643565968531458265075610705724702998374090486017243764198142270431480373845293687436005363912672650761568078131420041224

1195949403929770163428124078720808521666306459230567032070981617225290269602306308129260227978177435716138102919298062250972902342140272119169380327132983200837285066796

1628159235828668657609990441599158720394183682247309036694969071774570121777144672683511194579923576437894693563554208606297301046730719822766077439302309468861548281095

1520215970525502361564783557941968817555609138577852192225964776994102305700383667476235506978823189655698241650286786318921131224060618093860488326451730840169501420820

1573260642922288691115250254993602180051041932609690738474883239140241558533260115285070430660932444248252414417460744488445341855282414037622464301160849294866458299850

5205426715174054544202062807034332941069337272642970668910815358484014909456938246216847992970706873787740628928325451342260408837187219903835552674722094123126765963381

5572502448461126959274697523814493216404446898951622400692837095862738068883362916372400418759586455204637462756958305079845381534715230664110072569236293492763936710913

1351270403054576969966845535847145591819283771246258352141244581202749660784771810569945704336050816850958945374630795183950034717251948443599494854468526812973590190690

6535989033569560310064479682631204573191011544496555112268417325152277386303081359758217229059766137627641766163476909125070161999553759643211557534396621688271084036751

1929996000886246321999875418094360053087413396269299820762966801548809052355871868076169855606043417539252404967999646517600026926916551698752652506773950851304080538844

5060926010964368810894202685615907389985099082501549463918220970064845536636689968589228638215971107671797678975011542808723039128878869961867893071982867549919743262709

3410002685103259296326881852414895791244327648721801452982185749771429294366689378364382229658591114179405184942073570645283259884591225939492744455714667765151145692903

1395253758046337585796415811636218722766151898241220535132242248825616774245967654125403317844988384111376073500772008569727762648736527833891636142496421063088349308903

3208487794286440497222682861859946689343794851334855905828206964612394649757412236283139773381908745580331675172722386647360336322942216531277635847730631534297372774923

1497394273391681398751716501781032613174550637765770978869597675742214937163528845687419756069536003933733202065164936970814922958343311631037274090986625747050514702272

3096296228767689369377387123902046516729559259396375704229682701372159784961803623480433790639703585558242210882352987239496885357705045182950991937634747320351938281144

2092546739336782925849653220080015263180291408646888248451312773062679808155954764828504172711392640315538504058354882187206324982256830837811447496727888333200287176700

0343696901126333771406814924856099222370955185473844571790541768739279171785165931986828875641818677687684101914918079399611289070751588545524699476508217675677211646636

4276819985224100965213050465083407275744849661976161460840590607410221444976223667994484837777546120077234904834680479955363749200222871127185135606616822912323921800558

2781418581599044062405262546367710924438317339847632851470653500079783184915824011759565900535027064038408707123341254904300589255268336910947101885680630974863894076607

3009576591462156505019760734488992170674595990100871084361834433277876536825877230822334434543292752645035082751036885278769677924752905435163484717783060043423068025104

1692709840429423261549283137612187925615980554961799348822340432228376565318861257502640780451529541971941954779774061554624267671433647051086815540321664452516231071226

0123089371047550682508603240463908671443690773244134398492841467882847147933299621527265676834010400353802720252754184352253722834489706581156085710599336667745198566047

2200841601908600007880595268182669131384173418199055901058235929020135405146103729259840892444057996292612098298196317512145353321039055651854375501550633023039398321174

2416388765814182837549573824813579535376415648600199102097635339207548493483886928161124090428605168897241562593757958318989500729245960467988554419093112332985271844623

3567773599333480407470372053280347069763700271572820230730576961448719664204255727504919503663230046513954181407251089307094439783121107332766875928037765071725136087550

7193764856367448558888257887661339184493021191882292709019500146842343636992753199546115671386857045074414321633904078029992349058144656520449623351543572010554182060440

2629891318181183565214801386554846503939174422768998320599947345326148515338090881844067372693578429319408508180054111250120345744530333139738044694885277098998056960005

1914114142279477629690392241524461598136812557513984940301589089128390396906823715967622826543755905616030020457274561527945965191825007133487156545101752772344850870016

6716220881347145597733125187680564852910540739280222112003286278105640243980038070190569773965297877909880031620066898403721551962433429799281009560861602014910881911150

9793792900369449912062176537748237195532458063615013009808650442425293478356985149428846338401965086282619268181813314535227102216178688563979618042557270940196904195278

2218972371157604600163990908887257182510468842834618913181202969641334893176005480461049509279897517631114740226421967576451984887245324003253302518248307749123252566884

7561695423934889778461915757944227708558205785410061634879826139855509192493376344196938647024154343282328276845684142699438372354025409393730738046343438282071965531403

2671584701694148331391499674109081253099843163120731085050169063293191018211053955855670398241962916952076571999272307176373081364238940148232626547055090384196166395745

8368747626735252558119907191836237358436871834305909221799544996922330034287082269330565446768376776558779608375718328106825559568543168045747689684479201244439748747400

5737572457408749217827564247332585933827183601185506372711682381424624589045907602969214280818057785601886552690999279221547127089592401794765078445144146455171727245484

7694160766247832606573541389446198858375674984767805369839295763266022647239920413652162373636146664032315518541515481118581084433985515043673480709801630625247510197150

4669545974914141011308661681636860421033880745651994932458611604871116486279983802481253941899013637727311133834266778532592324533344853759664716208338537412263555308937

4439019515415756799424532380334083072168419947899681242688846165970608333973371771443649867987671879725126150631972885540631915126103864962031374005154413457210449044670

5194515692273936673249768573916031054311481689222515925766878095346204881819135120128416258147210278096579991944160344384182326231448081673218912379946597469447371994473

4475461447290698588542010162523064196805972372284233732451140654028375526303970784269804597321019493457002525055959469814542210702836906765111338008271970430451924780925

1178562729103526279169742580684026273075841902146284409178984425616298673909100537002978855069858231909288554409934577175078784925479171378543226314655666153587021169160

4317209842652323139660654889308538301968340192413686714206972763367197581477872361433017261255520558300247566655717115576955972073111366164764921241000743253672111718190

2698647349658130101367113732229307682146982330958626017552721672584247759443215834483251667179538137449644406567381383266457517449057480046505684021118589827064600025498

4222932072749822069747698062260826601109618485569352118066229299403801572610103842829608898746065630467985022929902092919324617716061603848014225088691237342485864417312

6718474663451214908085532127581189474825178594017584472291425938199664790339087510366666154164686547007202202254597534809842350136484857947885547459213375096376782165 42

7914135474670062812767442222991188279981357269073785469091101556810429717305778842476385374026797415237094624639434605121639245148210507976032685986609400873234156423535

6676663753122763480101077045489051032405795659667413083115792797480969654347260724741069201533920909334582777447339616500935752471124170750635032088677174121934951466676

3871263737284611604087706301583760011153364921219020331806850032466379221737979268046636376195583620471957455881725511200080419911461363395552011300394239159753566741 13

6702232551901934176455731469482202225229548333297745560513730774251770977444659807607594546606593103127348715553264389083383702773822074314568944586437175412106604933774

5424904700149039594657459437712378972800584293962105526750395663214923767298140663500612023607935950725002611882988170244093230606654009722982433537652437799415918514 93

3904172250146997425335378050436214109357723549008818690739026951401669721483626167233238511176389727375572281168991998039957293746032217281928453409332584411696304062383

0035426887429021182429856451245795207347984627346195104552425613736299433014983937291110870529969519183533650534844242848046310476522083420859365173096584496130285833654

8195435981867562202941613843211632576859825629672018289067811241868687845973133871404187297007119198677012931062930969498993548139218759288048398451387407586430276561571

4733518354407350625531234366496005310914881303238446265204351362902626593330681205115888510435856994564993886957070611192357275711029419692899418765543688425692806386143

5130310638444334313396196979733561523242666417066390240362365593574944251038356824938543066369565628381349757831909024277049909445141111241230206043422328892597491438231

5759079844235177845411287242594935333309857109787553868398175482521217518103082181765015556345242994845390457756267446578551759148916225117807053288117045974059759 86458

3008005610322332538430075098113431012045083319042286546858779944012296561656389081595922394034392226001019722176573621711635990257575290003386602274925942193559612947551

8513699393249692786635065238826768941167638908982314619095968338612311858671495757270575686399745427113036591818384717808180841752010065566213047632675249466820599746393

6880649990764134258089308204090236323835178306322176172064336320110403440994091584051468783262604775680407126154842603468338802680904047930891937342390363864402549244811

5739076316802664679967901755671870641363324028870508745716587139591642925361440259784029087134377444175895665581130737629688934752717371131377900031008272938712248798691

4124280084502725122546352721992018762423580027844796784337023680736143992859048111112541931101350959312726611248889546891945355636218793299748845867834814486269804 70399

0091628952883689113746167315617024100451577570016037783703395725392613540532501146574192248012182431698530353639459885750411326222400541097051949162925743792819168762049

2675684777486034513839694679909435305586150042794448311206309059344324700093753671005604492655603472747780790783639504518762448392697987313535284348463888133230439792774

8022015296192355486000473427309507867806253076736433322824196132427056557794522660656950161892625626948238005695033649150471162428579689682908969058637582782838 92895520

3632239309604204622878106376581117827427625323405055645853432873936828810555823020155443402566605611991608331245276376749383219523349563178625842134166574826288 77447179

3387032908362503995945159111325505050600405519331067854022140139289307747103178334105594829183713076924569795220641402279584670586885775784687252894700750851850045306875

7078397466638450736019535737073785356700362585582066737243338623835066357323527255029874737157104957827619393563501675928367423085383857351276128826173200948780873129781

0994013720875321879621767507234310420409830054307545430893475609999670530062600893958753980300478168171271950939341910683317924294515954385112109091722673218321181622795

7059544782253612682950954862217231236316650725721863453174107239765038264726120658627233900281950908857409600576878745913537314615158976810289446734608601099736780315612

9155718923796941378351231627407516945694908680544178579792003655469263444618177373227167003444266776046094924812058797064658497352880792415315639324344125789177935725901

7863758256938063425044305458827958137410756022875844478109654276517670622370697241979180215229145483395625622846078366292668106052637667735843824873378258642885 50121233

2924720709607596068275605802893569806800904247794144805224614080192982744535914261300673153997422039808044537501195397261944848449519911402819066003262802697095 448716577

9231879915526503112323553639686684530243015961973492262352274323053604223375606417777316238560718050405591823508941421612604389373200349753263396931768396397408340876148

9660534676500201801809501790152475738159051446684042332046251958596479602895315590775472041875168042210488273979524827374967425882212290841842673273540506297473956251860

9784580701322766115173325159280125006610307845655227933906611955467698467069311445434165385872999911772553407583626730720598192318641581968334407756173587601158541062885

9025148597066302566272781881934320443544059426417472874446384970887538924365482695425677051955005030485739671925936918313224995238290395190787500747741928834128339315143

98

60545655499274003400051475286011764913688197540583518130106428191854277239798889115565442061329521483403974655953746931767971260593262273578893694395281403715549812566229

18360289709884483519995909272476142927384761134273942707646596586099675123050324376092528374535447598558719279745135456040928446312383889192976387224509486946643316458588

61170878840725966344648877283898144806975933463489209208474793366411476916948304375959983029844851046697116761180315756704307486831913501510866268148110806626764448818187

63734119104630148643395133316943864867191253051197712222605129272066288828365146022194470819695463197824818062306429591934380319107075672891130341666939068376651437334000

98291769177833502351137723780700801344937512480505007321978123851625063208104966695365173928688581397749077190782452979419561688697223167521045066713791920865389367499988

63148664101700444576299589453219559866395809179418510560086861372835205544642033275428459604284332904254398613235909889616104911040370538129550774681638755768131629919022

50815702719076490775784360216977227904172832152483111546737798675348633009709840718852917729363837896867045449380220171574243059909227766330242974417045007458994475150077

65742782902593896377634935315947832002254243267251842300506882020825497212921405633725581581127104741570185389317829398787712798274886567150463224664385555348487878687664

13556127610485824205891865197234353363957236099480816817397403119030930451000439618069123447708555149247296678508952198610901655248983577004961920373861655837363132652344

59824531995105916058379000397996951777069795246952298367607431499008548564143327220034989033840428531242112340210049816242918589242224349575360808260306762401546957888064

72961926684527511160095905378548316367124548486778820172520546398558650035823500202150851642366350078776071519856145256264951591055507477278537981540632716819511175410222

88929837822254533675665190251216823016918198394259899975941176958752150927946376619633608719250944468590221362185710407220499663863587129905305552702841046772620212744666

74545215457723708236444533570645225149756787051719480050080256782460781818548758389225888164176204903611290431281674832076693485228577606793484645865766909520661008878790

64301681675339162115205681084467894159123374146368211388337577326140274624666451509892104406381808305608040137010089074171937662958444216912279089328319826772333212792000

76809166102683872384324544206867975603948267365820673451550426763088181585580393191617327544295576436516201516644668557593725940971151878369217329993841578827660226702925

29147461823418663672193112269842360868545204188312801997803348787565882710587023709613027479913234822581376961217046227422496101673375478801063408551486622869995691527162

63350178980339827364487742823153814829001343256788322876071651101159587187145010578347629752887433927458182768600045277571772478802098965643852942388173879991874712975888

74116541332391556510890877525768628661680719210903432650301463542749204474931246520147323197577036120450117479398628215253549720267179089505195085909795621538912180564887

89967857306884996390327781322940152073050520209607654688325251752641013550651450298021343533658755591616836749428944121480486599822561119879720866930212849950166479523811

70649564273155258468462892001418813857435106708068039340064939802782040845015597431112977796802358220439971891827408195202452301617599479708280586003288928867410624713222

86909469466843904201205774106699462157238511091569237592785586844939773606880019229349147175801631652547472728540411178250031540494164953211474507549665480231018551844744

20493368214986786738789249575147069024662390473966302006615781093579204636277135039469784343591741426437780311902195475226920895479688782548564570713556669850232754612399

88230278689013175732191716784930163650138955291315901174224896073749460202948512798559245077379356908133386974703741573965582588573741349035827814340746255423551643968900

4607795838664996649444507475657969935149351359732001330337208116628916765098694807294403299176129072301114046589113824272726329685176094284187939802605239565406645933 0078

9657415766970867192092945253931552743735282800482345750914485335424049182058302117366937604973842272159134835520404257619992806428547140193685561411027658280857040342325

87586569465009202308498308267843272774869277955040661110043597647686786538495502722558854224986136573631739614809989360847409717649547154334062370732658697527645921337 42

97141242070511225644949907914419244832839379138042063623800297002820623931807561922648539739493308325197087313845811985701710024913652522719868408318404784693642902511 1

2927517669939886203438789175172675722174421453857611002834874321905243332939928094883906152940641101879375394113430317191685263553167352985363986776163506611968224723486

4034096810385850460672960147613107402400742653402436803792106844615341970808088122076807110643890588634215785282582287743675470140831003113842985349352824935850403623459

23926688311448077546710591069406547237346587789419248732364370240202401751759704682954025507967730984443179863541346459539022765743203518843111559754635233045235505529 28

5063631152408117678464547141327981340659274673208354528227663430595305473430938175037787231764648909649104425587969404117052776629702325695246367177584587882202846053 053

0546618172209655098183967544401193567027827210833556064466575228098058628703330757873821082112921021038961652312284879203280376653872019582051856182501442409628366098463

322528411015741736148823414602873774984914984590462988903284925608395739914059102412068455762480345904723595368839274619006353006025286844345466005785745687250028879740 7

81269950143596964947641493185049625552169444528557452948072324330277365557456104296608835697425307361113031082380003158538736286884921091749180339054828505462786410001 31

6947199894889420953086446502679610071504843926206331171186086120818715732734551370014186768692196704885901036242331516776017271742233050495306158524791962889675944421346

235651931221942844071430978955055985886244201736282607350511638463571118784683471588349062334258834361427479382954834126341741688260288873554781665323832154076835909681 2

0548845481718926920620367096536056408987592845081574274786219413671699432294586933049535372296799780048719102034782473949179717197672625388495281730513600892270524308556

48745041216033551514931039641863808926595190343501979503662299371070610980328347286934541513596718150629766374430994932369570495349591419352727151418737658694512236367 10

79380115242166244941198004465572125975358046399910239327977083871847534799854360306617746854816743025350911414225363563383883543544686001108638480214792763158479351071 06

22722123289000304008555103460768066519552168656718918133485068069377003906870137625405524668497671014459934961842246898041701860235101964990133778650832343686944378216 39

20218850859965813111265889482917734897361101747392459963978823770846435932564640554466354564506278165078910000603578958471411788114001470455657924684793389859044356095 19

3955983048411907085968390826969646301033129195405483566334707548255971622409572456078995227513743172920432944214571757021119664927329504589243511542249892841829309680162

790999014391000651606646547657233038464624395113830677701241619910309399289403981160616742076033829175930656604400871622422948641209276800548402635460460212864312968617 6

04867685458733245904904458154468976001291767593717828778157463447497443174722457141782495933265895990123263165431809785578257753550495717798199024612577728494463535252 17

033433931343645138304258981514773623679199418168832859918715721722876519947031424824878759641163100819127666028856647310961164666767118936763472129096300869239623420708 8

6293386312556128734065346790974199382881663850602787328004850544918201993378711308294933902544408454452831107906717557815800938675360038603557680639810642357510997331840

2806850770996662013993830215705749043949115438309056158300113014109913958329032586958376761970744891911326312164098184347256008766211159695133389046258905040210397168875

2976341156530163761401724174127541567757338515633409689244301406179749753717442566473493454579246967993011026194337636448667537561087691613043613748330479441225952612415

1489816998076810184818957800383257633878043531185697199517408404042921573207178938770895931557022742728265816149740801332063433840086890238893918333142004515961400898611

2156292436416947254313561701811792005595928543918333436945191945143171541350074200639050607167658190582421620289769684409055245447001281886858143549545420556689838786244

8320752121741094873647410915786040991025855586394013091551756002797658653840756823639735834723364825533520524772296088193054082105328131130488092625504406239557590391944

3171415265275408089529865726307163407322439503799871488151262446063128138524107414359794085766578525169438992746556218058869206623250449370292128851274592702137674834292

7777412718922155406268330082860090179218753829486291501624540724777886669988900709805056925069858183739013507639436541872329303118365816863336447793290070785745106113914

8091109971279294384939136701584241896910938623802725764521667814022361322904680977248778268641881228779934329767071638705695119901896323092458999902179757131345174259360

3251369020883958608546750477732292906284648172031950728149118009595051258134475715705546784844362377691481917998405290667715479291174266933488585670881394734484862143811

1159547544985280405742250695146472008424624472299321579226467217881452064820830276936119217385236785674096279108523422956090632666710480288192263713909073768444809318444

9896992931060368700953087272410262136788706972698569734002280363504755236880049558697359039020691931283070710617913556767421587725759822191294318645745477205132894587827

3757752103071194255451752885836334459358005524008931209918825706554866392314480862374414595305590030658095292440426176752542067710335536849552465464357701017897964163130

3867495239610828903257724489918622996519842327827499595893408310312932676610272410807379546281880763426986176170331009346800830651834771328705425499623260325011619282730

2129934934139799634261512648506347348327591363178682472894449321466061847965057304067187482841519147416841690897229561606700728197766108311410618692743573355356462725414

3794098134638322862433641334789788909126415318730406053172477381035370826694106544406226612937662336325906138381976319401303989855820948209539186024819909896648627135664

9875701497953400676447858682049107627096977570549150879619769194413901594125683267942639579018278254556896996803223178158977822565761387167701321480211552727673111734165

3118922038451473699023164776475938806016075537483382854582595062910239360016544735394298287668207371219275299907369744081134237202013142340450525931697277559058565 7003

1320224654790866575834512665456826305546193147547187391279239721184692177837620063262466814563224968776560280104550069682815286575425982348340092647101283284306430256976

5264469413502706214583191104917931941434930177034870263334601687102918360837412532758776315402230395628877684472328030214298729494647493950314649180413351420239388152487

5937371592838326600354064289535416985061992235303829218610486918255269025574988931977847206199198050179722622476371814459796413713846207982090839588211870480155631852081

3430516852374158421747814580115630583117678770897709425691536078780911362843548240228283945658041952013031197722796595983884093635835590118574270448266505634384216 25360

4983408543890402854304926899306613530295624400202826734367287192620794029735061796925410129175376266082539682218062164181246217311896733474309422088767060630546716996314

1821559092924351378436435063226209312068014316916384978101227107150216792472056263468706739088756753944244827083825087882653655581974416635778492417631881483621 6441222232

101

363549938942990784092165099611235325120110314233835506549388909275961105369398083023861037130475667626055583092784943578519785639053474326745056049094077085918865106545

300819826445252817408774654789916151163487539306741930233427583650496415686353883449297026988302128385075011730722531947412188603060660866565477142449026181091509558074

276082948581103520110703330305870925795123533892215724742478567167678835897376761772117513058693433553676490367437388170444054369873859392664944715350381525963250500204

906764624458522605213931219629970578255032704577980448589560702099021534466953706842845217821445238667104694081075061673174785697189794278788716052834504290221243983433

906947676464131458180704818421658186036693764098943364938142679991971798152335108415706476165027604538632999522440333893844869869487964925415099694766465468976921648142

392356827531850965408813353323603150184543091811589795300960882380182295483646650730834615875373909467531125542610804096592752714562722552735012176578243220128221736864454

018834091125636960586741735441883030291042295409312139163251665271628121402003934852462926770253355678013212110007468751049272921506773328246340543305146411016945801239

28106487165119374849628471452059228602216430922707329113148505514626087654910369689708578112513994774893832884265980561212183161153030419155186977220696407051705583074

429556802055860647919856113720832021397337158971801009341952992408631804186229969441590014844534898620679645053077361183991361144007305930420574967863953731729127783588

215629148003084298917636403425821977585909988615422915064881854893919464924424427073460653662824110438068746346424452489217270080026988968400977049906879061899434905987

912329939316865429778734053637410860411682122328032144301845079668191420237115246394822017709314272988455433627298647398360676197355206465441486062065827311631505717242

08876556248722268292666020371485112146244149649995152029735696346895793491184861387323567372723628672295693070222728261893624209144683434264680026607719950221650936519

4299129890949673488691609975570841538563266979992872310669681200387350435700139029251245561252590033221973074262855770905192682914207816015920684685189496026803241645023

2179784549350036868531120074399878675353188044734785139543177396060055361114406353642161812602126386573094690593255906397219492038056287678828743091909615406869152123189

36278924987012453882907430838160748627389678774537823637094678891160341754045508916975405922739404926662391204064685581911208987613022144456972568625452950130411217198

09106691154157468748584959986619908398378171263310681533641332040836168589505925150168276847524016347344155357829189499213859109112248318187172194868838657130408294774

24440601099690156383523782761290959674815107225829181329982874270549873596774842208562067520671599421809118851021464388153607962345415092134034300612997868682579602938449

765918712706805152707141861960573903188851284338726837664102076717571266109274582233522905129948528281613592546990669160185027594016816244982303860880138242498830699639

176230939613485459945178006710782255193242050873901267954978440156989887507912316527747212894791531731284579690612883700197518951091669850679856760516587307479984144568

49249212797902881242410088408847783731591936132045482630356747492775654138559987303215905407629514363788522298669818514967083486052744652644981763753210292306259036835

856799489149248808960412651107338909663443593384412495832698268242760102098965945460477365239871180175186513673393394494426767754232119108230942372896686802203474395932484

6237694374108666028201747655631949510872258892022152498032458392717242795866773783806580166137292977711844665025011525812407307096301804069407496556849930872630421830015

7041201379224567873194025821310945461093699749222618583746519732322036812782167978548254038535799586852096711956323580355674470587232686279282060364871793666055944117539

0180898377902808434422479687088565952716127883634043276080005804509481613762417573420339477986303223673828425719406058354743783890446415406579099993185608124308463533967

9917228812637887991600338337885884554719823167938893336283777320641146617025954208508851805129008431630715043310753954554199079646509023164428834474068719718935466735 6494

5681235430496129888453760929770893199421463048220766023539824321576162548761284254121061611331336488224541324489754535662653491424082249134020661750072112694306124966341

3321878554680294591249172174641038186713621514571649507321984896957276872688364311053604427157154289136681571274428332133397633430005081273956718748263504621039314324319

9754919580909613656257002432016658070951887305899197870679368295244048534202386527588450776292787146342219301500563804572513175940122856113554368542010818237158690133941

0613178329297920410991327454105471307174533004723383956546187163445703401855604718138157808981594703684942578682192636363447742822186059642474579472170150025817333302273

20473865732153469489387723270055569460064750620414873316115686228526907416880934990174691276069271989385055648400243370326559752936860238742696015916450956508885490208 98

4849887625009506966763593812316977903451178055406673492287980647697339913892073568086405247056322186738247850716446114309809549488092473731800719660589269647674380197026

8224103288656111365849274866963126807376841336743444843549156947581728533594401006332752305783993875303255072167800795954106798161051287012353253189048159356172170323986

2529923914117857155601316674495414313623225134309093811713897244019512308210327892658362850240295904940492941532776864235979738724580593951390567955604890224296007343 17

7928261847519339555645937150468756199757974316901535721440668758086865545248500482735713085317312441357923209935819400847803553026661789990340016450047094910138334017722

9462992656423345478810460564771403017078618144781116162479625532392440680096901179092285135273147955036450161301300382806788453356339113517314102426784397707124485599392

6430774456327190903208839351852366033052369976131317334834526825772849605743916715539515852956333008194184226563499761732066839099307221655802854054586971189012219660240

6694608294227814321543243657610175020591296018436712618332914534588474962392709765816939404028638746776848598592213820703486498305225048439168577534548195511374947189521

2461814152399021533658511335253541411165644033399101481716030559606437342868032391003140709411132623262323998396995376192271369735001483985838849714481681517149745907 9590

17744927744511130627284201316358437641792093329243167440055461231142916162639760706635603860065608371404383030041343474639895484594511269703337758329055336412225359 27524

9734553483337232586257096338433420359664089728564431789587613518100942754520374008880107278848188959516723126211891250019208737338613365013734348864042522520017704620726

6882007044656184738968235564714710782682630045214904324105536355425591842135419686511373978866630975008142564319847403775914587895906098457426078857863202314402576460 52

4637248392024315470427111900318904204033381700098642286434180747777779834425559893089229069745701872020468182941675249134855996061980098948447489166287604198060025970012

7365693936297540932085945466756234080461501354582155086320722660389340137673057625340655169815277785599299882419642665167687761191736222702092227833605250770480707590 71

8034363357075638283659681399539076072706818136565759198668375105461152180837811919647554096709582495601782824567273685631218502098047036246417619868271774847822246349032

7810885463141517371814329792883256249937115629715737390115836310870448602510300496946914258386937065120377046630824216489443358000596868730214852492879538242286100073642

0364967914869424254773064472810425550872919341960667052564506409608790024404064247311413566099006514678880932791384938464806546101789056276456355644526787973176600856 4598

59045759450452936327322914034062409343851631402526002102085325002803141809837523389639583076237367334254811893427718926930339828412036495177176010034675192081583382936 32

128206631310891456020148225230455288294429174005143891311827980981984843229029838696282514873944582039109406532801887540772094907478611791577001719038791280637623661 7440

144045207022924252320454057628069657930850203981218378402067202501202667529553130834943534719363417273406360262579603136511978554856693728464042046848927157780434586 776

100852896073693144133464873773525015924521197659754590876950206056175781935910774036258357653600808937653281370843694390227229865322182884374001388258111629715534575 674

032149860975542868865798743690009497050979860937702783572233883314539804939892101714335826189674003122527997303364571061607284968264026682347704558301545855748271713 72435

847099486137265871302549402449573855889966053537090338925114540555812456929413788827165199000437610796725728059987482047989567855938858499483469651949308978149972776 3473

305857071790270935682275763063930497022966339552876337991307858593142078113351114320121026019873042167062601435758411797707904580838088498088166626185358835592420063 0530

246434628992308203070806494107304156759771007752398558686759457317447670945568426890385311284949880181447745665050961489899151762992416428780004741385080452032953053 9184

097689946319969559127867694931959273366205430918120556692462152740786651432352659207070867879558641686045277535750207487671433377060119129403158574310767777795213590 2613

080828983248839483209499884568307672417592994303402094399322708275483573885074199171369400498798586194234462796084144473566520379282953170163351181530293127230254356 2910

554586395777780221165886661126933574072944361455749056372007128254481135578340290160485176052432969813550274714705263542935264813662388695848981951679047612474744680 0847

725887139455273671088784750842568825983963683066764766451330823429953840637149396551260259641269166395532942221627797607874955291748568842182486374632474778324492983 2354

402571567607928674259528494338989676434365754823075754784033503696537687365498022398780119203544049128826835941953971843647255409053142105566632073204638848382768379 2610

550038057395379402151364136624967493537324104404348623823362492049535442857905306545277265072203465929044320220171632423583137835125210957641527412446577626167543609 4709

743356400769041436221806829935515109138556573734119489032184562204438771527004821101276120814078245264988636103832650848085252951495226355426460671844543042653382666 8610

066557716951714429565559054236819339387175320386411552242884740879638726559965035453160178728429959062489756943146572532979956564427538102595666725587611303086354595 0868

484208170230903776010731371062342933780745475082378560549479876902139056655858928600919904560260320637827290761553970383110180084490112148119277796748391027288205755 9782

053508834615002190348376576463110568401425042106378331650979093472594994266170452072326910171868068931598950080623997586948389705241612230171728940390466998494272133 9295

681261610046509028456212675739414392795031958650235048110471685635783540426485721275402638812871946209203813254648116170313586767106436587660551655133113317022718232 1568

773621958482168564652846069706619054395401406510630973336513811963331659490303921642708535422804979802671491189563642517489134412142636155478089214528367082216940259 8711

263211438852993916963048048178929629882011238074901305294249294801611435330239008067065721378167971985686130290301299399445124984690100198919360598279169730514759434 6496

028833289696608150563450566093781292361334905857805509456421035309073601958446371216507319820156424220132684566877418323310247319218685156434120327170305730660785175 3850

970691717079172528551174362787130160095220892024240503057564021537273695926679974781070727937239123557770934682847560107630127913119953917628186159430382077839824326 1731

9663133362063793496768750895240236424692319045416738623583604828374392788665477594859028920402019395937706567321194909910433528551798714035020307605578201914838828809464

9648208424176699245675831226247807030905576531412632602429224362037195329185547180915964431856852057882350103091076128060445704425147997589608880281259978623877435496599

4929673220844972443458243503689780365184909951214229401566917453416838309035284779643067608611599763678720495505795636516693845210212057124671890236358379083391190802066

8995968969901881223218552528693485736518886301604529410281797360806895495240360664889446834853573711706079943054719216487594313141269759525166102522909575375509509337185

4490007290767612634676529166464558037153306020553474162055566838087233101145670608219713601991166960117726535124144051093620360100175840533446898756534900244758018499028

5112905603628154372796762883123816577437517662456404578370496485690904281846741434107660754984114657421533437962825237739351775877039942552131816901739901861642141354392

7797334708765973694817101033181863768927283763660230192059197929591791482244163940318041477900282857125177644841059315644675363309241579702126264813042808389337706723982

2865434173173648142456296618079313695325091128754694980155031799451669122841384464630874102798782095587734617666779332006361614129983611238785269844967622494946016222419

8481882844175972508965043238838826776211538694490722314080038640966747955659603365865500834501574668100371549812154559177082855269058782746268018954840985480647767322593

0833646432666789519813230343847805542571189332448803371027660806642619768000401457681926141234214210908378826034880398715896746918681275950354190406896727813951321988421

1832561094874735276486643671335936837371907167136153442892072527305707780561606591615442358910784646554736956343970737221781859123010944369231395220301011367407345705952

6133029367437932120406159970890681203507862354127805416826582353742593856966435762710973540865230333395749249771995346662569428121211926674888665256315169706607240021939

6266842825154475614963579333658452377240996873579532275919009797415517213348453335786814228739938519020936827402155999142045644643838160009990650537188184849381608655035

7227064177438662975167896665549998788957217902623090845448064651856930925569645317224108945164542679676181972883295841393513384459604167285457399141508049594466135343984

5014276180542209659848671099440825081513239252136069510626733736792233221425995230222936409047664596154505594842048813114413172046469267049759749059935116920439027605157

4466773968708032478040634377784167250219888494354098282116000727729150507598693656847220169410461894445826185511600415494510628158872485140345190055563466615244737496076

6113577874837400388629388488610195028128078179274503495840575292845298389091576491324731010563331478134640265046262915675377909213724782897003196325968912513302152465612

0543583762268609282030777416870045904352635817494636724551789784931750675390464041603363847240546498075003930024576610714660605719495109140248232735266912214960160708972

2072205462881003873076229689062152629711142892734633921437857583816799570965129751212882470762293756572134890623618601418995950002939343301174633003329729078340263825278

3796053000047355927546848718929972065613653375153747792196249551796922008557314794457428822592422876773212885980653704654024619938729649935943563230213110848242495018000

6757189398611897262182430778317833445857036118160941397634465162725658288616878213013425589073818405734222752790944015079633506963068315858425959758344133931666799730480

5147104205162135621754090487773022739698065649590094569569853658432083562061593452925424189291617305222097935246571227066400541353921262095374160702598813126795666674617

0932371740523629631960893652984442507430228049766416403828292571371636030617625967249957176153695852486644931720109608534572342362545038544414412716384767262833330818958

559364760061635249859063288744503255113776818130533466466995015477493242098568659350490106211412991417730998045997886539985559972088652729738821650877480019866860316305 6

1230114449331935784076334183313859772732345270212652657729626488462044050323775092702644091599212652486267716599659132457154139254001538116996614014497922059852865463119

88145874191873375518550958118710196924176642924238937549451631594772453110198414508008761556264407882172093511259342618446830352107379400041838289360585440706517264491 68

8578728545265072810491172241294152234684844898973496533155693932685540211665594490751531039708324623445957019685643267568038544519358687335149681959769600820125379900 840

010546335233641891279605446876357037106514135683715512448361849192509499414144624632178459676671911648776744489599464431583958487181884662742027844189992880327512449 6669

6486793458941329860233034829288762606371364458073713401017269924003140999628987593282399732487871382262525474190348822177498195455707963780042780145879194411890770714 3580

11030266245429362515054346165151986079342385623906645515459086899700987275783385647691033468638899428963619169533138310635144431946929978952150427343027450548912822404 65

6751683738409173741484373181971188226411967029514001048449736868836048926288540745371246015784688794778131708392027701850083959940135078751064535614615484503534678749 015

340275140901834645675419760454833086921693902489806750922992294071550692377787826669912301589909380813372850555299059934716784235078673905803655389520181111477155275161 38

37266566870550325145683158295906535700608065726990227214337914923752422195825551552739047664151524230841309327935561940500532444145395061094916327038715303701528100887 54

080933294790986591783965408974119198714373411365127164382405244158428876975714977114147142795082958870299279246833213370515267564394231135026287768903446466363218444592 1

715758792411319963298754131201832522267869678996413293411317636653889683205119163622399637364006506242186919822306441981351532197319859101563625698621817485470888837822 0

216171014912432492165323865576908527472547859682981249480686060644449351918303748366550817554225733526851403898786503007040289933438197230196147340864828347612607301982 22

6861441179898436755838915900846999131405413831939181656430884349788299151717429748649096383896734306517121736027545375783431135217215018269592914932278747374257245721 360

25662638413895262627930213000996619630032325220131382188448222153852531276767630485518700683146840399268185487653840563848319210024722319166100913439507678555138370482 14

28491510169897533907897563233992177890388047633681374846516892263571623071840641566324924108669239676012160108144560923213374291457844880612478637738826410208618024951 30

57338836941585087823197098151586711709517388028679580151067880449339024806890990529195328446696828825455292078708090501661485367533081336907004801338828585461654064133 20

25069383559631742436588406472615757600993478411408406299823664823574855435335590505361262742820018784805295304476986322636627829632741637011531118234081786739876610728 127

3257785139211380768154189444041763294630490061864780759891264283257299873528716127741833680517563794195244023212888549117741506531116818362269895319004959229250837626 080

50033174333856378486749582231058631889407398076144969201791751393353298858853433644997913001657128680999951557636883579690344998472342604194318599122046582749564413763 67

77021611431270014347716120164648321329271182571328791058413578619311893745953236310239127088901391290916652719237745868641703648012032953287516120129170609592709077735 61

6740193911744124471246014178496797282493661458990725500824349970890968096364168915696208984519256267193430471714563044323998155688693543372623026149800352837166513591 216

93178382309796485222062854188473486939359438432529987537651192492335099196668931068393430992917742911260879728304331663875840237022011217239456114473365412763340270584 54

17785774852486316499991704854769484320531209292739986610752631319764337653802952214163742360237221850677911388725805767775543742535744238997963358197140322779356139744470

71194611416517615151238823627905648863589472686057334479728309257094391377951656305853890416816898769258083650688250093611926107891124270988122269346853198517066637174204

68096376655728936417132493864433405288735279025508689950937615120464984509308482094646064177940775927273518750614934528181767517108450236520442367768151326743193252109519

20058767491849302776955965409812253963577116946711260602360694394572136480764649901663784374841097357300987497338721557269597603311371288315838030624903283304861952149

8262358667333635943608153309620435231806990586725316679671989775739671985056332039162769296127845043250930278493655757046636650050423538070002104337905436545267691563211

6230708153866793287528039918102228796675492741413814600656548508777948994478550508894914805288265876884445662729390819614400683983080524037256950641143899331811663770163

0751930445002156616091239778765007387433986121377676316379949916035800294253941509361182877925648901997063611511833437732287805137817006854618939770780027545047057465744

0961151865016887216781580801618548641080898632223340991247492258081118532699879573620303636012024863397052946712401923986888626998354310920045622791699844169883212018095

5945505348853261754254953638515063118962530921766529165824315900458349693970687658654243281945647647853735525153068989109796668878100256983906871131921425419886641928666

8754537724431761446354156657628714055353642878518017662196471266899624319482731009397417104259055243749687333036596872146018899966928025823050002504947431957887361774398

1181941294800460295054273686692310079383355277975003835915620816472862995161814151182430486495587047642038715770645275872367708081805904084235237577575754008846875611166

58139251193567319094020961428901109648995715039720715905042478578296419139818646456980690038836467983679810123769463228836091585443010494261487603503799034417768599567

5993305022532347652546889552580628931781172182316379508778459343728786415144638730344373009824804924955400332234343780588846442656516711172540890242516568074345760293571

4812822409478460558543210753456382184804837562589157517060137146819642168971776054953019924695302390199614826260170628481879635799123969700159444146867753185645831272547

1743944500821629828304937569597521339743912031065212616962292811787214991497537254721293068703850875650615027522642033730412161623496397880994352708041669332272183593224

2791116573630925466496712499296143990750000976357102050913521721026748783818004596833106999955907782554648912836743394515824781528057461034151134356381056637703548798094

0326526654845824868458290675564358632459180753460775801695839975806488137704343413010596884392991793174174742867125143313255177595667391206113546736483115197935420861547

2228327585604017733891732111418623652027644917630825994309391671301603555506396640063769917744823839840452727764172011229122906118559512750665065064983546029610296574744375

59069555507510185935075883789469234080884424210404435178451018449469776602243257275763372138266732831848510419137225727990030230195198814570212171657227651891902273755803

2398085602854179108696330380502830582415520792215467550998816607126379666906266962228804105219437555359934786322933330880742944031636339297431845742947964530484811266724

2960554778937241659254754257272948183035852407989706013833901921881647311557850526432810671078304253827862550735751441780943794515208769644180392945053371069580028806009

5926155424042953849392236692825186545787154155054376817216194941624398023669701764585168224184823494985612261220525940694686843355828800042360426716492051983003040063908

21604948418223179378001878971473265413916596941466285446072011663385275508319000328193491650079157574254157264221073192131424033453260766000540883242243039536428511701019

0719596899622216857242058270080448845866160085246854711766440337454171697257316643571299349388718199291759466318132864866184797194493997567376889688942187412505181286522

6543689028478952394531890759781837842937386927112332452240270485501136807130149912753906376567761950408247294577951614725178334058293631438937472774496789651364811407002

3132746198982924674668855898433555455760427757376276091205804848027199984955395267234262697175122246216969012780081098444425535915175083609503915428095603229402450125344

4800135907132943742644076571567194141395791922713654100901617029910319986570525840925799843414159079094047612707383047817895510947309066257504993635142278626507467665960

4091872737225477947297521215673568572609788566464575410138193018478212336515699772636151699971162308147353868744559534540138665599995625362468346296316834097975504564056405

0102772610837878518203493705332792138943408370417286053516274318097196418515813346386178640580852899326162674792028229007798061487378774117335830276131207960256078488189

0468622896278439020499837036853688102771127764916480789399444390409250759048332890872832414244933523646308167981840710198476936630553895402873674490337398474907585950356

0607200358784504301688110214426498847297177449594105845200382301253166078708859906630371280412152890785515342213121547113947843863137802569372757622158576792891521263000

6697169938472634430828463068278319532124797579764030143681437346152646919895021034218176259365978453266760250667448846712460966866173009706625245017692278852056906735900

8431604124592847480566894697818276779475588470103788134571631881749442899892987167278685725466553677372831124446176730408752350650543917283196198305056380900140711890771

2213292090837662204930496794578667884181003586373834270913535280939425518198079349785464480249642736866843355216708089658368204954333674108689934436229704144434396109886

1270196212361682942389358044719702097389199208230232314642350567106499113603577319563884719181976988071005806703870572501530190615358555901464126696691923629059503854868

7337612191372174427144615555267452282574000580347074835030814855107153993891683837311092750205817019512313117767785184487776872680421393928603421323899581851327766693655

9818064557155854038012159297126776864384285371888091790995730806801022541165164298547985978012005303253225752499741035431088380198432200235756740827330608396642555936595

1758305161534971344382636799028771794289082660425974949114770714229702525112586060390052960240254826248325755757561216131279581281215685384085591627805709293724466124367

2058886815767907693035051789576539340031112592979725357775442005699664022021383473566037691695441318906191606144682518131601766098616772320663497456046509728826595127539

7273336869417550848721788938864061839846003716868968509185229834465775516415629814905434676268421144524839941312303852580518486801957088922618832522355364368736458347914

6903567872259825575975596021681452229949197379268978806572774996020618312361912598422999744591793191907004469955405843606932673167089261261701339842671888934212498755699

0243059505953676654027533075503094906893359337655573217538335664890410728780373174267309618140745156207212598314724574843012110660947978796294904381332427061165526973179

8122204673427121226641381929147327894366091827878882764146146976422205029114448418441381849237635277149146906974336080814504292765761587542170524939409283863732947357784

2342407795488209531262353402755035057102890394313368148199515356610924477427046999116723789895516390463749839660324274199311394429039057605907415555332506554821517929225

4764254508718962213113516699330454312000074718298006506387389892625648905439739686679432127054939232744427995717336359106333484418514695342981280896868740832375408026260

5943298620562918175441222290018402100592584355705001162633413891116472241032935430679924686315539002795139232997222766212995130994097950530207390559581191512433304040788

5249710925372417474301388303179701844108570451357681512915362442949250375261611011837321004651896146782697244426178043464984407081819464885701556647291249400183231574748

9212272150548567617331055173286755555513727525722807015844443069091168420794485271927516752388469452014058436541244190068829957459005435743080561559465242288193127203292

3409249033976547146518111314592519057258049351511243668916540226006275545801761174352814314894999161467189352381443368464042767295387167526081339509875962077857277892498

5598283482328917720811777733834663342418849227919805296830926375673184709704872233707624758369814717742114804321363094148725478149263087466784789524555610053498907888898

4592118410223597827373165212801974854134198697705439538874973901457412211904805390085525475177691827060709796771265972488441968233810582599435298238267236317342426715578

2964601050683100461379274890656030763632598102793661123570622546093038459223095699447464899594352803595728122073590021484674876096284970187989807161708718601131703969843

5437109684351766492479554420274224706377160003575229427613743288102773742437633846548182324254585865229937019088847477674002680010096731972668495586454467670798787717513

5398088398320732770178046249932786188807671330925433892842895473399804678267914598196746901983068398922634392903571857330959662853884503112265863257014951784436813918535

8392042964337583894923848122176550210355406710582772688757513784359797904449145269179059270350881467187781681401490099155462169014256978035035958724739149761619033348045

6499169804389482848716057330970807205046654803487557123331222486247330163998671379512788679864381025425604253579275162413162454955297310236459199301134961429522185316969

8295710406851480382213988837696390758142551957119935917111875508775962547737751359233870322994013917636580370678440085956246876399514014714572246854340128078585643043939

4706997121975940645921442901071293191405742653347134164128664510758564588123951140117795508073216367810160373433760157315563492556593937367165619355004588107322335593024

4825696996555838830534131866761689980856668282771323568706812262548462982103131760771801239058725534724204741520016176660218058824619496648746064563878996315071244291538

8404232450756032140477624258036526092051914829101027157745924142628710445597292699956501128606688546274875718766506696977960282304138110846930878716848357909254627803494

4264254586120459971996080031663034795899289456325531254253443173998536945980583602867451850810533130476475285376287457097777675413077142430023253474040930306942182911681

6439039165574700321658811000624207185247579746980527651709727451530250946186599372850401168149577814259636124014780968378688851125147127622317915331487044871205793776550

3040166429701507673885504773832888178731212460074527612354176665768817010114942899257349101235646676396258065112713396498422824502730566937089237358164609535611643437505

6502996315167974534093823532899970256271802165624362551179869721624983230956871968296025466806500004671622402396653824185505730658144606359055955496419820113969652844

9301573931052062830891492180634287155353700590035047086460963541097884106096566034363535444906170070789958318056145339065047705274315664176097215179194892833648128660460

3530718819480534421773042360408411915761644613091367593152863992339638705407205479884795798861699621806442015353531790044765255822727670645936350572878042757348558987682

9668923357242880868062463248979189584169445790029592863228882938280091603240663532023822312473773678740617940901041381533057610283008848864941592255754380946063025862676

98917318461563936799577053825149341530483493122434806333316884026997670244732749061897363345433278280820774402667097781782078312085726344560947085548521426584891018495

5031866421748229856734063285256420887466425340533950450231027544934290191600084415039849520215325600163927669366477809584126560429064525680617095585204187128148134743445

1539374647549634422053892610995494442891463675389578760558424841902583118172885021588583788068989305945215392813746540887775361004235101493108460765432909460867969100188

0384162250090888837411244830646123775233176545420148684333531525807526673468582177646255479877158003780737152412484086925690704928900666781140279791636537433395145161411

1072264561762822599124788490992848729888705298083065245993551737408241113367757972723356826803378116357549983476669908238378145543916215512856385576818804480343213210293

4216240813548026230882224191231192566120525501613498997753721130333183980936216624718633004541360033833053057938607902112932032887174995145272045062395800542689317348

3540808487347093887388619010213621756502595911351442127021011648093093037494167291593624568786003466022400339535225142449151606042997647942423498377961134986659102717829

29931246528425630398244060789798697342060395170075318757863061612647145001003208115941696043578824718476564441351338509186208978574621805394727057491475888584634266518

14683879858080482488208055020192887763490205355158562400189346433242686366341174030312234237172519433745140146909287147290589015061523895622846152031461702617566758586

9515477682835625450643162643745807607141785780027587608048332596758878853180959596581481600806521298179195843985925295184458469272466477076091516851746385647187622149

21562301012713105284450748107002650244641785872435677992213039965218382791418482652816251046147211630479078224202348421762350643914152982300554362570147636995887845014

51305748180118673087237596524652046435093431303966655856293925156810381694666620313639439917344896713982981562741198826272351095251221821704750686253399267537797846680

954204214991134354070102205692681999404251665893392849856566734733257949343731185204896148039890749110350313946833407778669368735475491107761126947029878211256476568149

5666919095529367320559577279295129820907151577300917884779633743002145803603144778906200771554904764805424270513167766893532669437359097224008315940237583883147635421

4086042129957701653672965990204836305777985457706702819058718648306138252752786007587907024358743812118409743929083622398912538931673184265955965954849588145573527311910

74691798283457437678588414439809369994532326065725996975882421929358791454369349104390422637817675573224987314756957013416041498872979606838574148390973894482340813010956

583016594628071917844798263328455536359745121872603325357817924068700106861650620278236810634292874145081373060954890892474450662277330838526810902090813243131511849533

9696898903910317086437143284268249064812666240692670421336716047057426553138242434220856097935006501412683886793330241865462230747202532071789380503171991787349112267

18997084782360811160079801956068213562664092459514059419198886596052780459618987988978124604450752043911951087425762535033742853443599668025487721129385619101220742965

2088842954685544392519090409345969976232251366844939593914310582200549776165119323633557844773870072751670887701833608973105333061674009409074876084857287050254039652206

98438780791421232920380618831048172321093001720191485125537211230915157190161271374065368354640345909017516919077376312212625722658664549644959123749618371939551519665

573149034869814045381745736975898227511377456307867235621697668252392452981590467562185285845589145301482226584053595263909853937176619710158783873278238375584724428825

637262108186657407440201307721398294034324598455034889846916089350460604602972075243052225050884840981344559999075868131547522204656145766233140452632696535279292379792

37392919037486967196463071806072478474739655051709747509660626023429037646844450582472220213238638855714513234749820349966040662117777729786064434146736621110648053316

43070546516482946603376435620929127032409021043510275954039706547299792721089618396731508470303622049840208056686992252465953583737496013314179003537005246456058694776

6889730944136910107442020272238575142052254708190896349914219443174041123748517288954308018457409103199946112805564334422622895793405173650295313360283143297024936562201

1880738379659686693856304036554244937571974210249374057019369451384825698600195321516256897140183511652549600636011031202819953496704616097462082917736572997856457130696

5165001622777885273834072598355967397082404631525917673804229743178528315649444954503695640110936985581798511502721191591763800482965813078985947039731206537802032276034

416297329698012059066824690143029493995250158989047710508025383515715842805178152642640065745840279170365562834115519991788499516885789808814050968676482560815758716892

1378526143482401986700859594918343799687274938039243990012408929267645452342219220757841378533424878833535474932468597801974307389167814296105044449662857971986849317044

97491550675521279111615838057069681965725948152986672133352580867678999959649515960610988893062288696613263121755757586483262796857114153350264721407950235985958527648203

1363986556281374476129381484774020225044508591218250956247790738896549452650203707851005311569656220298499374750861361904397629959903131190080431975245809400009145582174

5452662580527403998131693257001827684217223180514068200502309296617727018544600370433014323452516468664561052172069290603672027333539615770828488823728813588940453449022

699453876878465724857819287558201702633487954178253554557549068250069979739505950461342053430926981905725531230338713302912850319228135696396012838534552115943569140977

4601581987774123819889586672711142729583108270898639204621803453794556243397185624971271409579259619230781884045355529797774221068893673388554858810722367687848593822515

4035440994822529059283484780136013500915158114532921442353962987554830680155885316596656339447818622239365806973126128512002422047411433759617769910633391463758052035815

4703062295277217035148212385304932657058446113196562254523852860644493309181674712723697272029500262669493867606739072357441326860714781306327962522521358345146175319707

3528090113543146777394534710240060895178155583075721182150799500558837744547319261292538304998174304023070223890981607615311099288865287256021794988663396420618209213160

4317680117781542966367350787241918340580597801941364138016285792981832086842444697476095946754659351392719202708606667009644691264257896976005037528190966153756618554580

6264352294921657908349843792574319715877064686820018182320486899825645633405013849665576402970834480618786696893121635133966878313623511774979419930548228986619040064715

4359595792275442777794396672633729746627797753573196084347249181190210119294392590380260264841748444782005168568430346614412500612254411855360366968299480657213953513340

7886924532705912914982801741121071884134268787888298002107119318415476906323213303566470428019983416257261051670413116849386770027750949884410851369316956444860759317083

5467673690177738942973154551145922770111036084305577182412122340329282298744398644640191956092300014394993453060442579969384917723978161494511312042048686379167525306349

0066523958044028984353925557848458072200332029250346597448132614017337334841522087264985836723648805643312830469305304873539059684897769410662489968164655101825562769089

2330654374747732515748234642076182693720200111288490837408415666378790491771579162617447253356921102796313636396193338303169096058563478651583641040952185421892539384536

5190009456821882351219678534912907472733457619087952770071453429642885777891979700517373331894256474677870595141670950151254363254585850590927777223574413690610705925417

96579407364489401336846212597403776943629267107864806916569414494764962755479752699975061123929065905556029980618277579232119869045159059424907676014494433021447538110788

6168394173626824737953620485786673661943401837539950788735707695697363348906096623415203303273664416840915597267506068186919542897295549678007420888087319998422933180164

111

2263918301140795970491267195672661938762353423067783745037399215560497316196545379184136237601366609873437405615646163459852384782852331973079137019825090585326929428640

1288966155623665336680867967626902193385870094706204085027017894505168178682770319342784307016451931313911485790961696844160662092837320833387867641488391352989258481845

308669975884128896586702428755687731235900349616499576082923775226893655707635413408265577248890243575485397525790911342017983026115347451748939422823882771044974234435

9228203662147297399136740367101215970943082487534476980106697699031419407850208010006384516220354274895328569552580166987140127909455465844685317297663885922327228023922

9572551621704395377986809188708511955501483450065354205895881728190715946327770613634760904731651841773200177627496686192983004847842222516625268124106031714365194567283

4889281095890446951076541036189885348326694340218479313476380613355515202360217636561827113154532531524831850160025503530023509981187456840139784132450412924899510635618

8398860593998518606626698374306821560893536408037221056922170621065402903346895715239006679969843981971994494884736379926562713791440855451262773768033692487909647451106

3094304810474408259752902764930190996182867206680083812477082804253485451549448267335177099158651397207444535596290620297896514822799643822846241004949253809663171584947

4649687732429417148601177579254648092229392563484734484973447687678972551867684457804193010435883847874498471915754661252774210651983403688768217709856479897496641796375

8532760889483399378980386935905003885915004182247692621391632221151170732974075729995059216141953417954539564825806957558191410105474085836697638897485443567038088777622

5342372523668586252868607011120737664441770347592390220540292118336359207682874681916357344362122584685518491173782814989331732943286878667341277094195061406784309559634

661183009377235593155008408188204299011125362549515865879877933201606023025399639582088857852464068389306031488155118818510633928013806882947755338685768700628738187175

5096202301678908295772993703948123553225117730651413774797053438937964519477623368044456610377282423774640653174719122855087525701248555303495842547755111923104174126089

6041764530473844861849598768824455494974223707962742522613785956150815270717355225140609247146628770765759232800006362366178185182014661299689732045067019387912233902801

9679284857642151753605032428049537412601702757403030534227164186394957860000645995239276691728895103478325073178136744237372764385742521775361860499615778451640562125120

075711268692543912705480417413062908526548019648798111143734157547549991708223661965071527972205089698520513649053554727844829720710781457451456846086111572956596317579

3886474842463316376564944811616192160462873702904040975367288613066251380909350984914309152013685940349903629793136440331259011869648801208786661412089289781050701755926

3159328947593335355585396153573748647327788469290051483756269645692713830108719298422825611441326287969308554351482195929480670805922268245906695987049805652022632188600

0458133848213383801071401193705031999865589124946066131408979946504502916697723288306019518497855653243552234794761767925765445820526605167994476072270550260402538069305

498861065092174655658887301788838412895999596591593227930457380841437898235779781221663915400107412133609571689636670057484589520547926161961736661796689247300318163307

7684291239817740029380690464820050593056721097047072358576276559715286685740580991153560586905724472242083498594270765145780433987934167957681379636208339039508329344955

8059536376048546223113625167923643813542477484194804358933214598819421360579294167412261599983610855016724402649352902692947624134258256387230319743507861606461769510121

854106322030817166112414886764403388729757657684439522022196100737434541485089168713374426783595270561315212526207386933218305935658930621030492920953553814549511601214641

984397939037189643986934878415890922090939070427578219059357094307672370243890053453120967003508961922242988160876864329829397148824960412446513280821124892241883314474

926880363423918296667166282241567818396752437966596745811699149281369014502279780135976931386539455472068457777174591385361784792669537103695089637722790612812365479157O

888690766768908193749340610368410673860054110026278700874705947106586451439196980559706950556505015123393666589057173313364476421307065675731991903938811893825307521O882

859516147506809468839245740011135470274786982168621274321497665188300319603334562235534421417368117005957663493676223290691848803237345192434912465665329718238417396919B

658713331341207105170736834172412344723718672494151084270496155495037955015287382248605384606767203928758686262616984382280560158757866830925122304015998975384892283600S

5980752159490258074717157735374806690050015349853590369767229710437159210984693912049054262463396085508249182328658439340262938268563715259623894744717833699966180201154

788249913323652215956653404857011232582770818862150130715779346002743951689275243551823962408391501234739246686510022267351541533981744380629368188431870219539467457880

683873304502669934820474093085095294008870695181863254835482496626070665024995646819410647040712273110042154418554912316340734019618090749872367033899749934392439916588O

6512836679056416957365216050282244885217573611331742947356357748308478330984295743057306045041284029711489736552333702529333285934154488313740058107262424647751553561I8

97734274294259684019406813333814161074909164044307805277729770093827687366826928083628346772591003284128128935787611886533037994391571099173130876841482981473167438924I5

07081233304663866652688517152287734958086507090211027130801148807052510861641816152556748171047862194001056128001346471000489600816181363398613143759537835734463470097JB

022397066733317884712174216623797833950508494584056480430059112018417300502241962981137643255508625993667297921212396833626254330792019745755537985778603339936113973147S

735858804674826631905550317147510740826575455384526005243911716394798545438546404774452744423208201158058940110410945383309204755983130127803405876733811345178470423962O

88493582933994762948927492630761519594758086352305003906352697550897371963214320279758088695852977316219835944256689466634986556641828527840530710394608370553195307057B

1433206165483720876373054215104981963982316223946155540162448192274889889915481816161014129111422332742480710512027849723849044026429680769803440316010136598808564229607

54069569557133508005356595188841456265319836545998149354703533396578804947612732090048553113590869747145353689015718969841577548117779410760419019145652728884040518827S5

084900826453601324291522888893797519995599859682889360891127109103099730675706218659729673463984306192842955042921054669160915418280351783236841552181865567963407111243O

6438551541562489751763977188126209840873609829923836373030275059418770626263573784813688683682933396988927744468696966294996896602769160036986059689047184171600019718412

362651997930127874191392722798698022724595229401524322084048175398124008257637136067030334477654114042743699620541251450482700210515174118908825492609774721462055366779O

07552008799892323465359252612737727695456712154449995014845268543259740875744450235595031758880946786706242795049681223173819531934699556151004209215360832603215306918S7

272259929860039539254734318064647706444154409630859833209894281173897950682749552487066282732805364792474107235461194594426537987749869795964322144950143520616632736083J

877895746269008896094162137344263820026765753391034189247610563598881700835396449558521922407698130525217319501323741964534289761658135407331302630302854780224851925520y

8323237454832679632890712562747524612292929641409424850290594741557599275124250850699663473537642940738391820741909162541609728445916256144727217059082938576767130454384

335990826160862846719632875160286080403534720853621937494941366551991508536892373512748507429481445196541693822917931530918037820472872238945587726485845658358368835417

61056472125564386401518568769577988275174222434711115526234369721170763345046653272476633423782119587911761578339722874708451279886599245183279164144598280214437270189461

2668035792333375174806410499318521884550682118360856897563251181387504609424195574432639719843107892775247235667228839552735523606622598788598539632877284454683694923676

06484412273536829219132455470407442166682067461265670796430120055351599047417085461637336202703865952273806423126218740626890903998167108615021958265006449778173573185051

21577151608476313142971006824824913916249289255612638476738621823334522830285701381554374765057846569058839482531341998139629573625019198571058084667923649110592908805501

68338217718370944062699938269197068244705320057945408351861003661161560945082423986504199873162383835746664545218873590261219345316048583392126663772506208519054623623731

373281142581664384249889798596679541759200369130045370371962753606871830137648121274105614824487959863510850054873490298447752975753493966553071505515441039093865374166661

5215918335499738256689326452686208235962140239634097122426826870025339558269322438059983242690845304781419914989311071015716947495557635219458745768629630183899058242012

78786755796934723185030744610603483914598794977948711391354629025486282224974389447438682661643606831812488598723268791076616230538008602921583381443284532240746235541871

99888174723812853596838289500620225246463573161082663603643148563553160504991541441308872154144331745253732106513961197330694878139130968733186136669619394025720049930801

99995348219427655878113235218778124571834239764680697306979340608138301890778537922762154846640882312642981642123941745697135666153982555543529498172239014522235919257411

943146421826647768886677250217123294152289784461629105662854200714538834167848913456674450692912906319135618469153315855813507648754132185945574577015648665261674662085

80107002355681687581059596769989019854727064175312884874941390708664208080627944501319670197033551810041719242513644274485219781537035777096179780365547150100641818050211

75407358363121007584869042036045306374303438876152382898355737285260279010965811538399008202510641500684886037785145170399362693832261888230189959127100087907541662874831

22394452285023918486791381569205686354321704163358637035324530386031382451788944252008048530148042326400652099629600966417769376130820806870201308834720999916645580574771

0297265064248590310071846989531069017432283784757153675098497043483482059931223224751985485354554519808422814507464169325174171166032932667762278190834897227510030808975

25205030246649351264612784890874118303858779985696639063450520022123934526686579920442386148475724289010511432888381458370014867822833301640727611403694136211573716998561

0584173544560803368889069252435338105397159315045259209401288682676195785113132383976315662128576482640972356689220650459488317985864101056438186988942899276314816131141

1197839485143896402016161447028222175648273420064615301617015673045186184377052127062573422587150628959854886176205229616886565215837784247149479866487707067324817990842

497441413030772781221727570393898531076626569482761976332874465960455935018021232313616829469105165367996131496561881423079790028197020753072041377449667453062511078456541

79246380038273856860224589540614393614994048484286969806483262108464380743336558398688089988471514295701014512054966842896674866101866876512973139628321441026834165860351

99114389318076256443446092747502825373247224396358231441866189853373236916808695029220481436237204812347239217482268429106661178494353167259238504105377931104908157295811

00821890410996537405229404620198482059471956546843007682203732839906146792078803130621080742887826745652227951039753018549245010806671435652203409852097171271527515883904861

3112946667339640992434926471623122346056816004405457214628656625616892869974683821633904331762863708095828508190969741762120740550851100181533020817181367176854348727208

9172606743181038687549909642874280410652857444783139948749789244343537320188375097795346044005281197852692754424896325741629879428825506076395165108387117664673766387526

9750883793989032883574021122956354320874374141624949105157682271741771193223294539197409213659086000047620243281888116116364492871555990829981454358403717095653052756 79

0061038582200502577422429256977373229660938546733692944968154326056858234085073909150459231508132267678106363250595862745514892282856540694522210563558028622416764055769

5260322183356239637398808014246118755054609555094122720020016673290789780007096384828312446295596545022120743326436571869713451446896888922951080200162568079951218413160

9833097021782768602448714433613144592320601473497481486913207742459736433949200730885748206722411847926824053304598692754589707213154584154047723222208392080363241272176

3116289188164339403686931713558268000635173209745052437830526334298205158572929372906514408225209314003833442860094371776965082924111897120873707840935527547594667607 70

0739582996325838861067374507926180433067251405352116133767896036538579850322184508372345725893974482449203287338602799204201279950628458617692934934748645046491968635338

3914495693293534991399224536641881006310307109505687734454230964589543614186308762419444741986669834771340530095779787207929146238962639088463929969063931016383363832 13

1953937930428049848759627949597793364835848003784161018633174149813364342738798030257797021830886503641810622157102928202158718144562664355191289239074644949104273596927

6652311469566584458181144763773752350128593126592576070550556501984335677543209175069011358786716195493655202913909437717332015580807227333191001779316519681542280883705 76

5075988375970217541177380871379853389628968302629816280553011075577589785348238548129828105014962885887182322180488826018274616448790291728418400029367469706652600862828

4388308813356335802435105776352859335154443801010244994217478189655318847087378155223644071454282954832131189537540818216074865979777622791023082336484361723366029171933

9926167882868980057891191156961286113730404222996244445645228811797471536635791461504943812179699911719553392915467985888569082819110314718381129967865651453059679813190

8004221827090056930448074830083456425150236093637525563319351742703436292684023477641389903904336235185135026393009596644489877604174378603817234206079071437119444 75395

2284355167788857090934059098426280075550024872583065055751125935338963151718260658460012893268297421617913081168782127210392175796983370070556004792659974323649525447 5598

6234844740409617147566288275325091737591546032508096011443270411987315969680079689360407597750180374094171469981885046628982630503406281730282319799125530918362807799251

1330655577258958054972193754768545034660721841155705309674742472328699207294954176864890518966875077362184297915115758515906698154334699735695187655323006157281995 74087

9843326876548387703704838867969634471044881036662552958628343438480462678122147642516601976422702036294508798790923068997762622023363411771182015399050173071493996 11554

8403715228268577998385109384004528263675761260563984391352940873945574277648499241414685824291385791775623295402016824786676002102151382666411013004217800623443217919523

6861776961388050065925090702529001557499487526013723194269628376917390123280026029927806831063027349010110031406707790045792769827931368494516308950918290483029650019089

2833649883721439618079320874232876806864909920444950735986052895721169951904040833898408356330664385412889007147543866567070850184695379003257164362715713573472251059295

8651109840684568971565825455725633355594576577476180201568378523640314978538098315662531858094257851058139046154652480727198329095772958055530088357643935465714579135927

6746643254483845455827309139861671777995943499455031747714365979668345645423457823495302203379392308962273442866717466798655682537325145377430092237212027531693856659776

1078235980188630750756040836771274187145896698345702753848703603042584280212778831071351132772777499204648303734048642790225836760077390083656159122196628426903636660298

5432353631799451470339639496138687167177630565460918856551844344516890334624584682912414317535865749891247595560896040292155393977446428877279516223061246477930882654235

7480107178091527740165487104682655357030989813108310430948667539415116316766582814037008668655476438339770427571250447492214229690806227876085444048588139957127475613019

0263751507387831188514410053710314183163474336751174923205317682583915034235450586022630586870277924322969161754595344045744739971171211920778991581218062470944155069472

7351401234550204629067783378176214191890146908807079818100673485939679726693487696523832201910820873136965799813776062143345608391307991994916188426151766201151633013365

2402686506039217254034345528675180825801046290582083646811073518337297519614561225253713567711887842519940235439622424775173733453533891507262882321321961008440631515479

9859952158888471159978375100106259724395444290564783724282711276886130872971766745038778447060504981792632630112179328577963021341729058450693842312785744709828199036893

3332322524565878486499097175032022167285214998827754658306559015958435161855520548978267361357962795749330522011191610842080236996138900439932835066345181640768389987922

4132306415886725004798621914872760913836555345671758454745567604129133224460955148312535361128714521074304363554789821369772009279301372617048908280393909998355748082898

0056130326328048784070258801908538773031881871273074613664205091087496124190176913449970774575283633978323156110356482124709035953681610393841375688337715782793829413546

2391627057605602271779835321108182513441990267728610496870350607337208935530864008142659916508722359726946248604836665891130395850751047449472699392645264605522665189249

7645308072131093520962315760848831138260364791668246800841422358877410461155693227078879326874361283795757685381840746778209846776273672446911189165465543048479710421152

3977928920676389663589638356641805894421474539622623175326557972526814663117697801299371109824534052292709330039962951371443831951350573178805666396104212257787252801325

2838483308299728818250749885183147176842056425852484751295545343386763129354644077734050007219968380140522695947091991189551884053700772316581952105304731149144913758587

9146562388719464798452652848026896471317271515010401260230712122287922468881136328743346672378205581117942027216479225653704224487763296970129276913150042520225981080099

9798855902356890432902314785387387240209474483853941778367943233256130938177205017343097962453249510339055541937311556781096005736046904087935062129874882405284974393964

6946584677743623702801414300304401661790116579760269790511369309091941300404399250566495624246838020340508948029000026376220057864055215622740155388028003468955917543176

2186131476052556899895094989323834728790825399842346397579120047011707656479780085662647461192149320181526856763348584378214733117677538206424894580965820041347269571623

7038372077935659762005227802251822529632003253692765319844596583810375507928342527858259174564350062608434416311297019553754748518566661090821301768897917608118674 52945

2988157107661286021403190196538623956291513139154058784966307201942185802070548054644775476955678720896015971790736172414061141977276502529013859928135313797092560361206

8498388821519225181321799809333597821317510704838060219422357710899459174705182470860381957578800281615561611665853316284726372858902442337220872462933270283531749284100

4434726304179900869416928737222783808028826637801392053292867729280423656912564514894549289320084006096841831783905934902376205000224339124261918283291801084770695 74155

8673641192801283496458206833115754626233916641813906198852245734368306773983610104642316464133135532352227375266445338203004095510321928897610402878825716543401817838741

0155412948949210975077746095597715247869128811244829284257197468820488565490956157685128610593526780265614393833585921788673566005965300985365462064254346379106298885738

9831835436967921926477578037976099151997763156973019113881696777038313617431644537854570073193605219700151735410076686960130543727588160704998778936501742224661743819326

9192862429301061984254163460485330613122444410053811904217052950210203949284993280229187520880003182408337953638642237831826693545467637857036064039922133069977838159195

2595555888781915527331155001317087949016677272842630344949747135658767281439426549205263006190800059109449562185452175790848418298466938419495588120747001060376835634410

3320545001615165724072012529860495889470797218342597016438475984865828098086905490367726146202153986362941637717930476272183136596809685002918031141079704311381186418491

4158697584986602245464927651310287275862938670400213000595178163586440778748117806522568506530713507342458944602683462083458031339214535422338754448630386070872785027777

7750867762992022240271074324591340040289287209026586544735621084252719222866313141801122028687658119592727128351324215088164501912089966902197373677201815546709893992735

4219911627316526245069499200284852366347161921013531794460181932739607212488739815750393461817020822302795639335431676555278850293516327345947265795104011499734482377420

6587573034636025051615613927268982066048321085542322084072024003091375158254174604210574559615339198458900273971574375380224264155677787593079348042453601445500804634412

4054408972659399663300799797259212922310941947078384204183075966255394324456096358954142256613673796928542371376474952633755697716260219922673491394272360917846761965872

0575970336763996591575636399342505878894376680142891641757385033888100786002186250388008817542820476784272595671960484060571210079656164926916996939208557744994249775562

1141413333243237951509364096641340808447528614157971242508505925850581064620512457221688482019793441153229915524820485989751301007559884796958676495248802794232312419807

3887359066664964915113962924488710570456434199781743710931692166476206909405665634791222087667353162014395794445762166386846601655005339479315808774710764764454783961099

9272348900285145797974434466044051117604558617700843387605314800885367357021156503610454879928403753217591482018829939970865269064553815917395418408308262969421422960361

9265688038545999145315531953051939183455018588454934955788237465098560821298424007951804176372673129537115995329871379480968186645546885144362583503666244085255859306244

1470020188212204642192591723459339896192579016317752923865551976562492371628465685389344727069088244110281325841910775750548159667543119569943978415163324068894070439326

6081121561861902530322071390646769773268368150661123769280024896835575713723112884258276892878231095678118996696977409489348724146614973972794941266107636940295823628303

4824223717633199718432399903981284939398586827244646054071699259408451818357188910943512641768469147790922528378375323956019474534909055101642338078115996634482620714158

7238135420324049317575753370793509694613482852746513771468342056262363219717296199917086663830438150499887260671726765724838996734934781864059973119900529660035329941

9326485263280930082210836770501971287076699002478130085130527296684406061016437823071913630704942204938341960095350999398713515879252197394280615135656431340816156888490

1747824857836606800403926759962908630937903111506727671941876882462830751887950689739054897988939998836266317622380542165500973438436075892142420468068102248730738778094

7877235273901645574306789841755860978059801159655866608180878353193002712597379133589578973340087634431188614869768378811215108771266775723101833251113919851666507968096

4485325566384175831166949388592161416674567555538237860442445571263339667721462244674158690129562054565276810473636826097898649630056827907376919863156012916142569306478

1006989197020226538674162603049633997127366767957428955566314014881184615445820075931678835668267089919455698580240906776542744450798568453549054758781082742772021979660

0382020997865951969944929335431694916491705544129448209296760925147990322588900495395499572299301806217926211240711016220957855270488860538192842360852957014065767422134

8313326026342177054373095357010890780730221354982636285288782655869156856346625947313258519565913846892446244583440227704967066534993872354752090977064881709000110639125

5718740682470471367225709217122286721119172532204691459692215612297524374555068820561736954940666273325260413704462908654461510442560076329179486145106922588794437481577

8912508859111341722330315674973819208639722065135201582800869824687653737532334466697837732810111452619599588666501082168815975949566602903527583321493728643587459967647

6525820860555244733912117129797672981403798909728650697649203220992792404020662929795608333348370973437110383136407411648476005069852015674472277835899960842557884272473

5306117051602021864568180148423771285876616591199011868814095160940245806826199051129611212754574119275346382293994753499107485956141079594010471612965889159165679925544

4422507796929870576970584791430401182545453561117242733713999994326721277193231933913000689026267852300420447759813672847437901286717096517647825594600907601267703866659

5545097404684858912713323990972933798358653508097267094853161645453079087157352995507516924250275201254820633988339034478284894862532638437039404450543436312024049568194

8503862902856933719155477929628988453250576597682076434462946898145244334205580449946585823750536595705362244038184299359354415068135070920577444979990516676905380209621

7663656068357815873203792310267605286259507765452905795272104204138524615686057656282187475538902520578508007406933729429877374630579256993775040268566158661680407615422

1619388979463853633831125958244187970971954952240747959539519422976858251390076954836547899338635688061829056793407651975802467421791025033221696437760044128207799049032

3308123176112379972572146928331218471752129580119166422739351443766983751998453064578736640777380699652918703123194595004164514120225826972469814614881846289049872728361

9238127204197791615571313447996875757033935401932546128936025654403122892178220340954344388369354640895445802479018084762970341979623021602945161124769165018600598116141

7274382667307289529784092440165695776572555572957613024253354790950676198191970142415166593574460395967361853255106100792443466532944181807012147646030124862963997533347

2474598343390086851604672402252161362633437520040663234254993084214185882609820943615743240845647108254431018830907726258570251121046551644620072039153746473932152920633

7979709851707477303806053628389815119695460644201711392956688531188328062643343170171702051702648528131841251634392473071713355495060515867136110105805046978358806115072

2926127576939521408030298283970431417265709132950829112534317694308075468655132922242764990417404748134107636819495739774643796865183044465668398318484194824417605986382

1383782353137304522859498310101762249439405390186494271323242724894320227208505266704016110654929802726886299503201407835500681783266124429473379347087040369339474448836

4014630143076654888691999519065406311664767819404973010037519857054167252764702308591992273998423414933499839097864034390769224272933766892660493494416947156971554705990

4217421925882575342592999565986426768451410612384687880654457917717504027762823606330079437584074366948131901830157091267666207989331700172020539549068640945149774097741

0943734521094285968471727023406450671007142874586355868678236996339079944880743760883533682031211373244515710404186429258794496377751457723511758660881749387826753877641

3037795190605040647630268064742690687941447448269351953938097017849673194645289143892223863280101218343141180980750479702438991935727257052706054704728147464744620822245

4074714169406520696951676001055917701364897132278420170033566432055511462217933132322217119327373477715778865837696990791171131728353740193571986318244470476154481845770

2523441248455096812811666497090604322747183366809811678365157004680187681709568893241844073623106025662747574796976374720906354840294917347274873081201950527096375836074

6254547935295423417127268198133990264190257633736116569738550874413024110469983643222100126772868218094989076236868926338441751885626283212965994121375179698905209247519

9878946684512057837358450431489084456881613840158803969872920098540862969471322598632313171273219428914135494323359122937242309660154616499698557962763275554577984454403

2230409049346593466633088573695960232857501802685021200071271420765565772640798302072919480651293602177611360939113593935308121773628431894417396213554536050024613878 79

6067604080531693002509344182571342161361444011915228953065300777905397200339662894179465964177798355925342228777465640003439352647351783515933055719947159812125038017 55

9547577745951520713908433610011537904901507890705670881844574114508830512299700209969611309712189372853330835942880990492700746796012809697182928394128198960521323361 0705

2144317601499501995895358971833690782831386342368579567965634192930497698152099745288783121459736846075674766360600396879174354213850348953262958205447487692413303661183

2868911277831754602912791295567600741873322554290824475967304300804589663453378198753699486335053772244515901054165665887069898901375388802432758515141185845435365298749

1572718865356408268327225110147009098986271677224152877689862267703698396995978281655794088825371837645374022181884118744813382879471964454937570020723469625336015922182

0280817972604912829242072671298002423760022464850087988988572315884221469461474910278915546521230594248610272604712698132414938149901767357489609411826805047717301316458

9468234540907505186166101541070179554175975982804229386271555107777996745912246616384747714601117896485867721918610441753073190603997309812718219106724424419481471152783

4303920653681221424655022693020656758812764754343855734742632815492770562660066546711848765207267335654731669662177069833645835826830207766337778647489587312552194785990

5262932469843638246857387232842777898399257375979040382852977873976846638096241754775914830964299155281451658037282442319572092772658269323059752461245056435575591405 34

0142976750824433474662439858343761487938996414440251411300246886937195215783044869344384073455908669094644813823535320421925347916889980143917862323760175521459465427445

9179747606361665240186206188847558648738058308937960029471475549016794959042815618428725158529382983632066919799567389478800199587980348283125992532975991050207350736 26

6025424310513286294034955417639910947632944854290444810268915259589521457320122663991033216800068635048620073570348589610617902776607181677068179366318045382379191259561

9843885760951802894321579652120565761109398991549688403137754939354085695344990800901270384017262618018204667379599468295436971942325792206474709758952510700203774171942

6389346496757154272504046938718835727756344487694459508498638877616835505884970028685726928054212917958581319142283803871372827552868306454333148688092151697237876013755

2798110459669882917779624109707091370378839304506450124598867858957886370910482165298115205833227808177991560826384334655419760315048884191986084840626490039958781693294

2706196229745312713653269444151463625697342752416087344026979787900214636269626201203775338836775496915720824154639242413505757494323179537955463238524199308940605113410

9851878055884092735591167161670187118155350582366098569193922700064682288358994486575226714800463165675219420196618307034521639282318251127783428328954142472172600810147

8758030212294751534403712723321670865724341810804779578414426518997788601140202712329482376233826892096411116301103221754850809433535043407762834159300300053391341804464

2352624003807395109513701115850953473248223743803508935289352165877809760032571212835419531225508204126987603541890077136511086047181941080884311107425363917138988135975

3193233465157986477504684575869893919631188241144154303570400082640126777953841106665096407904987522171722641564904349596267065632899045783760113685132624683460213484557

5646260777352188360216622168327326752466009077868093484338791981564661208248486780506442969584602102930453724633487602599622427009937073587107459968463323761050293787408

0790673626883143222400263425418349464974943378764251078629404943559708895096801337624024819095581134913297913613799767020614698043501490706733150637876240206623907027183

2790480285148295606115474069599725619492966830210211930502506226826388089654062984556197933891284197588113582007257435616857677236659365775185363740768701008248672732582

7055947358759499152718354765991563039137857167298544040275171738826587384991774630988127769334111333640702276667861050716885045092417681798125076910950909144115833770303

1224085067230545812454372712593920137937129895049768896410465834850748194540118577023591724690129665114821500743422992828122279978737566992007129628455472834859249657682

0307884679506470219473098014949774430994977910414662850090667009042196029866331103011001113278230180588984555896639346087537285190743600228423038962151716195641806562700

0999956013489692855739326265725163640020407998471412336038886746834087295761469762659715075739705145216740388083854201531984940935620834941808448216518439071864548579354

1198865262240327191774938527578225038226935970540054945456324866881669370213670549848865275526796375925429531152739433899621680155290688351370329024845825815071276360054

8356859654294106517041963038966990987130227585345904165117509748613302711360435317037171841680987314181958503156487072149478854628003926277518456650430912603414221842320

4210133140285691502890354177509096542281408332982473259563282470188681318906977341224009824572047781021244257867941962179708245494715165738807912130559425579829365910185

9295214588919897364105748908948877461306386842597756423252686346343967798383912119284815590463257268442980690380184893738786626776014667556930214317529099268057699957897

3176289166009235193383288594182243444483733745857062267504976827361392921256590388559093917896293628679219569856722077114516768711517036384009901803018928779566852973917

7054160961453069937591596227833917112798101298430027748909358748945398117353381576846139508418563804583298672849362869287801275472398663892413528204505780353885364847097

3275449256800942755965160291034577512451359419321526895090145445952416961136980647081102315333388276662298702841007248866052876037540430218578429642231635313350340680402

7844414752008268900378635478415653636895673116883950433545631780101661551036438448861498271395607957291390579008646612982441522813477864422645334901586371744886261087774

2837337586277208196830696383180588110789389572837677830118699970089433656062457189838150176746551248089349447700083734530619482002243872523604954757117394662913618252446

0576260094886690256581329311350674437793232025038020543501852994980778105259985304488755072493125506510038857190824854783899326602860337393568736445574675696601191916014

3880775298766439451357947607144709888932776605307411631153013911049232821931105880973670474323440589088285640026797594353464274210784149061540929624083387910815657899094

9881571269023662004328021528958961577522206062145798828404918233836573512543126936803798726868475523692007777213811796492400361590234577003494115340683355738248739131353

9191475815852762541337589984203618879551380731734622447712358536727905300835514369174708516100914754482240609224557576103986096635678828945500333604333720846556725164894

6232727869309895094558630983011437878258832122990671329181939765220257245584008140241309321334209506896941907787020261535758021821051232908135283195091572599255500671987

9687514630234571315295363312886696129887510461508593563383915070262240450567951168869460511094662767237607237505288455533235047477489980158288045753005046091690215964523

8303826292106303225851732156752808913584883548548712412653274247165752201579356043381984648837766055270265526784708356010243568608832661178809770313606137790936091130872

3407720961810239015785329203547127196910567479470441464331213143956096751662611289444050366388193402864204850380728609879198298048731334990048455219406573465415553780146

6230578392118511432796651242618604478370665694246510969746211106625572671764171963200006067946154444738300958784936312325235285189985719809160006814191918616838901518124

8046434719012840007070411738005402484159936710284924075593039634872403013753519911575118044416261522200880897896833332452200044042973126122516176556942460302075359589324

0208528924924352707396565138857291831649476344881044603214390876483265516729876219997368370945854939343383325884620594015729844642978627786282332906904492327659329244991

2404613383396901524758560101968276123720343055108979231752867773878201326684509880718026820392202804849215743424256804673069717712187345608571520356706217008389369827536

1917197371079407399906036839520612792285497138600225654467398562769675644776992563009495709270131191929824955775763750859423691449068347634546043944371318679563025883826

8179289757907386757743819735208131268068528419443653155109713356812574828346979017890143182775261808833136399736028295990100005197243770152558648191061236176764114583236

9347955064737525034060992789723071950827784970011960313162475551008728852173812918560876774785619606389351240925078580628668215537123624648969325920348921913067134717109

9770337362895959783509532523598271289649198711044815613010829185812739292211017926948642819683286246929210846438959969288165507695228987337238615783294284880756219323094

5740630838089146283641381169292290832336756904067876535761639617542666459380243601974719954034836508568871288552924293518679946556844513208663020748995316987888798390898

9545654675923206058192261778837661222533547338637096772959302399118055662280926495004498775277334625217332638938949491465403821243218073615845361140460215552597536672022

3919881050284681073770437723293121597536227505157238764529436348387913528988787905821550111104233584518310227611334335804789714149771525157789848298936164491828979317790

3264672287494186754532691836408798282666191059531425286630926707017324059507062273866705792873388271849518797606853228061480881864593287994277974612452954095267124969044

4104613107390919211239694068351844696993047168661742405051922976752985296684867348368116257663525545752244031843922462332556165225141042359232732018883363608901360504720

6337496205823706184846895299865267466363021958716136936786724835356109940373813996989004401533889590391032827945728151139758776274744531842930536613637951698172746052088

5464692320417591143500135039507534848961569249509002075560557543841702901460781818699249239391194382411002570518213882300980354585878116400455303112771792666147437748373

6623746020207353840348230968767789362305616789380673178566313716353870471587401728905272491252082716893043739822965883776322658705827681347353299478638582774381654779360

9856131567793418000201475804245593773783030360395631349135763401350303484073992757865953439127551602075308652365385274503788859712486671650039877419265374145011160495071000

1864930153582757306955302684228758483403210477930534852209540792573563977027573732045507991015818436798097731303516449897806847700924124658901717949986390152345242314460

4073314623970454745260504301874212994382125706449150735469297093206380290377591188187390157910382794684926286064935770579275356999844748784544845580050634603101943277180

27220312508097750299519197688024658963981141533713506843610011310996298251092340076886268173814285820414110607895701149506894308627965260288795353976116845807304352996 56

27122092196855252288491697344907527823409037934366431756882408319294978354314731449399148449444634289657179655332850400946759399635949907338642665621874810726324787243 04

06290494679202538485915398026608630282068371068192586367561616748830031493430128105299599888061886299723689641652045680607795101009177465730814546919258132731296730255 92

05871656588042033913153974415917287194875701426222147192789110660275076128944273224291366994665927065725104193631023859817540798699438924388900893934634121382813621947 1

80798111450300020433901579204393125539951922260908899971256309233027142912501440398187050042025960878087081358686649017724447369521944670349065022384096930518259399680 39

42103858321601640076947789192421214895435937440099937964372856130362894311674370894674473797167618208641593168356184240048454270762185606643959788021421643166519009607 28

17460641376846555288918618196034882585207484555661908969318905511599162690872569882176388463745955064737773914158066740766500977834149348421618440383883109376015393889 36

31063154871345972529048370368834150168607515857990254115319535607077038149712274469609689409530526009808174927009219332674213887448974859582609816278112393479338277056 19

74066637897136621101115062064178327309094386421044341001568487946244696759993524704930849927057311124823975923359898645282770415482776290851650379608150578350982302758 74

21577959779004592060888004354874347352982814703676657845392604783416704594176917489468082537728362160668568691135194523833890891891110912798514587414076502852590672091 11

15279925914212843554397575899852207175909946241446928463268626340215826602498292115869480107657666054916190877864527586671942368344343143419381233500212897771314250005 29

22497167310777100171685118882299889208472946752106224245534716079912083220472626821596886645081673509616439339924477511462369670308230209426462536749134703407424587665 24

08826191033625198046411371161227393497964475670145458128842701067719626329369178482286279120561849828090900732388615464457885782858409709604774411265916188951776629166 25

27168721718762516513631708979617646484309698655020383274990369013956616772453347773594284105079107689335238793554981878467812308125958197651326354256394023461702890360 70

81664241775014401517898713394071954068036840143777538230232741686512733144671792770675415259373709342131975970409314814919948898722867822674696519315663052914595713618 39

74955247480194339279493656351149799084354265713102019714434595201177875945803750787462518049950522139898147845950331786257848999775100918367587992192282605503802850781 00

17854415807312470071381244813199018918287915172974806315354184027249731986676050304698921400122077849186541670628894614373987362169714665740339540354657042071325868663 33

62792832121561488105450732644643989838002417068492199129748009428508166943564929578343441236526169848230319409486334102166069882846231098338701418928078207374832054307 49

54164289632481395094558993370449182618664265415081900582707235884189828099801915103517706388164396642793470437736480269578536810367986653193903768776582660850368718103 57

28366643365057717227866672879309080997221922188813671309348050105257264157381340641178162104806787143023891668115663107287605369457537784780650585029952691170323330371 67

19601123523444074752681289283558682868402632571605697450019444495170180661810819064410545524662199746055108376658288581298144515267490380226956781150070305237616097616 07

51647435839334104077980557722934775991343313672159786610550734227837116776183199485930275857261397058658470986866795339609462918867805571711368786100785500743881857106 41

26920509669914928942274974254510850292688489395447482286306749939427340324868374491455647358303787886059847569183700454233012866275857927095106789690533359734293614200 32

122

8824671488045835314413863079762659740489308711076885123230786079342451220171545558018416921640760895771328756451994313802327207500512875563762033975330060539152081158801 0

5429443178555893398246356458310469241582866117884901004634140760949982144654146039283751024735058944073491964954037027224534572201244636870031384833984230918324024901579

5186593337859928294701598579591968598386859819449054412980974841995536896369963547172061818135855259262329953991693759170924286136642185584676677206230544414885826253705 3

2082454374482290302780270657510380017108817832818651459865583258719554645852534197577697821772923612071333641240281634970014373150088552712917589196082897388980202176082

1048775450013336131917131976208564149148137964066304065403542908487514460930454287978881572479698005656607995943883648202831603389466790445113655307314233276057871082079

2676043953535466422536781850268292288474386393039583740097385870142651086344620470517584760549904572725750315085058585681974684292904170671661413318850438469807973560177

7104811768366410779396794362026848191177314824089976925551288263145278012643813913756963948911084647998499767790629954544951573178700804071248028333711170264272301544239

1360206618534063071130828152074975371931468400971062250371959216386582428228940836492726282455960515525235059120784146707605675873673361894111742761679155530171994367568

4792853267322657804559329786442566694443389254568991712253772984042896325472825553367989035077223103164222255936866224029328970790077496213095713496292896492938787597564

4766633100100120853804913657780486947710046538421970875553324956543022798404918623196790625317699314735845280277782058867692232477965323935598599179926338256860793129715

3065955394878327773888071369814974640506625175143141064669754807888250621080369303014952360247668717017783045944939821354666812018915558951093237791066397432171715286101

2900733351801133111413566846800791039962245315059620716207040285515702614335810300780277816502377517200967856240169078815127492674101606302319578673330966741953326317915

4869008772125173072357898092522530225632265419023399141038030993435384580270680334442716519277258234253690592492756444995993189337302024156133921728688211168862565852709

9872906016565710880738948758361695217062095632682460903996183788978720182268702378535683441810012149346172306767528635990112808891008964737792459795688250073985023807750

9591298155530669248953573727641323856684678177814057638772348760403251294676471359436659413455310631406958562634633689731065163807435534312610069096745376501440519811834

9235318871940799399909553892895779875047727780327084660936154885596579969702595623802846174744164826320487241282989494781029882283774618561858232334667185868834958184 21

9194388322204129256621821361627794490041022380140242165823660436391323122954434656949827222067908232880702415173532416231227477573032561247042685443385322470127977785997

8351819954302148759477816076188527083820208483499747127552820257969469966553819395937982896137784430079530185400369661661157628681207957971615356066202735762672471209934

8262911458725856610030202616546006848143871460837279792596145993239211017039733769873495439047451665419547377441161880916700728524973472353480092459473691441023228343634

3841037207710013462057946640670753098240855503576886997007848143675510401545336522149188838320213279173749522508591040941264626158726646161917696317331416633744493848851

4859513586790187007958173647585040709656344453006310865134015456864546098577937682284642303331017327992712144695553139665054817276401972657906219538813253509632509474 4598

9891087859016969225819860487325161631487225332086811525038778390309210963973140870146326277487361999251603690351640181322863841015778913834548208695394911416541921224410

372588235373528329662254040539599551254188046955347016927965818084221691146779495770903181471995224200489058855937464416153490934682109527381906924934012585401621298388 8

236067112990472785382510932212676008635678872991110606474438540631307026915793291147168357489308607341762034842422557974325100100386436881605524668277328014816697873496 3

320996341723779237883035166054871671792874001119144726245671247002497622458224027397707027023503237712416913149130844802726400994497259572092359302306992730005224907365 1

419777811827203052590605053664793101848708283976243597654102454719021296462440721878685429130719911560221343598109261278099244929883532142115168804430524223133517203668 7

591092061218171572150132304915038524312760160267807102779059878363028490995133233325655424283233126970826048284109116462935530797013471599928564268898897007674423346489 6

500451148248944884381190520316202395612515508011429345593840225890445235491606770575264177519736090440204483367881891568970277893660449983167550333460997434539066796812 3

798331363984458622049182698598018506280836505817585632828672397021647037926115754368783456640465906078885859361572607337647947362839418949283576050483364494011349865491 2

082100481325506837562819702568696995491876219763972045027900446675639511937613156006454486485525074979942085002895444499533574504683662276687208248316413599480307060161 1

822309156175259528490028995293428737617351026742418815937159948909697922577214401390912722461788838743608051967515303147911414335732073658590493042773379844712919544464

543044009597518309741823376186337811529128015178663600901647697454495895732314679554993893375139495643626014549592646737347216021885313265446860088537207722134527510310 5

952625307111023553788569164995911692208388770780517354383985678670150963780888646757576698252325404424542840088268747777032853826199762542581992934205912179778082778705

118584523089729856387686511275072763410035994146601222794895749559203609946037804848385525959910817123628274420401784980211031762787788750303630226361609766675801060395 5

547997874569981579733423399742444758845313933453664591755258134755046344267161094890817996895922670464402169176805100590718447352631235416442486477743878776517385347897 5

014025204069329911353255614813604353296831292899129535615290402759131767734127770463653085221325754863079335478829963834069993715154224943808824064733263123350461388215 9

479516917592545098480979110892331137789539664074608345730097651170607524143628834660918003849063569268532964005516535997857913060664047455719084866250414627633572044250 8

766033207641200276837147202583957757254830817635228170657759415327083266255391096897305850456225936898498975622702158265265280620025184416489819196909558212078989726717 6

461338008395664877220193204636718817239547049302092798661041184695704868470049638641259530673766660389421761893875423752240187458159722844252290797737229255101808867399 0

839885492144913863562538863789161591888424051299819365171369259169577919884949414977115194315755826307059948575863534749639756855970386267805400720089745025270519397969

812529688311916459045209756305283373094860238329627213225690073767533911682947198127705742624375221375825032008736375320464500057389179346592355770836220414528501390794

640672466736018275379854478140769212560856944810541661956752647507452390254517053109406626366824745757346070406527573577743201023913341138135775033225611439009760994695

213981770284146108841360856591839329939313580819527070692709276077217177790987638546344602405169049949774757748287300933978940641473694571985048299051348428670709915093 3

044579663919535589147801094434509827017367724097904948059288384657526908214068409367522488842466125605269509780101260577282287359937984997704573176117586941984674742904 8

640123199674462226863633260076411070293488972360126214984596841603187424531584352054891859045356941964460633379884931531169547583611157497667740703154480578978170504573 1

922588154943114390257934504998955003727042361826456804158997001369770936461843182966060907313547636451203468088044854463947988105468098496701317954942866813882765844465 0

5851797712368142674585475571382290726634603164381375014654291945359934360820628279072531886575179734245746612250274430190922429627769366531587168094442427862498407494665

5635268045027684357821136273406943561642515379232246664582371079321459044461110922107039760456510101286976059355657979972303939386839617991898691799159386947086324241601

0161030988737854395677314829723489659476714272134128400437621066020055056620233395195755864512830241771428237157769328767978452275710423728172468445461622447270366618640

5467249250144671204565478357269214470538798054274436506654919003697852300703720061418372571130911168102172286721258959485746004353295033114695796564900624462269539125112

5286803782637520834620109193920529940673531650175837826327641004899446489071950821478410208366996441555489973118544459531232788198843519364669075619174436387489862 63627

6503027206860925188097236088479380162529503922552102831859119520952160308797092230631824304905136125178526678309896484076261871121193856452332588015635845663256815365974

1400635166538527833027993327618187316429284343663516156632552808774054766967000188148312992649497560617449945560484056516920660629441907471164740575561955183453874064046

4663446523332035103778447675885666660901728752098224471156450455672431109195734666267791950351111912484914064555435257725496566392678522191762385475381971083198322848394

6922294953737493172577554258096479498589102686359453576135514660375139149684512296745547842471779306074999924871503917371133105984182400457742575917866712195051742896119

6738988890457931260778363161297825694113684507888434449786640449589578177977537633175038686569214328456590763877035085454922881774592134644790364096719378848001565990852

6276175097739401643740642321537883413005026201713197591248645879230068500253381354891513199847109339798843654460482728915497107603731744512609717170621549152309355807845

6283921799509366499044096091569942149387112429146631205469132238606335268740464186976497513460031842898257475120258445983103982014060948468707691618383086238789106146824

1972832005261503851106819905835176956323656969414236261015215538172503693919882029271368544401062943672645120333442448082952850778650223588668479872923363345267582654613

6476448809927763316550213007449452989543396175266117490169121300581095542449122673852851223438369598397804839752464510120588267428306399960890765573443634480149048829835

7189816409645101192898539876088229692264354208244568412368752332589778385243845422759205676092607958467279386639764013337529817410380083971366987501792518889005096162725

3029063851224481562947348667180792167574395023354187971463703413595070688796031015016464474223923286729237172565384290147065906886729406948425427159052053409339014321081

3138545825970434858093148550633941870543754820191702268823175109013294136171877005809113345796164220301259802204310738296686490703750445386844284408790945807138389620954

4920759961553665571306844046952129948783097134507882757248448998973497039622465110498777004179371263198665504248264504271322916123093246164135330391676894514835 85692702

1660319065689993035272966942540932985850471428540838671088927344310873634570026911192432496377298229400944020752024662066447383917204425754834014539408035462734099909640

6907207562290737468413119858657879885591425214708035178496735893693353500754467232461312525826870463756343964639059479070392896125976486670569071815846029936042856641652

3782711376077456210903179808657173199934312817969129429620455695201243315446083595765650783244672112585007760996890714299062146372250183703199851692200508139102053789839

1562299250064687356797954066278295022474592665617686886293211625656050084542945590329173720098208588173537468783182064470226861276497609063433941566490264262194495999726

2787988742068653774840012402712025274733443616681302272047545870270464098643467573641494455604018046902564908532516727167095127900523934217027430328861413233096169647556

0201838621599787236096212275704210996873700071225799429518696300412954188779628724093278461798048647907699890577733886055838249114228316374869287853814814314622428398496

2337855773508605110105092612932309949247418926751316188186207532574038684806964904698279991221611386656638203260191359684025156492479064439833300318078952369616518405614

1033843479937617818210047435190907065480640320569117166432209921547124046169424710431522220510488536638270472083522207228179230772437862146298606683003944318403189711959

3875880719811504839775086249284520537660993570351138469597842959544064857515050620849712281233606395414804523183037590392304193129358047025322396391151279375936015140870

5351460629899519407457553548868619822693246955382102194720212488490442636553953053943533004050613359968320166698794772079525866914331899092118652260584960881496268365467

2349840481438109431985740368377466370936580637339294587946644516826114833345980894813230142344973630008177003029015416740179574436162018422076053577654231628456427644093

7556897155378728695987224574072558628891627542274001393924890937205554561607842415395618839990558247974029707414788143572707914922104355774546093490057376815968281968019

7715906457966057549425418144532992479819966571196455188565565681215133490980465803768476182660137371756837274130617046443808292439165371329638261653072310606823338899995

1025530732744120942699413792011010466320874971541058744582611599590687724094286978759462694298805477860776458005229407570308406385160436059312555658553312311655463865541

0300726993447731181690786628011955322480995617127881556393519012560771128846947322636357783025965642326973847446967479381719183498441201028423519219594794118886452077960

2313430632171743540692488623758813507950130195421256853896066931254279329444504604143997511059306830758986663120897091786738144310626732857291352473357373341399649816225

6056419741787984415602133012302960042148911478485752643862518142961913689838222749415586395787408237809038341408792976445118888023372009886401223617417643050637038107692

7834118901729401329462385639308531532244222768818571728893885168650075904844157097366365393941138141433007084179831172439743212898269308828179084546083522761688392367706

7018193779063875186889826789859261224877072553707746809091738121955681542467817582498635908457752822108733709471005344704316264532650987541870482604085049432894496711511

5564045037599221831531506218941768079740040267805916373594923058021891056162459013019130734230922011877489054176347523478720509575083012458009760894490201422832151708178

6382971517875410232779377854072683639518037227273990724613281166726202385362958044768744894558935561294166787711514445952150728011245289783898892439805792611339511716337

2724763220356134137665549360910760775428863941537509112835509538910175699273481058020524474988606569414068805981571455349610211640412992071191573822964968790061577717959

5751167559084819296904355934449028936964861900866118483640844547092298808201869643501024165892806728934618921288399805886754723382266287467489827600721396917908199418072

3645143829894301735530322511478911846799356943769256545351140882462644178781754058520462752091194551715071532232290031783786392997264579504136224307388747429588420386537

9630088782849749101536714044872706052239832746544345293565280769604772215030240603299608201442874066463090243695582017514819021276959194854806500251596627667211562658227

0387072145811474469794369850676576882350350632790855260275245784521393118971219595602227780105217056312396344461127002125867235737090090246740694925766696530479737142762

6570535957448790876199299797549154709567456866266693317816026949924563786133450401286482682564546781442573431384860066826797672708278086896403276259416788405714594114804

6291648812852001026105619845278543167673430194180028739280946151948185468663109907951504488133650125461221413542791228391303694726258978002733168003031221980792227233930

23339699024932381363017812136011313683103886773390012252524103762883920011468747397830196815162344845844170847115405630671221502524006009566845020996544235778554788000658

44530633545865197485172711046869237082103742383445162144635916426832646976748511916117839341951421626343457280693936967164403452311179813313674725581030233003356470901931

31184154932715418721453955008962151995577648095322472817056846283525646275430576200001721499028964513843833117116585841776451061858250821047245469890855260571543347364077

76605980218446223023035764928698353547722866911672550929264234953591129778086706764009141760472511018422954289469724271931395782511177137364661288610993859615446366855566

64063242409655115271244874081386339584032906830166375051593788944700324467944420500623238056780758614051617949156758784854753798036133204588672013137468422277739338300101

59278738266403520793857372733187872211600697886849105688767373546820717934420516918691207339060230403595314402484827552498849668639056151812913549133306891980509583195477

45243476885587738310414344539933978264379056007932577903268772389335970408929063798623591326882769120482353226853311130949276614545789111470718167829879548739779664939333

40499958044511144263787805500618120194285658011511604965975147093608482447840715101656881915154763085370899177749500315467791994055430238455568432958712280703393036550800

52035902690222574326941773378330207785088164599774998435514703041899310927886362855595498666452151460832243270293986535471171461120978644646443256180023520004864169280900

12519555879677770712014158540324292727965139982142625427852383618248974400837646571991833956603365094970785232889745881493347730994863978318201932446879191922759819287O

59190059075406735718523481186834643477850541578282244907985840870058612052844716335877723304380827091373968890506845553523261822111023731271368814205839838547470490000536

09386576177773048783078559721753881861727200525683687300867217691544126287754950323922211648722178615226209112425923774670550161255823546154420874458421496477602559572744

02918471645315888525882788428685588949650842703942989935442747557848840925342436759646126543810920255629982051541480193747004803579667745874100884131811872582385785448488

88166827713303205091480499027063386694690993512487922710050927746141495291462610517648599069299238175990244161879497023565718095144258549957651792962105674460022141336733

75480798603596959468569833797726728528759250604441846435778240362377667029872832370604527442555068151809555888683589147533045701169217308190413682196197924446042649934388

48452732374619113711082446433501694801550253284184628733492266134568944139002470663355470788141605485679668656432669813031783621646465882030050364454812922884964461641O

12502375768282189455118458059573209093946206486775093802437912896018008279404748174220633279147128420951370105619432717629286857909094932180555689790662862891547375486733

19869373846641958561889977082799001692464942519034737770398863014920111835283723279199650777155816340617619698407222318678270123540349943842916855650515174383773539203222

56115992975687965574856371399849841889818723863447747735150135350791910818082698953601252478254704321409778469323294380147255125843760860998146021996986315484735762920800

09967022312307799997668926149370948131311761267250290202511250176953858317579332482374758716019598930968995429457152380277892235685048764182413910232691491655744448451244

60830514578741750738717674417355110130764553789788752222185205861330107591272012841581609901291829118566671573929980929179914961000046603332922577608675656663359653418185

42256059884318442721810942318109063106059677332784039505960669772102777361411827206434298538442465877284718517883633752819325425830054996146148373504141918616198629100670

6387911951762050556410548148530782364372016539996520950423890411674976436029024341956762244685142089456568032327777818280402353171727161613847452643425617030040388071688

8884214757957281324163069397719048693210177484934395220889779774606413216065912354287306634985649513842991511937541568268804640344074003191648752422812899084960491088752

4818976629544394637536219830608530266297677622972823708841064030679839570455079864255624132956970690312606937272270497385849063548119466535336602643153554561276200224543

6501409287070103845024942299682469894819311063805848963834966842315468958573941781004780274310243436170639373226671414047645055320685254527772713379959897191293351095335

2183762378144828212389833183922695403629441944477939338629478208565391602768654742216021974541625034794865733087486963621460507966572115585682647125640198323687221801627

3702740854612331531487465319393626134387278498409826789986196912202669754590120334748717152720342378485320199773941089399183239935343608383194483504071701605001971940795

5692916944124826846029055460248055663749256769023585654963056305499094738338682002253272751014765382015718968917643861760384669147127050209010166793143538898179926138949

5820568431525427173339845849677842195542284969377538214198753983628041513853270951507067783176184926867492451878878256523880787559339874189923941064660842841595176800291

8996744604562347684174628436149052409961460913299078250776717299487379930049706095219029159187881565502948692499482638863968238495608685005799920812591698095762500632522

7429775722353244646312991916029527828387808637437948487816007986691844944952033890911819968400143601937093305472219047178616199521109291127917001976672020216396691842224

1349535264054150343478493730094810750009923037091538820155082003301200760403674004750282381217235331054698371010096575548811661428174197142241111465396494088106871353167

9967489742793722635135810389988254516055367251021364169290032107312234300723995569142252464301262735666817822859016903399525588647085175428439793423274925339989258403923

2079835993252157435842523037436661389938632000800645747508807093799846274709665708094369299361070027345381568480143981800226497961654983924247214953855166490613388699479

4487640706256601605517178911310515789812484067441540438634321808049603577636933696507502496754659653517150085997507640004559542637011962683350423969409324732540732174653

6577121897863354556824170391037818242656724415781843849453825620349781174947104658950823214082047820539992217083096379247191435705268927378829630172045984163967659793992

4684512021673155759406108501108401501493958481324314326483170638352293389835732862962500645396532323409016634553497614539777543545510180022729878166610572423124306235039

9126692725593983870446822440569021752720890597314031719499393757606517044308178435846890232264090670255825631565271039919878744996005669653116942017890333193079128764045

0024529260777573554483085149912160462604079663570042929414152107851793951248929311310872340368754933321199716941558224225323452699165148427080749649824320910870913027192

2073605282398890333776482440248216436744892838932717872463012952137775840656766503422548447952734389296263521706924829572233723726052121486755901243751068863616862068481

0753252551908087008239375667993000525640041056868732134577420110043021274796404626772079602886807545332844611639636702961676361061209564091590392267597725612770823369101

7979324027600947790504939059499035509762328552456920149233803895551145369453798964243907753154386610796172549357971644803446126662353804145557367642625144590571925802222

9306403304943177399110774599480518484341690301247105284001145301170159264176031004668798434006763661357541593810739490233845959978566490063310019258076179659274890217308

8186545124915610084564921917339841849364007892425340052885127409782607281844993623344396777834164303286170746555744709588710661228597984383289888277860899498259344457062

5520846669336207364561329951753464990966070993431256359749029656718468035188787644371927343228496757534874380605868393873208710712341196033089306023521935023796475301514

159372862118229525906701857585590486981033613106195370441077208602330006694355982298997200389420507124130963301247398989898650161344604163697641299185513985641334802440 10

903820420980509818816370765602535422885206425047489586808991794668611719325033248230241059805584766380455213789323057235020097155747602593772076776087468148213452253163 0

2087888239853556684084620188776333889382394005969382347555811966044053660170851554314092863357092159448116017532965834133347177302711059709051781158901708660299001605124 7

9450702430123230670262179701141510682002268139997507258321303561667949126100542012864532298006726890009482097075854102198848852954596019747306361328429698553852265238161

308896659145091248868131259535362960576603197504295041188439397247053605789479862831714003968480764211909414275681202732454233195931119523950562906226110099643989483816 4

6448745866830754857785328740819937572748521974137180942967777411722239364135603321190933440755678783811304519984514862898000608483869420621852719280187780424866808029951

289703473294463170946003859512545338683557968905846517230067044889684061086304061213515520387421392844962022257546585820866984060498655425885908145530994843493842733 84

2178645051398542739742909585700856146256183495270022814173253676539794691275297470131700638354159654463424496835263505948534474472107805610781082964942647881002597931877

56392390432917853276342037522975657527434082950845479470152452608993138857831239117512692255667572885133404397696254039311749337139944952935680106037969445956859759249877

267348079073267618245233552121621496802344929254288655145733756557659455570923953342814246290317278154039983415564198377180189821124760855955518999506207300714034520815 5

0332981497507024426772643603387375397314843137407092665449229520423199007345946393119965350568073329814865841109199443946272328453677112848473622460633136028591059635237

1938716345986963644390685405322319315241354693248757673046338170302944798352260205181494445850496120326909233752716235513352623432072194300935881503393598974493352695787

45727831403967039691001707341493253102206326301692523701801202442268849290981955511719561208381550144858365716651026908664871732381901486099246991315460820019927050473 05

68876141893298108310235264828108102485450220875722128344134379484999727920258353434172044259846932740914178214394928017997456598736982874282682674844212137154682327511285

3416523703165307043258283372112371376096959399375495362232222197465961933252907404248760251381952426973910175637197534300447961782504311533150675825627353434762525391425

1527570478784376788524187196634624199270080257610839274976226365490018653206454995155802908398513272627321967284783025385221904792786938953878036869931884660310436335 24

37327154699898811186443670114028426202615047388235899747281493343257065474637022451887289061255030273791902640396577417606898976983456646470520471635921448307099584306 47

17275076337180207145974456525141885050371637738189029696852844092581931744100555760898502923271016008982572238173945436985296547694901873840646564373071319590114076131 7

428220388331360297085261451234907307147624534050245423736666855757401206040598869556301144154431416969860740322078860031327905539178696741635552402507676530856322427378

4714974603778406460934681299187190289275976307015984088778174519269014769103003023497945851100018088606216286801109151161040983270880899543375511871831737865587764873

5885449036687907416918253831330636233820582015488544982788382174243758054923815972964061963105821516709190326318730933813850113091332770930342511221097215505610491387050

42621980526024761160750597103794366477152935349171861250661163673833049956487877936569791129319587827777406010753337243279000973240725607038880087113730965963445709604 11

75089444791191621745516970964844762046174128154536128422601551301589525862806957063924443547802390516861338549861926656762276035823498556919224890690011645986609615679554

15492157581283540920253931708073781465078832035266513457070655368569928112426567084480177864614974045492118431227621845224410884715075701908834950224375698850494542 41057

26624609458320959625728720309947765401073985580699661980195345005065145010549240188610891341730607592384394612465698616056090945417729073640093913219567683643332996 19997

9654243480265694068369867061758741316765050602713588257443370721645881952261339705452052344128092317306505198995450285853835228737872869829172758078242098261060609 015092

5211090898996438929786294301411140067628774992078172379474620916899838961894863077306027491026703883943352474243421236712177024610116024061118571687050824404916238 943435

0470250813649155155104327250747941024737478192265205933551362553018121964249924882129926017740801037198842125474472160296950027742685077521758669179938013625842241 739856

0766991654327570612304528672807076318947058954406188421315713800339848740868094144618458542303449514485210539912489324554866593053355845782771442377433853392082412 776321

6596753832665064306388069556915620262225946941429007998769834420914699989795683266709041413980686333251565256968787899225743279613964370265431446393337990008198575 307978

8176133743609483928302233803279732038653646242498055114022469226478931592944918040382983649034888638697564414855608560537172287371482282368642781136572039474969069 37239

9156100368280751411784737825622127759959167758140385329868885136552323139384317422585356280701915440077630163476477812613243802484218656698552740076410511901095452 898382

5634840704558080905221476972204287382202064451116558020215813722046634766333517561799050089778088560093208145417644390323524193973154721892092020247427040585791354 379260

0976807636298756614782644338926121213456594456801042227616524183764018673453391488079001363603528432174780401683146726188067936493046865873646311483282698049202278 762554

5401785091774977282946756929354516660986419481458364447890960383045828536989560806656661903603100453047200242263938815720686746900619298730822484900168163489212554 337544

4758038873615865885924859755874278385999295417099172251153402542222969223436617778192965331349674775722097843019381844505985075080564948318967922058343414789051025 761749

0997582001234724410285079412318437601471421378295110218735899391708168680559881335469913794537831377114135491971243465810931127833266217592227589369934770498435251 921073

5175737020054573451305873132214384620876027751840534893713237662907112055269490530038882280420480821085192876107677281454892794403227236284124794319264419243079681 370382

9482005869742015013235319872051992035387731421588221148597431823879098919932539341503787689959695804327251777689866125889395442013917130641886295106652689606241452 084480

3403474113314880043138647317158446815754836531670512601746640760728095967213272942245361042667928094697735836383498228114766916959695573277028279120997926086381701 765421

0151115811092683287485950646262362992592496943781822916923432548536160415251420636925214774598094198892681477644390537611191746302607417041449224194980017062386681 69466

0798924751697185000942874444662410058638035586306506360957209797349307096488907606301907923346170197456656248712884986738267854626894512094622905482754233025173213 28278

5316517586934054111845715104834632832008101152525413119367954561261216103197427832103879548253917431409877065257006037671968383047869291003267483442066840667351508 858609

1662976572277996926003868273349641875833211808091686185171345911568508931494044819961072502337896779899188725826867053775743512465081379862813483506643132466270924 523654

6983517407042651369882414330883552631945475425324845884093993784631867338133661031005811464551518773050141308156660180611322683529639711462401049831484646040439520 06163735

4258469826052125453859150536462014165930615241377433037246441039759850015689210279093786786031851767240469599157234052900381676072424770169781521486151947462705461 422112

569127225381590502039583051468585003602420054909455026826185095875426210973991322095029548958289174232981343154069257570588015736545351789274152712894131431826947 6902588

854786599196898835999043885422748469473118751974887712150191591908885813263769981215908089691823350590797148746133368039706350036955539422590734536933263896961 6238425410

479462948379374638132461240825306902691465021517196552425678847127229752667920280384129661575021846811088693939925268794395032672870087581886104930085975019692 8095204823

757138945438844241621756626373518033643277401734125942745582026754402788121409286880412030002270631808259318676648149044725799446704594238281114878611771247129 9402343531

934763897789484462560417454780205295530815025429801570903923821472145748055026968444380960380164437261953004403990818426587487953263601747734396347413330292755 8182000846

488609818329170786212437034594042815016041477058185644132223188779392163988961273699240022516053512829372365573731931938983417538439708309035853330857183693113 6503700819

056545043298422099381490045445400448618221405506052871683134142627896441695333980296879517536667847550967256739077347181693399759001189891113962465206160718856 4911926642

694016069590616104437801498666998284332465257608825710108991959111180835030365079742812307093951984025480169446262059236363510719901481567744000123130725410256 0560159316

840532816997339070715372088090132654154084084869464851337622748284991612747470532225505791833719782998021737591424344714866343808459910139255494465966742373011 9121569104

847330698050817129727313806629229678493165707003108305567889791232981303751031746783401120813530914660733675118762187223478968370382795012395444953967588537384 466413472

024700158852017613121548143721334792656722446917956258633003069945836289103812648952338880763843818809211807397927683141230868809469171765526719713411290271581 4675294427

625213295015430115336903892213841013378832753935920209051942093938979412893807659530870273550773042219114300432566846240722890340217115127316890533946785187382 2784851582

247048412773926828056517260376445844006828466737354482592496916214239033889318117011389131501922285410738181452292592185148575252476183848077583829866667556762 8883100860

152635893586357127421777289515445831293470980084717828967273412903707607159708978389934279262557381889563194050727520873259994565456085868553582631337084934064 1393491374

917907721822853116009498270366792789235878843160849183895336236157657683342228959270239391538681195815299606645208188618438969720018935678548949422165123610171 8473760460

603829616442733117304025869985215372164775363206607206123292932396240780047428067808124542992795093505143976021192911303780217039508295099069944393911955003407 2157414801

775922997193282716303971339380034926791334165089113946424796613209472508981957237504911331291669842737114895866286493361339703036247163012832100708415202627689 9630681896

952666694405167511329378690500341200428417394295414458444754124677185087103019497637306464566186349406566159555902220388135713617682347784631098854748031416075 1831305440

342811270350719599014203058814923214884129356944529643170313190357793773919016569125771780698512180115203637847182328949534446551462678468202630396663019906922 6106232408

625749366744026581397192480133160604128043094030117818395338456973477787554965052267198920398064650880987659599040268354582908404887435901210248570719056955851 8229964816

324570157037483350716256177878402602480795436431427740975204629600770595232159378392656058476318316453946086912913902970752932457528311220631653077945599709358 8019101795

478468222029813794967339008709934835236592521531747809149296359576803641645315741984082949704240390235340265922135259853080332431785857740605690039517496624783 1146154707

471015241092709901906946055592396830561480772653739984996188802045261369572230890736822583697253467329304349079623158386261263444240863141034997403655086921605 0046280979

131

3942595737077522421277559627928584177313693090364594363844939823392918548955367647356087898997197778070386620927463185198508519268526245382587024957380501099381632387082O

2285644740218903503702051586824571125520231530878090161943705387176816448293944911865897681228582282863058898610296052832544386045241598879043704773737313685021353938226

5394902342171088978371005174981759702068682570272597123999047773133208906742112120082183986777870565629424693820042378134913066630287587520925647732731674636966474625706

12003467156409034789631481868213459806127296825654557674985919728673797538106742173858019486939329232978545561007900054725330516708185495508295733164530389667280573878 28

0582588338674395672132016122364271902399903185069780948472980377051720009130562696720528431664964904631103134009037815217492434067144662452624788444797032670809204796602

1525362297779841303529182491667754733386109757717520892570062709539841927672468102127146446516361829038678557363550700118367241542895404774834328434837475643045274450680

84429432880032635348132660476017214226940034962856397025725628371882800895476023866630654974045349843766403859846458144805319509199259451498233613653904412316740712966

3261870424198840786560346889440980735651117190884162131337038328475271580140233374087460550901973370836047200902016509576637603786372188320038639008631870025211594327839

328479264705620690613940364883094552023943727600115564487678447540835616039984885137729230343235300979396783369830091277799794971704628531410043534933382267484965817 7523

53127196015906172828462137140103053343920270345130194910703446174565771792191324143736496939690489657328257566913276453108344568715944990050920224047579942148514139 24545

7253260782716471305699581372348962272410151358146339359857391762405734462715784143189580687674488008034901047315958190724146417291159828969702593777767500673203833675999

2089814111989857577054569008806673510534703721329348905545549720535301055732469661053806609950070275475226267638316527252791548331453619391671955103025151941717812391244

82761114532218866777176734174064523789930150371042110232238847769373142553479838888837413246016948494522990510710895587932162056322085156530074079505592494459 74351034919

044113379156636661772461772342505771351604426630343126850879654103973473927452905140551241030298362589336235834267865532047780753909090946014043806764382911478 3029646518

21538618920140071149455186269299132473572992206115252478291665457569566383251147867252560975782740644380460707154246177387337680719306797191303107259747170951 4887374746

950349152024257187438661942079828728831057102815793640271463453279847349413902420822973866014373548560908029571346922653311826406796525443498999697808629347038 5536157001

0379699864618068028932857297252901428290197040501117093960401799400772790730059574064535923916663881144597641874269200937829705225640556982994851451162539665867422363083

9350814877821146971451562642183972324519725639392618648161829774385324106434925254168626435227043137238046625985568603693929750779400699210927268702532645880616669225572

96352244204858671812045093427166631890519262450149084600688052235094443465827402258290512950304979961401660772556448746146270978688973256721019292300982454 76543800864374

0010975723666092413723186800774476268629452639172489702676745679273486954263296293307799383060695174258956537533033198251374669746258716171967671157859591293894641900519

5942311625542655890360404261172499890930640509531999646107199705169382815884249161492363960011138453259171654027649547661729565903127916495173602942136281872645926776782

768960296649722476337423458853255432357319176321653972435014685762391656504338768002101569443582360863484573501025317463807091205815681879683858394510423679587 6705668608

026594829523255942029210268875291154369222894717898363306284089773039313699219114714816289438323424291014432554651237764299212082763929737715940641397405209230 5309640784

16316554355424023706082178182299215828215654017676697661208991448060059529906543762363612026305988951007723575565165921971451678367435512908451113629555953087586708377780

01919961816556831004557022827891916130240361835164122316270904542939796741282350843309452202160626684268919718081496548307692214680927243772183273970929550675247929811311

5669882315933109698993912254344793577456066112690464296564074054192882264785479797974455922998213530029315531271761722186319708982235311610694068488157552894816208209181

6672481312890810462376990701322935328445008140894151893110987965407214618275748055858024353816151391877200444585065798479202545694411477966399229790532027713023499786440

3442182496163201891811804326478311151594681114816472645477617634978713086688756952976260303166684121463430262440794686978285987418395031713942900135590877225577053738582

0548895316960180686323258791510705889327168594220715607638907859760776550843037319077176693145523913518622363114271160854458040847500024799875823005870882654914622078726

7268063637453759806741647944848149738923875472845934387527928676644327256479556156983137246251045428885007334895130250436750092419290343543601085669208294261850388397292

1380359209785263348133849959272530132582108608715627994119122426533013518546830147588361727164882181498350287788557539659788906106832228618619529827640257722566057247446

5593403252865816014386518537361611416307557297037999446651141140540794271475377811114255133972834961152875330388644328160811651900971960558504031993456527982222342807299

1817130543227748759230282529480265505717408619925237062968149812420426377685988195395996847506801203170391350760070019492084836887963851424058353860689680377544207064271

6519419088448177490953805258104482473913499622829643783561309374571147615093589594579265351844458244073484489100939557560229105606244845612560043821227900342790703931871

0752380856389211364039358354010582073711565980725630307728235851731361937077949085709984740133668500208412981911223595969682225964545883214339357119483477892683458897400

7603069394540213572476634284866657596679093779764676816301584881677046038970658327592363081409299550394583781385143772011353236289264203778348412135950821407271208953333

7331687881000034208405761803890377556602679235794264582504634285826405824664704136419947470546611833543879107633145242000537808999550347417398247970896323065778806615087

8820932715957565448426168832058940287967211719805372836240656118907999138028832094343311246912681414321820670235390665254965636176151324015679568910354880795582582532800

0037407243165059153059069416552440264422265534657082709714384284185626503226274931962958902475416534576128240935354062701440070911482095483336493865766285662502730215442

5979845382948191301499938168390326408029823488774147168249578484818180055520669101561370253273682398464025891880120337650092064772911586849770591281425393646332561493888138

0988193472958921912237302958434157505102177387600274257006713072132715585027183263165673229741655993878969323828886660475346398736332850561625487738543568830057262974

0754685155054477920664951596322925802293266172296221390772547495226462179754608682099390454206571222320193765629382978088618030619528493132084967387233260394748697307938

5678500759409394786742549882024204154706671982406807377080366075125464173710459356950619561033855527322098109391227546654430715285653966082482186099201307482570769885348

9469154319716569722476616161766706519887344489662458628625152225384290156087408995434217695175410335336340737450381964569616208605048641414584922245318556939210460917

9499351230970681243367449929819627438739313209704016695375731733639240662067336606643546095104780339053198170587577123827378025247349903133500461177887772747607203425731

4815970711634540909184231751982744277637794025829492137246161424412299249358946143400880938914467003203465218983727289742397437971531983072961242850269656158846639448405

56777668605819505314833709768685171863066764959789067888706831505251088021562700170042329256525775535714748807170330384064659421313222560288183751881800277819930167409 98

64410627199825274556959680616423014905837129420328352608681346582068661120881999425608719239598657559906884794479895728471897115760112663523321187199746658400358021813 40

43653557895102240963230644670362396496910274014153840261860400252104070107826599150449718939463765389742339594447809412596446793881059540177061710349317906010285848179 42

86927768025160634698186487615862337015251602363718178799903437959551150865145402587972642320057250444419433171201374682541907085310721520851269398466280177111026424456 38

07915882495667434928755342215336739800701841459689064069363816415346937335804134862168236281248551962820737251711644710048433926771290470954498217712459973537949895925 01

81926670099192319815139687203924078131733288227406183489936891596258827794341812597822374682335655696616170703773942455538602130433493640075319533984910869963335613308 49

82004622052379421261127386469773435298406141202933574769854283798507314230296086006458390203266832971066474859280280706222894119846852371184628817825233780181757362738 65

63385765251145897691425173865797681334553312596108921679675992162004777773660173969815700518930556320934768057520823924709307505648653540147069692586332447535240023579 514

20808860993523690497927164952973307312176270227758280584330992778199380439217268111765936516774634858681152391360381723993804361913708704178324583687073268792759151242 13

27930692493680672567164093795811543741008447431807984798184562295337831592094370859879306743149584212809177146937159988393383659676333285508452144871734987298280228722 4

59721110033706788519570863646932674715906123201811286192207037816689825406153183076538439839566907198048513952994033311865280881232170777351327009709343772864506905524 83

20188640544591213111790441279949017804606511934621383518200791941711869477902086478775088118722447763439728760053349119723576540686820193680867718864154368080099068512 36

04632696099408509903969483621715707186981568085994319275507836176310635228624435829757772397161712443900078527257874425455388265166580154504944151640169244989042683457 34

56261734020571401179382531553966917865085328616022779776182844821286729462937149110911635104493201078007237074496349256912189049313573617672737075794898819870814803294 56

08027014245740279928915695074977187763259012185583211683324563832463542641366304744206953390687461804874269573082528928549775570446553072537265354490916550585449050090 80

73608133007772880591762113407812702207157721799184691999647647655001009776601727964696655224526699337699312900316835211790528327287495447512981340250629138109549668572 8

38779529050442889661515542282376017449137351386273975389837428783645708577787917772282941837862509262813945141402838181071249937085703496006797685633696864767208107940 13

43326899229203210797152854322309368783898249784699775589777584162128896661183097189395618055898887347882238586298360614828197063885376638702369971905688791420015005893 27

81576679554126561532407775118370166626802333634507878413669795089586402858492504048493379617118032804557079693217083088711762591995943564413971495282622098711793930564 12

25493818743814308084727553083891726934629489967863754936115462433363099267699468701922308814839034814606660954452470388376120831651420100763183337589349399962008248009 986

04628306243249049275733571918058260532493551274110735860302253057698204994600214598639757850583213910161582089381025366404328385622156050481874565965450748771370455025 81

45886878709415841423636958253905004410260672794345106150247728062235096633157361251426679434140746652063534546583922611463235578393993709702620191622134738856245398302

01655471462084762972652574642687233863623153135535568166360519035718952446642375866198041144802521990130259589349990701834644312706805990043256138711210450689055750679 07

6762563050384262622513729065225654942654827321349984841605395004718249856560184731449074269196457452658113570265107162387733468833725899778643672591741155701655292484206

3152187379159524644831952515980633946037421867350241980390342917288205856935267866581582011958760698252396492923489146794430847865749745083669623296079018930796244587484

3725174922985957167812557484233167303580613732387207900933633692574245789195386758667611341258247160307089431687496491257167276344430305543587943967279839160550213514833

3240544829591240121878313403045929441621292750499845016709464849534409152471837782557600293979425014893421463386758834589638702611272977948184530344423147131244078498214

8467358627326536286255010397360301794632125535434232983465042878299269795250762117660416274969836359499603243495791210300046219201462117222885844133017429091519974561741

2214254815101274882630094697483512300721702238025163626542457316682915474442181002777612760729789236825861162066670221065842569816638217984120387841026448773556619962919

1877992434897623591813120793421262858085245471431809776763721364584043359927672555158766872926818986466457737181567137271273270691744607683702302787924243821403830439878

1254700621379655600744129064710835313042028929121521463452256753566839681362781899954496762854591729513324623126862712073070316867408660159570991663158646522388745623449

6451599074979209945903431392149406798483027225457243051435304549078550468576555985031418630848348650105751244928756718289912764572460725925529643384645472945207641184438

1790283425846409298456812175616938935965975740429295571529150768297078020239789169692521995691674357951072811013584487803476120458651798666327827480150636490227523192057

4521535572880248195865656936663054699706295114422061255504568691480327448602817268281027424886402616018934181365408817027519036327982203827038208983512455779035361653446

9844285133454838643525481083128499832894181729161208390713906848687892138101036796504435251204222146561436472201498942002143834420124097014662363438582726929610862525525

11892129071483604566795595168653305310545101978321657417626828559346918714924724254791232715290930876743532838375056316259536391996859258022570941304541646785723973762668

7111622820934741373839530974538562283954012484780817487662218271515701356954660145145276861488219451341105980195908637689785102904745476657438215181134510250246512821307

8825731151825135732082734220695504162056314244492977578541810479961944429363946697398135936691118451577779926435064354447789078668856676365556751013302039206410994467486

9723800189018577123739333851891151761529179033570357835979595544489785949895424664715432770670827265189437837948064033422143377240478678694063066207205650277370025852374

3934557949437175759482223948482130909053923116986640964837350683077249445046879290633705387963710117044090375203852640296257030044175089847800778547400690983159410928873

5871895339473582884735120247775144152864143330152376020257165466393135151695802886647315383404234437057963764866980429174989729433098371512156861929654197679087543590065

8851911949101426075596946536884456188099014507149283559365332865878073446563488103298867967164678135738438860393567548126731638346326081522267743361034256279904458071024

3729086055639419844097513553481936898944385856295365499862588604857134920887479532609109770303465012940402286028361334223505478683117531048043788706828828150004082704666

9007244607660813283486164716381696784460515517616537118876351523592785897555315812735391683028019688722025518660905434784145591634664587131734675240423205137846651454848

5210827433728001348672450682567941253261976165988765239668887392524079984935281412709665420506976947910909196918431017573132250706433908790786086732900306398353516515980

0074053082689602417037023267069764470297602604698157613952922096431184121990812443297453496971403552504286138984205987355762655435347212310996418075573943049847583589778

135

67534082777525594534690720294929013861369117193813173601646476625497435511923723190337566403995542447783944298351828774246412327916622487261166153119556515318379303744833

07105649190863769246430392533328182321026516298438247988940816153152609817835820542047575742391907906610890617263336352753588367285253235669995686270183081216740528018000

0255368992933693867881167440777299165427880446784135620093629755456915180767133129637553148165799090729310413796284759941779072489099883994658129887509883672688417262455

600654681144806371742950845879829312533822895706426355589519574792100474878009906747134908177116597027874004848558461844362188045597553613978187711001600120973806585206000

2274673984321980195069095331623045898291761625860113602182893530594673628712855570420487403580738017522416010536449207270031358736274465470777935266484446408067183202372770

9420143447234741680498214353219066185429925469027839023946936756585504715201141750008863744375280986335535666305314474278010855994616123990295628653163830851747979431177

0261662110750667259113679565731261079068742527132102089342043068644265621908898010268788167758633391987936068177968001520738645811186433222300276066097923728491881652219

2627682090188547803894435603204243307256873460617370232452680436617589619974446116911030486059055319970563600863935746768254932730261029297265221999701378474566833692778

1960326841205387250396773387410094302331065949859757562894408874945023244710451411575928385431904365099794417786945650590867372794050181035891330078022518367047129478588

393258945203373711540965251726143245821365109492864963727906756677570290788108245219873727940381999502492852736340743617223491460131337418826157775394499817582444937738177

06204950516722942428537471667761788064434756742240965920803803416278472569426829022925212523848585337347913671592469499736083501009084159991613780484158077661793091915944

71885690996102325600636796509775882954357381862985180328592847704137243664895146145050192069446910055451753162806533052264007677708933108986747592363114249211964737983855

9542557785447923057506806386126902090132506307939519222133860918590651259496621115675247703172613236327036342371720466923617527296437171794845104623856042582227382057466

694163997921781159413559646469298483067092014257797423800935295307642131187963032500338498464286036340724983128555939809627486824431957818590388875500902591367550437589177

47202605836213776664200909010705930533871909593483383302999128166144933023488324286278094503744594199622771925912129739618715920208473155346980822917945574110692956227700

7464682922506407698844047590191734774146649892635236134717021650665605904732803649682439836851481673729696986157231072880503086554205465345998321799681376938521879386471

37415299348482991880895775760669754960742573489972049994459247477655065588130988577915322125348254266184290083358153300425533240532142124836043612819221729578384448001544

600298950501182389646309785607091078801026493793765650947187932258338844608895546152946266400196083891130912856890749524410992955918508708629877492462935098430541631892088

6518901022108368738507686044955836700884497199414711807644132023195029043987298197405881795755592493246941547994050095786346491078935958277278660074156011277589271569744

66918567208804615859568973929234646086518259734809876278032735783122824239714918079349952189648914987489119544660184648597939334327981695217348104730074648312530680755199

70638079106679559876645300045068524334584450141943807605732159724289332940953577690289281327309503733441374448424049652672649123074576028003390297467260071906965178930500

68460857048624192192189552950120649938297904522189071154499699379144340897751170261408258401444153573912575268782112047795676433288753969726244731879032952501475268642555

9374561435487996402355502556241286564188871149563905477942768587250411279911978525278635555326054907754163728187879421667567485785614162304336382407650378801560198809577

2380487988087768681885315507155213567545824862107453652890422991722156412432965485960404760340108960793000197018816034018706687673493011680676253758341019449395995975179

9294529013819672487642943530349403840454890206908114517035801938191590053992538549904281164432095981042972560098939829781428018158679033616012883408918242656076965727547

6319967224533077566356514988924033280178529096715912155922079660592005166471511308401574719857331559508149313274438609677289684562866947981365031356609445293868587621 64

7238419472667526482797803646166461654600938432009638665105879383584087496361419706343732440744061759639504054302308420453116892618690547745512308142928671314547872493672

2627140194805807208956072954026603573000165918462286944297073535927977913514154572363668056135103373594160579869350944593076459253643501294918340746568221068024329544721

1988604208897044877259785222935514414359586661710840576191296480249689572963361908106473429291580124923341595072790860874734729777928310221170360007855457695093136478 6905

9130192897508136820693747662440979739060858195703694795621733909224235687871954488870765540417538648949470012505326595490659115857931270481659345687647710613834516572863

5656730814481072241362520297715123011521636643617555415377731353126073673845116106084151358374964999144671831238478172661278322910119023056934267889747688453059278879053

4903105077614255429962938754841751963997912067148771056699673222538913932234015644585015038358971663177969650337608726334162566515216857077373530402247944039875456930 9976

1184759503630211288883923449592670037904222571028065129805454695532114953758276649718670663197079169226050790892188740761776634726929343000028544451729601647189015641206

9943099861693668518983340105486629594899465690353605541570293708695036832258132391740011335789122838643818662142763911201676285091701580132394079716586787519933592260677

4097310151174226890239632200130165011170089640856609159641996020206039985951675993316164608725161053738306083013854701146917863353802341074177249771296949823893347460 9781

4666330889062394193197079768488622039448966521855308553719676675096359880517797472617930326912831826748620578092652565290342801908827059014054663747001383302588936160171

5015488868935910569440449721720211392456484749888583660011930309334050424819094808467898760038276692181719086769660182978203559920718977804292361836630667945726051889605

3823472875932823011951586555180178283133565830986137316525935030808523962559082865729213865472195779312243359147323752403786908253602164629481320488457398705136724638246

5059987311460376225349778221030101454298751138224482136735267372107046667779742422508385846488729490960052186147747715199660246552260696212706208541687872395159978268612

0792232886495599471294392000996760846261445296688181911180944889919558814713799314815293295825160760107116624951556593720289303451396577612021582500179271658211474939898

2521641313983316292159116787863114903503708046729113467566457367760316704471187228697777493720498411552534228369364699486599281638902933332959670502333106318645471594 61

9892117469604407479301121141664974922769663879341360554362801356521885916919160126501703473492198577912155996554507884590280024197928361468861669872940673265794485514 12

1029331647678050432030238929162094969940946268112688477619419298887283020453254835559564561049750752610524425580938627342063589277682844228129581464734736707362255972829

4372637748776839928637632345393663445058464032220274973415024875765991021519544839141489452822383268148398051997076559268898315447157695118315551061873006768568040571230

7832433568627813031698684340855208154148525567720882895194035878583720245447816863460841197036059946920420502238950509990547462252473281286675075605926657023819679854 43

6973843030363475513530101422313185966598355017290912266356068396272443084046523628652842323814234990867660915808115299856731882700013109382157619132965931457997811998 23426

3059141198338941256260215214745561859783940816914555513000175131020838255631736162154590351291061800153799212994811409282642152270031879995958336372414326471243780519752

44525862986660913976536312050348551968701426164273739442553870410409340117869028483173244669643839274132175880980004949886150245568393213699722603951315999099527837014548

0127595279256570668760339430121291197850593657439413065417846820219003381050360714952969682719340824995408242161639655390093100714463887620313491340775627556205877748079

8439458119495188200713205259403344213400124368874611015102666890106725161058423245264855139823924004357219566830736578634636284713947718760514239870777535371198854467657

2934635846988478103391466784785470873151504290874245923946132610892919503710118254923184207210437942006267224493626435095955033758381719510544311687327960567532982916131

25754356825245646500812280522636814611597532119917066036435974480828856219322673936865606254293822490061995607753330652464844307273487208879672929269681479748532637460821

1517122721743446121372624336324011843291889863711234758307389741059168181881668955953718239958310915895586515510793911051537159282391972723851893459502417779055151515726

4975708724287679440973145302040959076917387500961326483707455953415351331344900038752803101337856441911474235022452362008865534205354859562097776348591737878062122554402

2592527384830776517999637382300909619759380174752578307961563233626646373083895738467111670927506441574763282421096818667021420776326837552607761110898894496467327753439

01491089616361841443514027645812228960506038155465315732310359157359227589113492569009356071497797688700731954102283427174875749070918716304762388723409696302534078247465

0972500272241452603350827917050952440893757333312319820302154163507867782655093217137817123137161142123181244080633098153607437639502654255747238777747951603707602548634

8948281530335202194134646696375143271528080291002816293441826410827591952495181736937113651514537697574630355039688155770393898348715498840132387691533308860839615203879

2659834262724317749276862696354131068465624484349311650896687484469347103405348229548440424944548029015008712091167917658606527872481257977534747880316689003510874585681

7454970704973599571133983455436889482161005371526145300563991242445399625803132802285781555675337051796143359354831260719900258511108667542907999170353700606436838865760

34328421346749363284798134599509524459413666885886365358016564396724558897521380576361590214842158335086871769121384576004437560818745848473053787916068543093840810194972

0610599381353773087430936080254374618360436914874112722209890114767287794789539517337789411265620467480077129380534584076556161636714032813645067389621785209078922246439

1679860438040198487605953575729784808053646541796474341632080188152262997279175361841901057253910235261006661285793861284828721151144331217481644040056183601867286327932

53969282324260140007259899509705126468119851380702881169297470365138828781618018114314687279332331454315102504802737697359069296971888893454531535369515189875968892467578

8779434750697095948389187091999856782139078432637031657987840807613470059542315330631356291972347631112370126374371698836456993918020402360170654847410448271684991511973

7619157080068754318896722949153519195619312478770104883649003430712202169327164487378907486971688905566978934955748268599453881593423799010669245538086603569309347485162

7199700083726662594609163691191866834087603981605345718730448837878445994564106619464928279029871448513829662277069771018679726274836234505524986843346217582535199489358

3892640008216042578581501014868869194413913566869688314459862119923349722584933741055324623722317841154042257919783776260572497686118088856865797980045007436559961684910

508497230735345599811561675513166841957334475132922964429136515799632553834794507661032896848698762817246534181038300169434521758808549607478490075673039704749799412435

126184213715027756166827369026168815088921349350732182693228066051823157630541607230367138987483779334050158419807105831531091345171439567188012503059886950631165440 6196

180647080736123767536839502282439875286051051135181419197379569469363868127453165281063503834682835784405514229348007993668210804493606862164600676634483031922495893155

954045816949724136805749636567933079196367152569807186187473119745962143444787453875567679230213614859728630338941823745049891769740071428842397668062831729011908574 0325

0338545222354033759894400897932960374120546035438418200763258809362869789359558085094975650679953503345837068005723967268132946214075401312828636882307453773754966448767

0662370586745285605262876843710312085978307077429889423331966599337067247096895252498544582098143949212079393436556802569455922608937043796808830592173549902808187239 423

500831667258629898918865217070485809184394517416456009022597384114577622103057688690926001961997563649835723155371164800780419034073499583084414551229428177602152104 9358

131823966276787157371499828414080368819714227388302087450157398585239725106689569962349628354927273450826891202714752570427050611184594647182853205454272334515940869 8624

1199063270746182667956011701272528350864738849204602785712850234784675339031318786296649552070645507689584383918676790667856947756213003776080245180827345252143205516063

315040369941548606130675953365971680023592160894781404681853314710609546833691621848065587071341055193617451037101599216136199509708650137780533897593988366937873954 8227

896833361381628700251093980230769216401666959917739312248390826447912809985780640532434091371862188532304910289011371610830868087486723811585974218156537092967446551 8492

928750040651580719202828052992244416273543223825649525174246715774826896121618525639989365694450558351032818566413199762669743911116778094871793696573690376143192238 3476

655463730888517770884496832380581866104908902147878769073115712614123453649639008426994470802551462468087502320398977365460713304328417053734413903893234842503181688 3026

913873884588479420355798916511772879206893386316827004321470084246757852941660469640375384119123764327438684183406337690865633424691994128446600486512177804327689855 1840

2277518098790110696764581909532313211287890901497686946981263359885419455671757126766788371557631267027615259777699256164372103010849931303543718989924700939575282429 87

296922410711118009672641573072618623927137157828061141430020171114264825848892720722572122032771300998294999206516489224551883614215959589128237524836062172044495902485

756809969155862344381651671463163310048739229322622194319536546576778120984486254203960830112776745283134067622943688413542209394960718259832600135575309969096916373 4966

559745037380554178792938203007569843076695439031187270674422084002598679309241436629052273054873320374993181703899256416167449649356926652153481507710571777303188284 5221

753638550909028760052076444558521723466926597049347060455327562960679060663231026724420595594961738355220588146676672261525390990922373470014065613975006335694487509 6877

152084832351897942123225929201006950720089427540980857268094505906415482184309235205988084931561998371348630863160817753099343135682611939113131975531178789183155992 2775

0248461596706646640528488092119285325699549573270133662891020185295971510209873181846394428005269519090433002235567601977870530906642648447231165417685422211916748623508

214055524163362856193729156026563037485196337983124937125125846685449441793989395036130563585381465349742945202428280473039319639196656867950978121194503715184948395 5346

185707607532969762060421264450505727237362397565744000051885064382554728183969647559539802630009732975909676681125019203508629139643085638026878054242269120670846669 5667

793808428879759066333385191831658073331286410779699802773881216321646181972297993823279225034392320715570655636931529218829355461902365071690769166585341141620278688145

20633343518244458692779598046627970127434881999983671416159300403854909347712321623085392496688136772265723665048007428664506672319728135654420431444220544279556928481

65106790636054314227622982051060394683907359874274419025274298058005737842467721405075868932757823896493302953931675436558263682933422786679969086201880895596168898194 83

36656830850488836176460312166876308659491983675260971893798508642961542780131573884226269097207549301238347693325994685064488503584484386598349300601794354979546847 11880

58153183483885467009000362390165675090985197438260891762970517516264463210468224004535486063995436349987617932954392450932753221167185125536171072025658333543526968 00851

148963267718413948971015796822371232377799817045313497795344097553782240414687831031187320430287713984126674284308300996375651519942564082353249959052809366080177081 5480

29774680469873018442817224891333669076425734983149530154917971860386848864164136682031527841507301603828760904785507046797051029723762400497326796540443757434614926 30923

29939531021385733403946546431380393284754601668289457039852028989705619905005856331681025424796122497994161923808081718695617683335559308203432917111861194066142459 30234

16858928325402281122375144878383804375333860052459837715219808574847409115965605101261021357390877587448522307072786751026934059525840044387691660888998376929711576 75464

44686686822582163159384153340218454302004455678786165353877900684796481611546147742567664346667861657143213956058051068524929302034863111876979067573730613086580498 05399

62498308851629744832342091870932486556631685998882180763426336180380330937080685655211630835704265987849396277425742722126865260875537703432427972131624718656218150 21272

10015338143527640446743840817921398524717656064532716663053443606587885020021620083222433975474388465690808482297610959890903615129181432469419017754297201682602926 069362

86775939441303998070048562472270746163817080272609324491040716169431127216932459613466992507264935833273692236372077527056795318598698939329745192340658032028977873 47160

23174722746080557162944886163769004988028768178344620062203271019135123357703649008084555442557682671091481559673655385825943822195135801563771197167936702062224568 63485

46905987780346589745658447852408819268830922777054604946355615841409745355423947767342753050083936054265338583429240886791409368144232858790092230529152799915905009 06519

24744141709100477084998890303385633941944104864392190202256628773984626697129628687375722161881712782123563349047154994707588680405946324980613245053389545238627255 24564

40081303686346872684137921937957850774989027824315838273235033109670899330157066419524016404802598619582946661146570878969931266619486819119516889978571382608897201 65022

11515000045916759183141379528429240162630759069281025915047690687227211554583699847244222110127830084214948985427598928757962681408284246202734481156377615885856666 06287

07597984538985888539735915067846667440735588962117649026388773046312288150082414864035251260700750590426855154587308199448361425175009691988124717769429055846364755 781

59388806623954201893749084032275102063489232830934102823789408898953172570727814669839254676150509011338562734748228265353590100895281786892385130548337080121637570 61868

76378595480214881006720711402705172446818076076012994222575783785152623729804437548974460375910029399148776753477242411230998381469118795725444744960420094875863210 49675

62261710980657037545797784528755506706813330119403702836846318300506680753062470647522436185743826977839833542129935793422513520349235503103226324224496188486719775 35580

31652147106199584104062211908359669078979633808121363989284373107431582309352969442335189333013437082432643853278990826149765002466223360134797174646420807983695465 13329

45377557662272524462050785169967201989999340101333507207580771040382655836481700018825204873862694727977686819841244971301436744461773836037845058867731041552138029 55114

8294138196811165301623559891014775948075941177157011650869218541147100498530766154450326362434342050527856128683711788698624228453711227804753731921404895699108618357580

9549838444566013084746527651201522516404600863882540892101024615835361898855615679545559434153937745029100661215587064016652981681454915048709104536360233696267900742045

2451078747671110858160388335049073019845459657543115162722501328826413273272004586475040035975388411508587123667673305367165176672359234741082205566207389714589673612442

8601318448099009966194185512659140312448268215050509682117223862123115265230071316586542736092478521373501526308736450420970868697527750651953873468375231671703179386470

7833381254703056742160675957181244817241549006779553950339497434406511014026688323398138407299525779426860509803857100388472248482378758702492339217271606723237837596682

8065587790533662110909733434199797844334852653243507225336039871946098706430167889540908433831832738009554905680850927913218961619966362620096226963711045923058598579332

1394571812184970469239746847119409031628654826727816023346641245867553153722069865707588845615920039276376761595539143811410641071280366418273848570213166476675524075009

4911376831891968759459457677975505054359139588476862792044160738994386605970487557360320761893992907847325701218938635124345040256111531106159816041062934721259033810339

6221076910135920641844272757256362449921884192969284504886557168081476678966093634642719375556300958026571667406069670450700559764523975309356692672134756563223105217097

9726788201175673384420258585856309958277223767012426195014021412415170170625748239199375731539805651844747814978850715687141423536244332730212268108547082779756822570093

2570140588369636640436028018657355839449184790336263501554324008794445528858414945115640942709224065884058157027469906282629123540380245799757493270182391447626286719728

0067415106461578426087089210088433757773968560088108136928575650801843593099633922487482273469794807840647767757743508106481311385412026681801462169606348638693518228333

0213466133693279168595030799246911478332513082076691015178579716958660185792729967194005523669049880576522883405403829572329495666610618163109789226399407301157002457628

3747735763081837981616381190797360315872121983138196203449769816206423985075228327733732583324372882160597886109873551913778558814037298650838175116326674547594083452969

6709262816998044889010443639929417133550491721853044413612755372404380196616706233382973802404906298258609513935027495898750528072067318378647302489130579054815627 3669

3420315336593480354522123759323198928884374803732410786208605622462175566914066496560325627015246939470098514711600994089128351947568274832967227217668299349444534067907

3655254171748865839010393192874133368705592913800277273147754869775541198401388960452885541646155295967306253288304118555011888298415869115770185373637750607133505 43268

7135382591328509795856142302743603957649606785980108937955654308759455387321323500760300773010791789640837991887086196320867905115889137412620483854115784882203959235943

3270452459141147350780545540033581903384706481014248689197256381168631933216497629814849539646083874849543969458515257951809116266545236733517450163608333529739655892210

9211569119783167291844257925757550040307490466642270906643224285132454408076050309972583021790257921821690823779438834808656545165803224926641396247339605115247595839356

3660744382993667731189712438143450714426349066833674626796514487576042164900465780456681089762373524663044865139464823196604072671263927405371480600202881411947914174210

9645063130559424100563987780307683154549550715730058201479251595904602466115178393678560161253240627524204371192506137964850891981909587705809924908420856001038860 23943

1557064092622965854619405990986964765074089090371020410737732283300019434413131898982911460768793479475637926076608714832586479309319426065037650312207896059737211942645

80743339322946456217152992869877572374421745243510593701522448697613048297775612559901751454759543095729786727400721774104428762419527057513081889666990390251341804618799

13963917519996106880395794578140869943557929398861400711417249894605120231528627326923240980270344039357213519034584028877798823381422660989331849667478892669887427451477

26958495465807735894966320431968422729390540234206993659346551759464401935696808875736513565527378275584560031781257854725797390168904462967021262486121787662647754837066

53350285918705939390177804647175942432362408204797958379349852045596799025405159404945477441760839577973686737690585564791827948871367856297647036197192772736875817422066

19342121583922147169397052507247831658368690921206606029778144229145154061195058652651412881871106491879697943160479965128847403654083999670832363146792721324493626985700

74070471305843482632931884544051058661940375587325025731566527539290215236220741530144657593072275088832670637058214884319615653104189017912554845787463405381875056140444

23146119784513097056545369126825579444878042588295007696823218986262381111072368744992142469150404040031332186682305915435954967189013559031469691373622407761443103122644

86037647919392677174935407534162571025588711228300980455394033370487328235883976509095216726564747258020548090639892790959497565422906184606794762581291478097037557429244

80898165417448550967270889203351426936069221576656346844171425141884130073463421177091338849728156826975564830840913326336717247725008190383325739818737436603642934168155

41608528746458725565350762115073088506880345696823291535406069272424136126383496243025222734822076674328244625195343816866329833335123189264153753902689802092306902469188

80866165342597137031996465636551784185225207740037755013360935861416370449092199891298081723596202953847731659451593254575127621826588269650230832748779521264161760154833

90384586984583193711720936819908328893259160168315237898417509554214565735424477030798689073885423655119403552999768615121850369118581180619655257629645824743954946388499

52305229390431920834700408595906247758692673900383856751959635718838147734081913671101305078535558976019724131868873517693888960920970811996086393237611436761070356756755

19545098566772609283078303333849478963026102904982778516773496566301792404581418889016861103714921870755368544212842300294361846299784470990238920908702825867759979329000

24068740412895878180015512623850669835207610152581029222069185058989797807610317527440586023149271114853812525022120659086590178202587801042247426440305081493400889355369

94215541242794024288330163560084548061142519736000794683676743928913680375452258525035643374923040611881347898649501791203952139272589351885032274797283045553777830648566

39221525481817233208066725276659643162054180594480709373376273851917793186994730245908116531708464778489793524433572038061469870092327006538735123781189355588073827614311

38710477321330658524292176952009865221558322707624047867472018538357834377022350425081424571284375657147968931036341808937329972201918567752838733601339441772638156220288

49015960447422841069146815777966019357848536075667691643346140295482150176278079402653275401698834354341953169825972572703237672327104388502145755116078107849036727341377

56850547512833318281681806879234839423579028304653257090996989339293195685942215104275316137785891330262383995724924625652092380982163100220099385178387862912847546474999

73808460988491571783874044619813744839068734059128423831399226835506173095377600773344155197113447780113636712930495399900198370576056147210774583344123575225879819732011

54370849321291768820404507449491093051786970567117701141965616954874695699554579466490245980581820522678280554400630551716201091706454966577280912312827303711793448888111

92757525099425680572045715600308835125483545318216700312020988817566560248156433398561210335218031222038724546176492771976075616398999255588448247125397544419957528755333

8160752382890749579434084784098190504995472234917678089989735554281691198660291050548190046634137155573699308168787927857366696462826628267844983767811270499161680058808

6998350764297274029251889063949221831873692325603991587304511774483342679125168538532935334055740695466163939922312566237332891308244660515960756818641377290495515788532

8899061541234378136169948046390831109473541619189442527956102391550363062024648582096751952056686309608898274241602664868431018535792010934148995913294012970080979209338

7285245068367706352271937561210705529957700568521654739563093401459950709068494399952610995037011491148583565840356944392440181375872950668205071295542099185087650711181

2723830525509427097018083673324604796439163711962481671287816880866722863275435721997602795855021249960429342072378556610335213428835403247208840753622747707672265703221

6145985308480701432711332327955896270335331133019138930987426335720748101442606041221949987395209820431409400117396015001163746493739333742500935967983549598081681749426

4116175890029744537902464510011573256811536605504921822907182084906673068373474135408768646880362810534636606363963705084006032259444236806840701090414791408074372432537

5420452945492320548854094472777330867728957284483519930624062333601489465701029537276931923980079137424612898806209404251071821687003541827178246935885289693683871233950

8180711265203231606942139435044164221180458659728470198747058770491745236140621664718268422765043098722130253292080624410135651901924297683194184224795323048872227619999

8130593434354096340873414940098361137992901040490216519800169653451050932011865288950013498554815797648691997535540019962392183037155704956611140749069890614688031627456

5043092929685649204939142663256004909546724624506038670277906597782558864298724579515518278895492937913618643354949560747960629939433289895607101850074129740279027598329

8534453665473738254430506492187558490547253663137453707765892998875453181707580339014469761261208012192364849601319145375873842088703277363104544410317520421681702024924

2468543393848183239720117336727143759140357324974219076628395187262094877364228903555507099568064481385391213949440769981765712217675877697741374769300803853092414702592

6730972434251473897943330822070919844429634936827345559992756543084921450058959498583889721949080480021103101077469457503027986720456029668290243300490195865656152338506

6010860852470512592510723689039144260448041910688569171156055467655775413133784673435728191355792152125244163426728793514579872623606847687942449245432959375932256803824

1243080404415170323303546805930255459808401814193883913099131378039516586688564053442504597686306846470385293042281253712888124063837191968251315506404586582122027033445

2515002455569718520449427126617557097790765220163140992434562496582346327419996696591909630315077216229597379413304490691235466289997678706396442754397053072010156786676

5146256209664985206977071028331992535522210179071542554910668909890157143522523208824935483264058254433228993388181433246077620211927635355605540181246640118064490748656

7924930457530305406358998581036430506691788360446337600725210593072101922922418953223829158909341282287505109801463162395736714345765411805422274015650441113110505863376

8086937833038419596945386133732892471602532222936973132249379972928005992675567247975790953496341417002038663419614534418285905495280534099776308793433997597891056860413

3901260650786125608232043804545448631332510663723910133243023185466526509293911805432574169027595407190841762886773529998642546058514480703245585088002374921985388998257

6356904694020369388274466818401399841773978038022542914925919574927731783795745010729161896999328831375419807672649180013064389905496374484069145828464261242049616787491

9002870799697918853412624810879505542074352644179471940403447965109102724481791903205559775877384272738131058143603281196234824968770668087190298872341728952793641870640

6948463980108956431535233422900314798101668431369479196147206097308580136150705516651333914433244858694690120492477325571823058188574716991463603187793301384759759861120

9617071626757731572976133468570481409796915486001251242060298788553533764784651123129289985200406105400658342503451634685630309405097898362527739341235446910129362617019

8997139567103939774531009630549010544836570216719918722034635756363824411161709378889385493935561250090479363717154224080610343811886033057556124867332684560694175279344

0949702599774350146701995027107914478321097845715562188832741071097653063416812979334593467427567072441346943145162167047337673582685317196054281712858870830159316048841

4921903243630580503307219660182809009404327179179905769932354438810503240674918606915784062898734473767092342257822449454348280812656771732125804023869289361125565308204

5065306634056149010369686585620638108841621125072437468721892420926302653347648510064882401875311077523879905191475130170115551259365095027766386560475993267772695534780

6327049303948934897959809117789256537443265439374227850627165326171004913012764765858878130048084071184414269029062542006789764961699662037240749990183839249623080167971

3342360699757087490626659373346349331174723891567344412867677628500732867428934742984354169140694898144641854134452472851022266000796138096075270104077275968926615167444

3660665791711951892709120693115700607846131048600959027972951465467723383197530039299820230677508614793783103409232516679305885809444971780201240607532058426904533209973

2793580656207789144551244404825526930353303513359014814445164701717409678054134220952991809060291260715683392766168926744561155320928000933552341934762481168750783337505

2144864123004693591272455168409493435643592020840086639728874526447641681222431979005740376752041131459356908294863650288665138667187098401912652087193821461049629177160

5611241326229847291819735019232646934760673591792047346019502146850204222725499060500390527173983088239346961329546058235596168859614438551773255682572004086406671572614

2874215865629365346667650305399437264337771175524863334654866174710294742570747114524084069357459330581066479105872577087039412159583497208014347320216673200295921778831

1465479415676323580901593449839343603113947019602368064364710155265483303322903249484887401748716255541784323459835831317368567411948216488283219830392937820890066686416

5635632999002589124253674659874275784244531350568116333665037981361610334287691399779093956537587469917820529512660654358748024953105462079082869238253173487093488509220

2989974921677075930466511581133383047832465845377999596422451105520528225985513579954331407804687382883091817168865047464734200601615944581759276487831751006154057153340

8027770017339348691597254483584995703116908905023479004182696112774389101113682498365112435221732941277060381302634181652357514835007798682610689357810835786811158166380

2542863945488244743152171446531882122656033716864388552794908372296727150585998390200737043520451962130066826186389711245753979831674835802866090243753365970379553017428

6017098228525434662822605029782281922967974950706846947014111471827123771794543954247545582317637072002093952480305502486152919425838074644561247566303293621119406438535

1261733638375785199093033778956350709984872647563288157718587782973537054845621840616651638052116335535956157136554883040230839048499053464022753635053221312628579847488

7120879742391971298065157156225145357376229649699578516189472586019304801887884140770642778982111550389340417299078428877913960319090947564277462823464504961855895718672

7089305050991750825906016230356085410026150659958294094188223171176685610343109409015195560359419762952715191446546570114627356476034166407335530107840661217068804877676

5829603445926238647585574773412285590994551261972615033569804674294654465241074799894679978464892745481446292704610242772494517285427202517075137258797350890288356512373

144

0751402162131170264048251331975042822098951384552813626884177326700882525431724357988989275269871603988463330764027410207254286075391463205633419005178016954816417949317

3488781224447251861194100883513137555490494181672864244172188026388165627539857333600411059599433600445109382525788027664814425790955484257632566676861865912705148389804

4159755020202230844231648171457820102308761368940062171591863696647566801150589395069179486179388110691357391119291194764571638424398506767609270156013876254738775551308

2161491786331107567699693263983636019984305639886793035036311014621259261823243292023050487397355510388061839630338392022450218778063418013902925080016547655990060390880

6917718524407509635151958193085485349436375269314283472601286321556955893475903752173339569860535581833404046834587120394074492363546970801353967029689205641532705761 78

50743694104216200281385974099445803948437171223780859161062547291289418501433733201139419237769927284987921679570485720848262117895374509959016057319145103301592195047 79

98616399735136343018127076636196256421829613557147414196825197784512923995180948027615779051505299624564676859410800319655247778443101848753683776930481208594934751495 7

5363519083664510383481178383040200872495432943598018399096116144252080810246154834377215900747465469707895682559178102153564440601396893159844515933212019827378778074605

279848063188533935925532640568047587139273617127439049444206400176046596009726741995011180194421994703076386808201891521069608033731147927545046918070866330683047177 1276

99388843497520361508386403092367707966596240746653032058855795435358985947775420846585446621311741732136381937611174526719845377107365956594981659688759535272565530 52258

6777874361969955452700888850431275909414330236429519830928170463636710577004082355626345787847852344823998469437807398003882103551971467793674394843979504226155553 06393

729577597612139082829477616090698661599524434740720825487274741637905412032043763264976757883944911594852617550814823643520144900684490375525495045287112775902440 2682896

602399138122666872871694391424233569071672197485298549065028669385313900278006222810546594919496576773408717385262225853195847657410136500168835483562369923544012 2359693

6006051229704848086706559828336254616249888405808056838747680592416721489925466769707042797594707443992401921713587689294557714824440483700282760938444366726557959205333

2863638211834362027746460871764560182369204997521426141119495091413659399395988849687325390545616898630962962774556931596271112913890687533714285816832655822915316734128

890277343355493447886835534106128230021846623652602520308299055735996294121284036158487698284476721665060508430933235779163412598672524107411628555608874176483498 2071420

90696390405828539182621622899826869597594938059048857536815235174514964661426965879562019976643810050615041800687076584704534771470059633072335779079437670642119611 92058

242544441864130889629668960333915001324327960992277835339589184662575993194526902421463659868461586505934071484008604030338552638224638158915811836335966437381856210405

8201328165698540316735563816301968056458734803967516057164490401683278201603100326068032668396046855898129134031175368012912557689000970360499259145265139775772983 46853

05855369363518247572333780440075047551435090756127219522846296067221062160746123771515371186885040037147862817884264613905805364750289469072392890947226362566212 57205691

9773693290313934135875697822879124283350725027285956323478025040789612019789216413238743692991691397743472714978009964967297895391487270489581227501458990446238905869642

949272303541293353238761892115645887644297136389781641322138439455803462655791314402914125011688519989228707998820333274585508787396201958428491699988096256663978 4614021

609505972997287096124330457625312926815643291803738394819151464952919885361976689649877753470040989333379271594905193918030312440938121636064272059749937430095796162204

7067461174085734109744287490240722240719200849118581815181242763385231140880919338699052473755179697915334836986077884734179237590002069645477898046544209616558245456575

7260109829279462120160358645909800121461108129748652676649377548555016380093639144038747044068074173071114912039559556476378636872521258664199651815527268261024910471618

9727921996372881405772954371894830012920612558250088095864823435031158427250447144179924088583160443635426313119988381503447473273977326572582918374248682532213362019148

4736976267555076004784747507130263315279144246484583105426179273255959789950216364980568016721702398636422151384913678946966518959963698189528929209109158145580415830296

3877917869354121830040998688888707650560675784523488371448929958031397226925002634423933729377836121998946004608051929181573650714060521324366571174865186510958665531766

9933181738303448325237239280960676905236851464558272384358920906669573835462780112429104142056474580713944479048166580981587834729983910310275228746947404696773821116151

0972471275609181821603213271154482879902209158099544671791023985775776600759370662369931528510617800162228001306895034828243805988974280780978633732375367387515639962500 2

0268891715608720568198038132159271334649860797832469882632505217246773232158505276772769080739518020633239202228935130743426597860593702510692638789504893955632192116661

1355155598132690575754094401663689426009267552040653336553951459594430336472986972522461302873983497304830196186945556575297910677872775472113472308106665122026618370236

5900835311812752978241047417681200547328540882448388546683741423365059125994228687922948350772627145754704620061650940034891292603995543195783268320040354268718280682549

6525383158353257730798874142984638739305884324111675854532875489997195502300338352132642356527110701750793748806830785603325414601943320967706374935741539533003747883990

9900702531462965980415264558977993948764754107248509319276032948979171741362137841981035068496164039387135610981878533506494822506753456264515252977403298927537561691817

4853755507337163704805113108209276849359945306955812100822853145418170553396237627676852364626589367737334280355878578128082111574306197915537124356435476888116318086833

9377582789315224641995493001697844790900079766476198783361464566192197575452830238998411280198621038498830157743708738410828080144737287666819032370967428941970934024336

4458161318074772282133775375992468949488568872590487141814602376469599508013866043470594351749860090523183122013945918488907530401736869961254394667213996723140303493622

8627011018302110667511115697441309369448508843086392094696380055670063404787656103708240980486788426585055996477627529334517217948195455073849381133042385946444639016837

2344019907188086077474584650233245520572489711651503735461248395335503707166335469558335922089003314811093105035625241575155460739324444620243895162945071839767616987097

4697327731850083632859062863381325773471767970860082863657847710142436557087371372940575360685199619901423615351912187818324038266010409932768038702518282689905013928 7

4943375476282680559264438064463585291569837975102408599405715559620169061180606385304794627810116368837111501855642083240988162569805452419611080501075913425742311627438

8612649920868926439355212150847906167359649534179203357299319229870094573119991169784226885366510539372307341483362776594610820275072013548479905377197752110208021488139

1072844348389583374523960791312644616573885318211704659936653431264959034724197008910572073105140310031420016078368342775492638478125557268114790797901786907065870634749

5144162525321346591354161159377354271127487844264010320913869535451417510456835940101622677546837090867791763832995134146804688956935286804536200975579858801075441759285

2429641027544394174983197584543691671545375831879858306467153427646260166170736520150241250941328917174724357727936423052842049153843136718688623786700688669902695498242

234826535568866776437975758217353681724178526139621292352814651019033040202959808631994332812220298985891741331294125482553096867233116292184678213100262026568569686333

898603114906825151840653582620284920369110801300451065828997688939862230200298730202666823959834337214834359411418680094410242394805971295162152859580318258362458840738

192471713075627136269742883333595200543374022971689775651438500239796312208322968868544151807687575048509919864160038519290649018781843282607380365794153750889224733309121

890232978391570165470989902590963377562583277115219769901272065427673643144359633866983789906914273142987712102809813540389905181965902575287171101772553598109188971805

969066534622525599610871060290385068261037365951903659809459038756802348958120983781845663284751012262558117615391139727878696566364760387833095845869521297413602123039212

623072758316201715327098091760294702138897544744047645354181384402323951927105008365411261449874776295766461315292730408262464670170879217673162155902352103397158595470556

80242283827027971494018602228872477449515019204840639089778470639368376384247027691843714011326399534905539160928436499378627081492308485158569104536572034214111838272411

9259960984403071513288390846139536707141210527220506102534051019402940749759574527174929539079385860638632271697588309131577548083427308450034582094375678511762382918133

228500723956526732881809023821928341494144956554284260221379058861020041883391973178632547226069678634981468979548112924564919562757485899108511676602352010867035720624144

04191113989650805631017762544678994028211648920629930993950416269193632825056590712236826429134597500011438126624463961940292261249313966460082178386024222634029098826055

7071413101340225182292518114507453249611798278098090904059866888739465434533741529283527320684520374228670618018757744193084575684590083048668952181850546205836400727652

0648231602447922945765035027161024023604827609189292591418654431079730615857216897581301459977941667168583567014562797481377628779120199707733760091548850548543734919107

2444887826850797672742474988775037169509964568506621052359813315597357709655906404999570137621979292143842319021934015133733714638856997560257526096919920416796982308783

513389340972127413617967133318021610655335147840122718050005605899625441087429177105963861488871216534202742021940010898234916321433410966455236456415744254761628061499442

862262819794712099533265692883575707687423148256547621396657615870188608830873520634213818055080953871062643310792183401239101558732344978992864043400856643324403552063

4294570835086745978222019072043491820981652741547556192053287163770669883912653893258830090785933097325279803007139032546111667906126220914849586424631374604742928512122

5840905884715319438431133107476804463295291014411788533608414724183078822879553889265428666448434674012601752783005323779504717394619849841265861788389973276677309259772

2363725112409369357153099344533436315957211000477806131956256694919026610029205275667024981564837479664097209386142874282180671772944466864229698980601045005527182047419

353303659476484286197418817359912109181105178317173557233620487679773497979516425829722861089343501579983963113356714420777512245221594458881235393183178984277679077619512

747512520272576345924105999269154185950946053770947153664423368160345377494478203803147994524854190241582254730780105109221383043888730097415958976243928516827241735402446

95335256497883617447651981462148737973733502013899631749840480314174733112535768108772820544027530157949921224828188315859903217642180857611795898305076310457939415167555

40135991645960889661120356360724099260713876870353530836023137161827587949437078802623545134999471005751616584083140181416096414848555695557304840323932205248542084091772

1499157966705053940949709130942603584424107356659675150594129765057268149531775654706723150313046360845483584572144624672088377626519460492307291085755171808704011926298

599674373996670398429978562924491578367945650193823228919978420229143846192877103398117953279196400870648499992736416101929828283644198702283182353696013372952696400314

3205504271571656300347801719246420651854607568110387948042645886919236548593033626064402769482209740683542342439780194853171920260633603021898499877395705143192427941574

28371466917177565362215383803955612588336253255619898881383941315190594078361441569787973390220266643667605661260341772385273381717007465432876226735779917344206401459 57

5985605811985204360990748786201063309505039894971353174758183494361183335852575639212464655851461773314300998747082934936630501465316745749214912742258220888494609209423

21143346282517160783182427482236806311975876268107227796387411914481207607961353984499878324587780855847079140358040322793321570138959365817735396784757753859198605907 70

257149851997929188620717554066504414367406195975690246107524513634966072493582493815286236865926413923632758445954235165302660337023066455584086230656244569711087919783 0

0610297648846110574242652954741764866252078704004090901790467103598496470060348647617110294936726514970098727032847990599934789281851306023690074930957379371813869516821 3

9546812959146498623414918326207550263876248950956748676320264693455175510292818249839119646790918239352418715552522863268318942087699775967873611749834858899300898246 31

185447842241011310191145821330652805811241230053589649036369265243691936406940486516075632836894857192461337719895892533652652570482026720647698022098371415108748082727 1

214552656540049463226137117556522557855785438620484397274512811246989303953851327557208738586136332845154980999121622176081942298329537528843084974815265985909596031707 6

7549866453741376304678326072883851651589828190598366244240984123976754338199564138877339025561910404340709254058733122719515004390733257007402291089271063985702642339450

72301662562178032650525080887920390398302390563040930830181301726145707308395001842861952901257381244218064366115969970222769336793770489676516002294892551841716903012 99

0721201296501333506270071422766354974111199921981966469870956664006653242100394714517812910000178032454064536894501473949749005669062242571460680569254946226479467048866

362893504625320978470128681090305962783791319601090907816037257598889091566804943193195890596973623783181042943725339610072872574632977674802262448251157855302750058601 4

15419087537221131528876724434954889393712681182357650797573755918622609547587939006855053792263520713017519988485814137391208239095529104948088632077345265344956069737 73

1565388547835754306823098580903306345184634352421193590099177251932732912298929823998480331430713420889867686491831766482764551648509783183127571966685940965467399168666

7380311428772560547672156667644589756821784995803697938800350918275358548373510238035096603225525565991415544417369194496215692433112650812479498775233971600989640432 04

51632415661243250145503431660567536064435401981471072977478011550232305077686429235572979795505513976023219507014587792644147392121187155759311788108567349467436775790 8

6970048686007610485539674009396682669252994853769134670998340658310623221364207499710367664880906366581828908867836544765605239961168746643503885454965793933667829994221

239057546757896113200214638887514277042848516141037908536267285432992826190091240042693001842308974194723371882770765364599634437675073059724489094684373350253686017508 3

172039515236001787907322728885243637033300444409278129059345368663141470104659341883474689282629988236301306013766926988217798851721245414573378488230382467191665951105 1

7463243127903156087414886070815548311021325401335686854055883431018870889387613937325023408807965938201480483031644811231762015402434502589721776700525987685752911079948

876170334681232019932311321928743418466125998701817465611791461186892683702520165299119898887494882924206169496543089442346341753064626206632041270524790465222259474852

6298821801665103773915209569257176760513915129079083306308913138467076780713608298991899449053998432749402438897106017627516486543243504174682174047720535790729788190300

6476217956560515937853174699754367850429962280685938360583506521637181437581203594638980135385789008753863779994425275139716428576455853881509599865425996111121263525218

35373754089383829940714767194795565653338103356092091651358796043175645214900210873745219407016607907421171146389209287184760160902492319110422671510290601789567464238340

951983591142408642645711070748530076249802206736383779844598841477515071622932192031026050055150907697894319437834821122313171979696873083328746838329398680193191653702663

82003482464988882800995308021917638041975946273043423705049816862663146331381992449951350409336852132648622166261430456380155416702997567018107991459837143013400 32034976

52952164385778342024804974604813565562787670014116764532765709159469878574710951707756175897195470146914052898762386344666075216918405152920370643416714344 58101488124590

41088366769369630161221404303079623341879278070741455430961219509880330732327122514307467437929490847000111815787217604725628436874440299990349072352336477956148 26072754

30475073383579416952085411858141421163366331884361393046086404438120305008737474074303519812587955651210154379618540181768351639553142978892109793350644218922063 82792601

70808596615134092310144550959805004970933341826034628222661365245786243689338287481808083116632140886018962793337969179670238926003951088492322226248791469952469448 22132

22071622818763375411744071764408256359777491004984411315866456552169347946993853458952764802986158402264099994210004334206449394164465158608227497279056804659105802 31998

14041816664689710703815899178259905244379416476766531363703816495568807841719706669088818711189296355409708944935083808672087408738589167828057846463873013356329005 60817

55657051868983518288538558189418761846431885541883532205586551491960840135051091304296438673726701769209462568404821695594224381628363176054907299398382901877071378 64821

95962795827372843849302107651701114120971271895136778113363452251194325640609290920398920303114286931102996162897157416513531226509765663872541502188189457696063382 65402

52017462748433137865936683535892728894413722271223237303189976271218756359030552405933440680406716588549910892233951031852280400316307779313938878812426373994576617 35057

80454864709713365612269105452680323351709465782235132634711977541566480012164761891538394454382707413571109880273825243581929452706387243028983887862373972699481019 09956

47633987267797438188240786469637206134575020404386514048145408486372294808918749533368453833291856926116001360905269807485077880809719922079054938464491299811604451 05124

82015768134230369758459793135250764991726718978359204624413558353968020439011298808208487927059939845208151455527716045554509166391086146598101094364855299583415894 05011

32217591891822740788585450573707541719807935765713476422566400785520271235964984181478091852475405178298598358719450900926456203221456793600320098036589140365924802 97059

70423893401417849408983405889420828137541084532719476594015784918087988412786712883469730444534633001134078424469746761005216325231469607461797235227518889113661084 72825

04473338769880899788249617457143265389593198919380945373562006977956600732920737598787739533401224262824638117604665529549327760151654443398797796575964213036484953 80297

33629734054095365660271566620956242040189725410026903088730688596758323634848608003136749337880469881081792434870555858612604433511134155068347210280388630798842486 47959

93442691070980705308289506513928987245609474089911504991593266076126398135041864212683987924382810639019024427167350746240245768175241297734377047211534086168041782 9499

67650685806251274752995065595324988186611187221699214720955606547552197614550491099906975684236752153392973559712252757150876656596645027191717182052938851093694472 103279

149

1299729978949595372179654148220468484710797133152924225658106596490768857512128315157570089156883907751592339497055571554396120342875806175188670398308678334081013481683

9437033921934197424316334687716755401028790595185546970244107483690998853159223576836050185875678557363748587714841063401334897590877749058334553977057953521359016826647

7382725085565578135488763598832002785770631642240468395161657169656331177116454122497180862165255308450802635618913260435962920096403238435463721295159475370293413557820

5609103465031482793614106566034544208285371600231136681319091964102873049208500417437038337446281046462094277696378558093427578759878418333403996601935420267148826128194

8862553950438151533608881983528174947354252961305205889894745297898192765362302146492716408632029235659299419175245477614408430602237931856760648303943416218753627041474

9761396384298639152870831145938517668485369224524791339701856796618981007020472212331804541923079943922150899183397221294664854266918247805798787826538813387791747999298

6271645433393042460911284741410076110542008971258536672836314860898634346456934102417486756764886499932016760769139511745616303273744980446090780906403046763494443155886

9897372150230602240876896280899677720082954009728621969369799908562863781892068734312434251912571665860848853313223426184326055836735173575594624744942409189135202074248

917184402231826670207364676861018624736484927580147358881296157116453077730991879061029888628714930204679227525167103708071639439123743167928668219344462247657260407054

5998596828789594818122960996644984189543550512697462222284055821601781563848932415629429410235472447440652982759565085230803988104176753109539450829566866700980596803972

3878307108887309916708399098666703021614657172247840852262333384257208168100733965346032149843206972663930918651492548013701038387054784958056923908090714701468031944118

8291677410010867607146367034609701658774793861986557251491603212619971997380349016484226754491259673931239799007483105538506866183048290644335568139253044901755675497722

45865537013114885452145575276500340012894742742237558340321677426586029415028540595957341787349070980159085826530220465780692136863441823833358550580440690789048769469523

0168242268953030195038490457409477237858413080942448126386762545261790718566784949159447575258904329859715562539168706640500338691147025275287746323076394773662205021243

1711197669755407073331126759558114307664350837766138393741882119872814024301959257723399249774565359917373704823455256901746838618160590685025236871722925582045471781431

991858074949168211910106141017546675307620289154632134291872601569145323392446783536092923925956317992477364265588541429930289457142976436732322262923602401555030564320

28370518644027032070094133089307407897145934113546630626365872857188977005569179639209408954049496757766916683128261519805386857951638874569333961269736698722204498574265

207857339345005521824959736438727810394612054451563797961203029165947657469934154327101407474577289265442299660080219143075163201211471223362886891100314198269762081161

0237200462099132116432607069198868028640972266780902380740359354214499157461979683557148136771420102843682700410344318799421436138119770538705702515776750087453539287747

20196545049062159447237705651061967599990856948777593914911594201505099136774196405319122353927497551027522621259332903159292020632274315631639883559894769491278028259845

0835836799862035335202068546055921678655283576498156695323158588572387298888221915594480378709089164856729907213738605360437121439616910385695176160284757070741220885574

454803861554929996011109008952930561509283466502880398315529188908659028176649338550360211301004261404612185620272908638517057052077500603308295180906193350336573369268

8723114598640046622373484736298028779881021471019245854937487774531159628979254055017807474919647784067465527903931955658138966925439286116812702860780164924717579476900

4071383841871022921733518989407640808971431883089221639365968753798701420400378491301275010036189355286464804238014072668778949947024252513956832936672012672774688760322

8486942873013499735546344984108290399024614311248528848255524681487627399427149890898964065884653827774882015498940055948650851084658197861933024860833800725503537057526

7261626271208957483857078107167903963214061147985758927316520665138741418399014152408069427164153124841465750736716210143728566150672804848209459012411539705704846 22153

9045505320545140864908348169336750662852070850447616870476424706292519842182340567119317597738507121384356616120054129148709109996813318550345675525027394805609455333324

2616500497427369923689595571203234581644450618398094463681201084189262133146656721599470819817686659148832682318546016554172883453416704493091663784465689767634231201898

3264383910341875841362419674579944649202221979834593056563692756849359777671093103041411307312539564248638501455500757943604266544947470225969885102663374383018153 2607046

3610412035069829100774024752336575842434925980678196106766125498936694745793203834801189180462399344020486054740053972919887064890835327384625425978152377016539340906639

6161418136993626227242206373381984306775264803874177190613456070869512882942134188943261411559837419843096506180799248248599557473975865979178350016251247911768205661124

5687897954672289441161207246221821503611187196038675940463408153405209319548994528013692392045582070502328159177110790863859943266252683370835162218627906963513461 00018

9272878972239673342112248855253794962334805017456457141696886360100538717492882149746928962534740324906591107947746995501662902714298465088391795743901191544231663338727

9050548931573371400843033387711793984550288105152253878558588527678672465468225260139414212638002515110525362020285088336811671179131453518274582690793621433828736367147

8550254061831507426381713513107673935765006518722579662135584845251998140046504966442936944626432535342270481087358438651531657478369349438175618439389101920993 392079359

1730235133613433361740937889433243636766210205752064049860033947626117730659790071733843508611904667283091919140548761824903540960361117175873842829531071297887413006781

5729007187202852534737368305268388208851900652888992067114141756148218048590301612699363022004245730365450630834445212718140481106462655021833491808728134317000593894546

4777807178007554115944795663687523130280968563849766467416423979403809780240068223930439751487761855101468074924443130493684240279796638069701072185944694667569 52631588

3828526261340027805651395416472679784720187392873431743195634271468612868703186802680513077833113336497051424345861943399376038313489195361652219857173400602626816423331

5262753256152699866044674282100016307871335675641760570610365397244034349964075523914459700042488278070090182478520476973060681827286895011123040202596546463916882653440

6245138943800868582630992637078330478363038980860109948994125751256140153446384423708749095624413019599875638910465209667545877660086590395215269307249475934637655249995

7398136870468238357822135022751562771743922399554134549014307806588871451328133707614850257685232363829331474280596688096462099842247620743942690027942917237589747 8932

7985624247296590853215947205332369490434027966266307402731316432230471242896578160810904602256804488197247067993494893743915075505173557882736746630113365128062806763873

8944351073404778542844945810324021530268892670928927343216222886653080791725525366482531922486046719040118814979669189723839048992144990637834224725829744875713 87163937

6603835331958221258389950053175670095529364850788840429000362324607985108094470411877669656985270024236542148408230742496591289909650888536308725432732151415989181628756

7811307051625685105581512671359344832178026783508960472580054261710332895188363891032447371674832059178733650962829745596943462409255652816656642813369025930758740440023

4673137376777924867261026258403688081693860941830435421605123289943113775339106511731742579190387744275577466603040662009904063042605149202987043184601327389509099811527

0306433694469041004457120223545117101132875640395937024233171029839349008207273903649597967324607011744165743432549961178069176467596474687979151557278151624730605833452

6364851289816778469808818991132100393955511186968360232676578194608392777588773560940755982917542808611454330139500455246551242910049113728859660686718953557118903733330

0649089756833516500494824375020133685157284996369674642591495360373941154960982344314351093202218097093597803295497595988950811043501360621642003040542535251820091558762

3321754421758808594192994016616000363439101534009403986138161418529659189582746862217600400754022405234914487411541445060350425636232969603659720823649255942147652077137

4574795122002325330757727354406667254606385566002002468570446003727540392329608743253281392448927596263699974608198030761215869443681254346476005823451709865886875789643

4602270548007083790041330514172192659415761568791150191340297485850517148608173156097398981871178896399754385938514812712285659202786935286076096100145004686282143308100

2880034237990803160388504060829762941823082783808603522724981023677059060464634773095240249025118717986424339190253045895732039085850787195225501777037652162664218528198

1740507340026663725152809340520811671011269698677937225985693349519432693201259024230765182777135271884472532778020551144835864447823011547118441835229325114932572698861

7491226032840207277788433002018243512889526264348504018011766921894003013846230392559573128981537243816953077315894785564602548901233598445260305842110783664177049843804

2272775618146361497082205297894046841964210519595297634427944938008762375274587365404368603239256812039681539780620318441175173406354964644946886431290056599239710398026

0552719134441217649315767012502328215868291339917094347218601990614994727041937223244811365777364784334202259996962798552988234835813451982181425675924349886313315764355

4985220161874700944848624572901415455918948887077304374958672079248383857434010982500628961660799710944183699874784439567679292388862416024436902715465276002249393490369

0547167448296577083073924210015283272337960935692399033882446560129800791917643031420219423739939643744425088139872031104733044683994406298819697937197577353254193649997

0332980309505730194490517681341165244535932990515291198614709570353745265578742451856889601351304465467027075880994609033018356953660132791718794449541015603436922866480

2222470447675869609032209684225636134056348368297174343949134503501545627121130706912819682638673322131840444414977037384509444617548305453689936068205803889877247411952

3892924216374678456249854279850314493299533158554300276671540262962651669580914607881017471430699174419986584732904016553566585762630805024149558847753348985236467223893

4163656532479436451005902522586321364641258499846796161843552340352324721110521226636091573602713021329448208976614103780709193655802622181784957122075851190422878000874

5928677362763323009690437803137089525207666717572718299861439365551183716692237254194667980821666681110395660439337503728075545148480681660436746789432640453711566586375

0531512081271327549205306822200052569298501430885879183833858882722616677568345546004203873216650375630854083599997383442031879253515109883833853900329096587405487398852

9729683799722936601292312307160205509733930936050345903955144350530779986167924716144327074762450851301978973869927099332578952464554750676366826464527152552254333880535

4836273916262392529667664587548946734475772733560138382737290053938966565922305985710484827743980497205838211155382009892096613694689317719911147471703733748269810596270

6129131399606088218772148525578898249605715119740995507139928669201545658383431014260308085868849327192298415895092643571831409247104705184512875869988410928735902874312

0393437627985164110324412262926311001109691495544503094533576921409803315676548064212577277675625253662101808506368182957928716083982340214720353625982063645520085231280

5800326716866834481511046373704849973483990721027211903580088432422211643344450800225977952817971722699732374386451794698445764806394894918334385251804287869326327529024

4789047593794042859845274992227797210002389112154893838239138287298993173119476173906115044782792876911023764755025225717321948181473706301308841788981959816299954108339

0244410692706737595956997119535930938496110286574076506367694490893018558649870372897272343345722492789153260922324770228772629642491769808030278236217239379885400503625

7155488753610089011456864982824376781505124828205504920676147252714652189663004968857959976775225939740603051102898580396262188197128217051926322308951746815864772494006

6347625239985417319602616103692419571597760197169490239932872743974658804365659364968801685286397751552247599976494185950268040500640969843511307379711044119791800574646

5493078021521252981008731406046947356590646892418148391263600007362471055648198258938088745764536277429937681358765419179735722961270008929684713696493683678963525182303

8913103992633758596525796164964499089095524355086589025530278599077553259012730600235531124137228833954640486577783316157682986151786509241374742372088701308805439522592

7885302394309216595649098407706095942612962824796778811133533262952874797540987883556678790042919545157674414867840448236392233509566007275479391401697107231858244127989

2338820023779406397575365725162501335163672644359159774750611925713016230090937345100474527618016380709677370094376805966714229413589600824755383245974803932060796044905

0176920705851236726198458956830937968062543402509574621659518879755057796549195504949286712332513375567387160573563800289429902488512188012405686792361892475560482487495

5328263873146464164205988538514774334331725912973171197400042649872224381061421103274992413637133754743240629667251815657913864370256202430396479048900504429852624465756

6236218820854094942368505732727377622836552938642131946178526062604999062547968847458530441305937394727793077535078193557627344106921558940727573628596944966388909215851

3270610171061497976205385708528120957527632949857667771947593521524216768778681734370556742374024365096351799715330205714311146401358640282902451517326107671692022525006

3376243107416178747624311018029013318097223112382400446652702557913433386482338478240836415091426303214665547366175962561696659433120665985127670461450435565056763231927

2380345140253542128096185364065860659586500900542984050046093548530620626770476584564323035557962139704012841450715513295891545516692865827838940339152992238823329025 38

8857260584924330742050480774965818966106091810585454197932480203796556828039992614596920546380587713649031487744048091127428165482419174572111024497423161561692475479084

7307516632660981952379567646387678825343150882208179566747714706801635105964756831889832497120460920855699713375741446504693478302243271003842147687932214248571356564628

3401032424128263276542081608944807016915495419078899085838997387070670154166653384195835071697145193419203744574382051040777729732736083932416374562858922413376538636749

5504954305663770843450836517700464664638153286744482262904960184686050368834407760844823970025677621323572137726913923930959252379422056676983704392607890348267373475452

8332857659917761012569555350192620551799398021571031241431145302306985898701030358942128852315150644414206528495349362202442156032850944545462874140740184508573337343 50

7763059426122501925255325129918634214765821403830797952738737610527302639241822426415421509064600988318441525643072600146861460116194913024036693824750171418942259202080

6707745491575953845423781388608702178664247860286824553825706070078528273322265105633445664908743615822952264506909608316956172605265253491502070413802190340057017883118

31237419981786872388251010597475120234906541684015733501431783733524819386198287179971086117048195607925864281956197702496700421100095380047388039200472454678730906 29279

68600542682022838886684029083133520768865052779186562901289213124031511478404650007571261779711587696003625917889958455320352877641847839786316650707375096690883613 16739

14766831068048300176113605941255839026184975476669621728534035859219034523767151164313372600671055941435933213580593431965154632317833809081857823319571680232256364 54354

657396538915851269617268356652954529933665361650739802987340183886461244163651746666698938924827378264542631427203865011755309707615587334543102676089168151624212648 7058

077506359278820073571778056908881607984334565970610092424036098417826254172021527883071915797667428851450587738133761448400083912643956891713569322761335281604797325 6116

480024364781339419493919981444634503389773048307901722189787611415267584913782767136404814522241700976380245927541667269859014203411158804515183793647076944899216519 5823

326382816833363255130234263516944400844577342648919320741277155095643226103868910385700958521921628484184898827326865547042366752750752984312290873054198395044089420 2141

667808219680982797670774929898497124238809335414495108294256297327826692300410116180647868542163393012745589223242478674916076157695411688302134542171596584609084801 97196

49487228542292491332269577189910652192682352097342852936279886092116791707629485847496149978359834308720070047710218568974412617231091035586226249946839024978024823 10610

77389080430317290598477045224303210033049575696595575909808971877355132748296339886457187784691064035564489612527351448682310530027781884310676814363488368688151979 35919

48058645183785865973102712078058781768283476422045804174854652725579259321275422093550670915217460741863450104795444847280432287590427853279892586453224298523386332 57207

854943441007130491816007509571981783809560002874582755714595912142379824103442011990429800083484667984779173667633916755981233073604499817833000271462079471539626074 240

190517782696828793073342737263554559682051321475779688516552157856382150610037574421068786981759087972310547187859794509341635317309713427557368480465493684608589327 9519

387805483535183845795512788897107538526481259181979522714467314889783066814412948090438764754172032883679315394873192784282061408378211112385518592573720264234464661 6985

206338453406008552668716999882546853183684501164335422424676631974561346008496308560574537359030332058584604742117198315800728930013561157562071742314893304796447468 0649

638129316429235235811390296699494680014506883857295049880031742947556236767437649942436129590188781636342231949340725849731738971847387427493550985064726969684412652 0678

050219420428761073628889385885038732456855643881657884628098866182032035782303338009930591300723334132345096025973746052004357099860029814550957846283200151357359254 6027

351596756441636530112264712786403244824007737996917646650602873396656356793403983565680722196540488511932248820542798091297110077004500227754612066171669155913980976 565

982271696317371323802330818946438128134866452495995445735996027340374953198103413735458599614954983609176126285395307873845707594632930371488225193038171751154383500 2670

899582654526381103725254887743926013540602522145491981699579837164535132550990520879677994407822530807758169956027111277585448684402760529394514288800290953802848541 101

22615778414915814077499984149629224019889130831785966691538822900994694745024784490257136735697263979283040328606345468198590148086774140892108904010576575031104192 21614

941874314587847613671477391830535314383226695453832992239404561336060178214118865092920792949664091216003590511538805649216270544641912365189082065327758919973892229 390

12002682322223697736723300393821723674653052650743914068309474732126032088400989901480267801994826858553514806570539140057693454713673203875772451306975960605679539 00372

6584611384511323064583372505805316793447259943055217500853177863339819472177438498394166462144855051887706616890278874741977750727859461678481964887923839242970123021 95

264384876917116929419136764539897530221318944274689864451195233611358086995256573849951322723448589323113867978311951784387713506482307870482998034471550701418820 5310414

266822948160081609502468235978893323946769501559475750223592602042472263849410031136704409745365861030801205930892752761072852639425752928436218637764253542781899 3064800

665696367275161697181990722601937571168925947974476124876288821798650136747507500638323479883964977400488412375666865716142158311084736091393450002732005130798128 157022

256169065526833030836656381413470070819422166484822104159343491908204056408595224038800378073492616503002317179993148259291180037744465950156593998138623869286906 265238

206123362367459640720983507701108299079028069034109175096357314561823190444770495486618716069228030350137359522412331696418348799080748080408689982217275513161958 7809677

521653989830962034894093683856539421196123081021103471051742416434655171920779277138529506026751864243969265536723344784100681455951149036782838817570535380038946 0027690

705631270232301414130663168017467973350972541462609599578815941072780659665422853016083098094827980877799541513306341851977872303012663922539995594139496211004195 4607825

206744250803288180503393892187565244451699554137647784167163730755847972338659392635224018226080316927670846826907128840619197491176562869996908497073082337564779 7687484

667530526919298507928036681821437679607305087380808301446429759825417007864397304961083418619696619596322018403591635634118435818598205141363191530912517440662404 9390924

513588519076270688936627099055946468937668006920468283630462501640210274379178544802485128618216125121145700035734674069253679036890950252923989154817162254182524 52080600

390600406155405892907732068730958006174997192036471209788419244666709204449749874824088659052669358894877525751640135436742379201453072202353576834544681206865951 3932726

355927699659957377744110379091071568358658465622085862107139549354539122567328062952751900754940904896394388806424557072622115936312439491645972564910184257540822 004722

888846634512803044831901784007401167647739561543647139523558199976935901084177521973362030832625761659684119366145853311420753211952927669717042067051859842499762 8346041

231639081227908900560239147276254723044656137389193482911984549306094462945096159115367552528626105912712122041446841774978186305011129740039411193508189083573332 9055114

407430444467585330390819867745858706466753105873320444866438195473708480984019014571108015111144466295074606523305173459452577257589307863700719576792849542202391 3726568

259931838496357371745540503873578054083223542866825098340742461917212410659284052811166200923282960301721363849285104773585298392087098926316984358857422063744579 9561054

144370524882233580267567460992544022776840093593181777507857673345320731185308379736957382024604745009604524055600641568355404686418106415591598692574489030347146 0863686

842071415295195399688863994416298501262198265478995063129214796056471849993133924449529728833783355225306656081139111557599979071382892418373574090519324118083275 3210575

834434078628766429488113359530078115142595782796409283781276316746885253232980285679247320453209385421015807147401809479461160486277678673437757511437592333049254 9945720

627684233643946932701733610844401875653569316078803127015677432921109546037466986463058964329961957990839163885107358365539735868580394756294040228635209634721703 9450470

385257108531336244754542001052596712178357874633359416592323562570393312801881979934876980850885387379015678885924599933804104507095668197806890979130475312701446 9119908

171380579382353672715797874399564789154906407693819236783667232181905821363990349731439811967421740486606919650658688315134834018768134679042643955738590065483758 0715281

155

8128951074144096045017043965485359053827804348135830772445160037860973737431472179402649530772942955247321642858586419313390462255573142876790225334478786885637977093220

7047543824471370721081728072621619220345167663856985421460029371066131784467434949460342745909707794025711988737531399123813260009563206368236285830789874153227446212759

1793546312149215856310068890958077806059372828374066045173381475694066896879074464372890324045717468931622799152607670087495794636552981080060563205359323461491322115081

8691711550065566655477454978755929060742276192490131286777580134210840162883087292122615776509522101508044663744032978226504795848394929088091378349110527018915866659781

5395224336302038194307797221007492952919014175775251699297714793500137189964488911520647362966761218218393084899260246004189915466997385219675672930964342169889806341929

5331116601520268906755263925108101725929474115970572467020836237914457657307631055046947966134150629498747616641845707823855574370474740572018709533927231232500033657654

4182150162660235171847267215331210750964015851018981377491426545299866692087088903694910230493046303417508983514679902569728765115044510267583560834962773334541439538796

1968611228627183777026264999954378975661863845224244739492492150548510122167075552405102103873300284593613184584427867338231426176973563642708421202831884367381928347131

9508717312219101203167211411109395899922884674801741656760993787819687707634475970187870115363507042680620343222196248189679109056279926872063157344359507897852923069671 9

5110433305566783849538409612527795838910548796848486208679717493008452144359425346200112410842665667586897808277627684013469829419295802033057400474913978971059122642210

4073255789131404774670952163373109546710714788243474697322536208971843414016895152093293728957961799009745313228076318182899433189695689530472370399538905839657483550150

8194701003364946075415680939094827544998118100311514311243716206028508211677160529015030383998177874986196350048908052208969068279491550381572239746651144204071213280056

0653624460196857385732581380945079434740660360543591168103854745541390100521085682696417436459269757301112314276156916406438293041441912580970015014762604508430299473977

7044340602558483155183709862104371824449093244999094123969680727355749909747543929025579847970934821903280805091023318505659585693410369752178796616771042304942352335108

6300728128713214793278040206646142623007856140840325983489255712085118985382238513620972879195187746506418610105011000152392140198811550103331906715391496612736381353490

6201898801186062648881416943529275130201207444850693949715656963700528104436457965400855804416248425718544837208664333866575252285810948289217257839158191476913646032684

4762022558337884307066268201365625670604291660967399373963337255981754023690188353530079901593967249287745723100178133888506294267768452361064262085472070806053673762684

7668768462104365662552545771558209684895512560427094838699004537060236388671367910424911491996301475646726002794069393629208526804159391655694283101370172150024613412555

3880321201748024661994057160259811420538497330990958586477131121900577851682135465676925436958683955395922697911981510567862427873863559696351596525780100887775161394859

4765302893365917624022970657836985360711004953475675722840793387469639698205275488541386380912804655679578673802477962455807493572388749181720103008919889932379533927562

4929514306391754175652362056553753374784035475143499180169624212773057517531727140899284179971054379976646930483998576569703889161802688948286649839647382240305236838578

7917654987361628471601522751105535642270930341290634121405037470653876110440576312776776879558283969360687974929924730557570145071286487760372167136663996479451681218150

8956359322145080853486264524413804231937636533552748353332155834128888664778013962249460243584302230591757415527544778471665151580601596831434699386022411670339610331434

44142152378121270529004152968328358142745720548076341739976854032114278702709946582145669614204935860051783203074959984999453677596390154433298372959877021587984045 30424

17236885395654311324912800166886143213359018145988153451156496930872268799815440163790362584744940276762231405838302463232783555897049122876375516099352286387594 82647092

34548966040439552829693496327329619453926341254044358306491272796994144257715378660212159628384800807764860068442119512842811118606563381627587966850467909393030 24381941

47134504446109962381417080458893859796343824476120094314750139145110290353458464233986653377503403288751178445621701907008268753712348942548452679529059672899114 16216871

72072527895413036625313121616871840029084914010882474192903310039585332809030568981619595841464035008818383544776616176408343356576282916036527855053342920173444 23999912

98215606563923309683123260611349847459047534817572479352289989350094349507539637348289115471101729844079071163848822988417921854283174985756016443562226461225946 40283086

47766387359459884245047099086787716750091393003821175198111842564994499619250193938047253373994593337731252524630640434299251006362772644042522129335363983888712 55865028

21483935195378291219232513295504794270774798175730690981398175836426749156756380340241635030089974588466445595105263773038875334873440217725854816570326003356204 90417735

57909734759843947599584542976546367412107535150701385121261710170943881638681800325344560780113893154572325876316875914141839336568229624660091462055145978337911 5646479

26663543627823302548582197820709974731091603511064700974874000731522287664739629127786218446835550020204307191420072846279018318639787025702772268782391036972445 48641105

88891669111059220294449329436270354133098805268800879346170956304846028827660119070894073002820066435986694309912883866952379298666281788984272697048886044737676 09420261

53771779179096775127871974710440809155990679190772341720808085990428600454545675142277138473782341100531182443063238871528440862687566050697234784773621962023765 84411033

72159043811894698293130692861159856453139893139948999833044009242283791251755521746218912876051394689884726771874665047852706643625748319169084915371258805414540 3632674

78695396491037240057461302202931995031019877506028802379750025521574996446424533498859159093695439580845280450049366398305637825410562624168321730113237466365081 8321551

39049801939199625148203485233603520297989224377311115009168570103260035364444751774246989158935734705658751497625632680396958169649403975994610639763432305422721 3087624

66857346704606223493784199183801309939280236522741919860542642497117922820503705375874271366672714855309460807796290809358385465468349840363555216845703430350063 4102350

28534877663530471250688440872326675905655793347845911332126078901928698099359636775783128957269704288379935513039269512405891998449060463192776299056460394768756 52776188

98780750820211548536425379197075472907112634428136059928119173570992152555519802760560371809051890207185773505552327139139625015942725393023718644501766178359500 53674245

28353346296600400846807272853318083527248634331602064968738739216160955927770747209186383619128575571939484457227933909841306594059996512638479997333289827244713 52363001

17319458797298546955749664148606783193641212157264534076580706689602531854014782487479728063120191667380722377639208723247542101321740219317052468831195661303656 70703052

19123786177939256907627223847705052392706222283749423814306103440377982382108877414396313901510708203127545660795464337135345992806287196946972555924876234056085 99760254

23380535602918699095607613736827707044286670464122474056996749209859838361282936506774498024522167809570099379281101073932308678935464775565148775074794665008756 9269504

91325561264280600598839499515516257640827798160572755744396120181474978004513217821297863743751099747673376313134440669322169897906481415224596057960363749293853 90580458

0982603556819528939522166957415642204303664372299146967604386441942130136759001693242269349130492467027077824818455231134611034534892733150600012302853833423036382471550

2555136874563216693665604441464245556981823192741194508279688704694176702966019507450249851652980618680546381347525143843327899199609271085812200893576622256979935986992

2149925464717682101276519959597824700501422174754586194296039239459092882841814687748419136141898738281264835534324101696493465526295345563461708335109501680694022867505

6776344571474132177751673062077821877064922444475208280009834255704577849191916884176771786886303321268954919764573940753700988371400487025429260329638779287543770695604

3733999010029485263815003262899728551130103698581932674488505222842191805455408227747527607465389890506437479816983471777490761037600628289061945763967812992880200275967

2944986154770652047321954189029670858378060569502688599152022861683177931813811133700846648282316429401940350166748929939240870575647942513199958612783309735265384335938

4167524753096842469778191862434377111559925282827313295136978782140742545163126864523275808217439154214688943590977318156580220579640441942122537014799189278853379033777

4328734095511741378352019197915279650013938688485569374748216129271672957278563847386324693858405292467490402241343895188323801079314601834291621633196573700159794273900

4500063841651314517655908597002647003221302219852249741391577952987929096349728985117601811374469220942531031313834496135599318178835441647145038785547166597698246724797

4403116606061989122504156909044764662457128363820616674275647027759689746278418105147670135804259038575306203657293784016491669482713592856927354306769178867000492202732

3162640407025502795620934962162273386194868110608449358956017870858831338441728876389093153740724000728025325627642840264865650196869797443042592225849580474179227925340 0

5525247449502340839265617239093094230060936630323480202108678868089659181684792736833014327146956844570493654212738523641976274894176037602975201615359389448762202357391

3546834272594628295090576514319420959551607261274135359833191841235784196421342887256687389708438311410465856003768822032463086565154107992964690465770652379503453459602 4

6149402061560544843064378729945258226263609197006342345695812108101880429485136828673985213253451985198680652792016175389656184118252242529689346349832387862657383224882

1467182212392161452163325275670017042893990524654825877851241251856125788689455531665497546430475503190355902321438128581792753398401246082389071705460583596058677 19902

1834652830571868277510762506653709452988830211962730293185889270847570148568998529665057338471038620599638943209413359577964476992214153786551124648537943925407362192752

4684823828499731257186457655515086958241513497981570174378273366379934306509060649809298386303335394250218225663812732097435466224458876434994073553886358770672063368111

1322942298365405268821561270245962885723548264218314546143319125458331181255979147364840129146862219867375818977195188232785208093327828052850343881380195284554650513932

4690269115602676843585443935076285667261266508394535898309320803700107893243658291550801322381298871464809135644029247125244100245252544508023246157822063586871671105569

3285438016246834619674923755227751310510012677605764387991571994859706517602138914640631735022338464345839483543502798190272879730320285084688401987594987037814617966864

6287546670399896304248334225490492446701329392472583236231531199712398944621765884271933825466621610382140069902302774264438571417574587943978979594815804972905977772621

8748279191542139015671040424987960383839873080715504253039320113817262336691434188476626755032586344926729416355440616416058126006897850489024654095367384485080441994812

3152264377783589280107705287235798131917642254407902629775232994316244056822840489795856204209590301675307700984125504143951737057720575555081755126017901812007351334

17723724762208120008604407951239514265989643403764245060829599666156088903857106846402914127173765715134887944642689107694108953101190999299956309309050352222777233262914

70140178644514635311873837849554388250085692730878394745287492017688644731178310411019916006314988189299906101527781687084216213818395570791840511980675977695998755313775

772688789108864591654468983133474235479298051910921514683071632385535103872718754467670829529749053459537652593192516594514793368506381679734786636883270789543959667298

32709668006279053959998294577731682383260738801806541025146172162886788358706619093672979664222559336908245867103212145301576140656384883204620465511573100331062717763

6327253551051140113729479742423417996595373489421421400236584408113388319761752550589009245453137756058842247628652387606272469903021267047078094512414716294955702704018

9986666320179842300550700844075332796256999177187654265257033125495139708649447191452729448830509460184152955625147404095257980099014633837977690212939408531024885615673

5060633863492368448950752823340100752025828306207113719059426781552141092186057054209610307132937255536822579473558746256777651645331092982287602837922593025131851658133

77060520921086575617430123342890846992234973515116314217452542671397892480500251722320908212457411077611635359168604652376411852083104556005139095894987309707087231111425

47023121673320381085480920178739148793448883726854568921487783039001654774176228126058072835541533136900793139630006376970200762535050726123344151011114280709368194022369

89991308247424654012701940113222999932048332874671355383494579635836899288623290439722584493817107725905803949716259506636916042428812825483869715966530554742543545559734

332016501747169426140864138038046659532238806099596893049398139891441778108044017768041263118730703803284078136515237865950551008740358384973781723210016623052721994787 9

90743605742314099283345866153030265910880284894388262719286059268854625261181150655431439186047386383201495201419924016510173976740922604325484294565925858177689977716520

26749864198907493364258824303008229914088423037033492000321094764235749370825153883596128554028571511999684121309513297601060622384467853304303605283324594771517521109 13

21846929689013599203990675174666377175408931626352691592231667585283815133095733518294423401948575999288757158961137352500733529944686451772778107293555066200111662786 40

68458347421220153546184274562778139563100350380090185222039972627590546827269914375360065865512634531653422399403325698761990327001829322904538021646980531553098829533 76

18967309534457130377128599254581802272613746556905822595786920989804611674009391732335754451424181559427904164840501217527511162224841376487939528948768911062083467875 76

32368819950650817234936818500492013953969311504508406318331697956500115163300837827110749772860464151933114977718620058172118357176588916463557018448873306567412167110 45

9918528506122196801107322548295187740766699796023038472007253327600594678695267905143195257354771411115730628379487172387990101107371970337951113879024422857661195134 709

3824055168672986987094588552809896555090500583947976816362135995896454669367741167952365593301962543171459828163763773483041585352887106282009286734513178670579055862 42

28776977038033586718966440076045210507780109026374014363278004628628932431216984895696926812699655700961162978104880833322640115844498657886919891551164987759500820116 54

710794954761627253597443140698950143479155214870180524406880531824450548615105575082458334830601530515271410340134615871762049327376822811793638223772636769508996060057

6457607434908380867249533034011936473642216403187735017426283830918160337130530819470054814566633422929439437912961361179742997959789822201838204339375151390081879567 578

08498819671169957798148004686111102029985597696284193886876123274515246277330804465733695463654938404008197776097066391323765425391868682035668542766193268439028859199 67

88147248350231950588774756415910641899124069125309416312561954109543530881464234340833160970495044930981167353983129373553934118732008867086710676292802662313136660983

64307561568243371003247612866087421391893567521305950626336204982646550082066501877463331840481096537269399354992508460932223638918187900587249238610783215779790260035

2226643917254444606289329459454295831001567300550754372474262118465163712077024599682774758902127077460823281087774656437622050892211762862594923337322306799176150246435

99135638162060740858439742513315938986338310272411438507532080538973380115912508795623407291394530386270706817801468194772402893961722164417584863020451648879583761092

50676053716776401041227878179550018233197260574617611883779468454732039891838117019778662208080181016483471431403292545031424952200821114330744664013624225319259875091

51217391324329653494012095392865347084631588215049551680144287064948483156384372726304816947957920355668445778638297228895353441185206100695450417704454744925970866988

60993447006199388647273449927912722231658528362329253648259342107355249952854844231273220467471078064243669958423852863743227324420182834397340003224185901923803059005

22292896105514993883061413500649369104739021291543977494536051080648720801311904902231107072307706242833919528937220911487783908790449596315222989682708220530489656016

59455860755352222159573834959609286492041366112049876816520941632691258940845282229036070272775509104234760715102608470372049953307356616520160803158835638796224312089

00709419217345047787877409407146870679225942590522751818094928229533182148904042084393337728589025366508426327725814394860195937648754924471152085966166588308595533616

17058520424797750579059521204948991346273733393517973537490955401805020862425294715561008799154147069653545729992240709325803842553897467763514808951876988364635894254

2842207312036451005027160780398336131700022763357322058050472099901287776893533759857416645852007639216878048573675392949503384098229397970658314255553928295919229698068

7772279663972939077790821785173247610873556418967084941823230292691324944913403757697788000085212068994885195011870424308197047767765647705166600738206498848571710483227

2345711925918065271167048969229098580751536275170955052842903922436500482488074481318574364656698445218053666467548379873567916422013296190351708641479733715775410615161

7424044495799580315561115791030873134720990353018949994658219299219647710568822829861401420194639054232858443817124083633226562432512383859477635673012067644108501475

81446342859310494968693638264004641625964695149031961110485447759191706584392706760240031145212717647083332009417568873875947770632409980920684630534774332419452200217

0004662282023808277804197794933893918898522440855068666909872615099934327559421951361489460327548540028247463818557430386722484814571204128940220014152688477096246122211

4999228876439191910899400076450042673636033595644644270818978527417707745133958409576044621143276559892571212464070497606068894752177888846753657731308884831704130847083

0281177925946670120877184128659419901877875096320028110237551436356123048655461532988282999046174517748581477601231334341731387710905577093670657365502031757900430672293

0314450241991977428096762212425199286327402583704007529728174354804106393450637326750684346881838874833235411216634188042412330340349096757794165377908412586879828861032

3527885733215533815198830588045853153469043130898096369406641813704154859314966711598130899440825457153552300652508228496172872396746508251900453245582357486877200667

9497121636028208523543027838653611711245321486487942413213317008523154337277460768066376699618895122880491089111765955157364984738860695247166847523751446415213365489246

72761225853936148416514385818691738416754348278131766314291169378556461817166096634029127205365302544476383065335504511464152247086512121312900999001968151695922152430391

0229496964390635521990651394321630365345397471515735014459156097003147953738250722286432679118022285454450510066868382649729074813258480871020887495051426964293739258136

7718416906545215610876157378020535279580044684913613691746825371728035367843503618901245777758338646770048718755154181150371412945491142726968772088619529031100065214806

0479389430426121125047463622256753934768192222192006351687668258215068279880160735706110805557861686704947486404202700061440097794418761478549764582395624980544495512570

9106402708323908144600925117778765206380393713571176447632922162141265648373947107451322905073720554233262086352301211192300993282164753643692379063250335682531355433034

7896291533044923115380919957555329449870528019033511674075276366559814722061218043857300307297879217356850001256233180674259887240109968969813852397306191955928336948611

9032394925359441583659581613839121854119515199265507043722245110633671266896256727586677388238790791336450938651172201396285944786544296218326678451700202781884192400936

4903662723574437287448563310872878958458482352508201562742207923922033920450828194622615291844607061975822852133877969632367864013313041099556453747740645777576849035137

9167325321826500440150064624163640463117827796605357566760371613744202669161721891429916392304849738254289422199854547868948256705457712083060696640151754701139843828993

1933635228885729811548286913596032428511636219222076679819006388404842715260865907155370497185933522605711810415045793473539632763998872325606368960841031544136421487826

1192854183849957430442768683851459149189187424006619028314479859226445963347995310286302780183115008930007657628088707768885566711361061861689962499638799370291136195006

2155095910026639429805842317789649650662756529783441514973388255236336446520362201321628032135004936195975070269172783237013837157643230288810132963328739382457387462450

9689508223833084417619240847605102724686019144743039151080377481923871052911279590551749882293907551274408036416932829212553780088491928702854675425466697357397053653 62

4540072223989562013067604811339156349727760567144964090640451148094824868511796216404428068971957629753562236181688850027285694336528800131844121214112389838519527851194

8146790166528406883218695868306612959039774599056148703612289809841138200615859142471286229860041718906453010082032794088580385760890512269876008642460648269485048618 62

9651722184875183556528814663127523768706746675269172441672973545695673316677184928439431385995773850485061097313805782920944994446323943060687595811900386026190492109839

8719699336474633114066294511147152055694804027987358243091859738263977134041141601662377526935772236147756347905552752166482414609981486281328766311875241074174329874362

7538504577774254205757662739315698404385677291438378359018235173687708804803437428632366590745228853952283577868134721953766050084686198962633005033093604099782229148151

4752577183795281384891568049192181238360412271358296116497247108026125459205952486511422783391156497546662744867185218756181629471196471380668777153608541786941838614660

7485365395525809016966237780000655836818847197698244458732298445308869902993378877952657197280899159793941934367522718663437829079368244403206246396018669904972319031502

4362504081305535338306531610923789527333914369792593690728697404262340424758703379170022149258343524101571864539834784545175892241236136735291362601712155441084930332164

4230075969711058854695869571172632037985137192940144871195015371587916332125383079389694412746892273986101183720851428693197150286469098732848172073873815201591163794512

3010101966662036445412956291903554810519125343871316001524124785504522454804170858009744164360840375963801883860748953526626950353281648096816794488017615939929356306431

4571114835166747565462775942167253782296133795200482904228816599956705076073487042908590849968490952949104632685176365224631701348799893768877984209294851296278523015988

30153361268342991776614639254947705193020500310556054936776339163283953895578869977697143135446101324961890191705701201820667021165776654146051368534345117330284374 10975

267518355759247185151889091689894865760416453321241702808114846759077301327254794609415509728678967187961801220443350296879621964404832788637996040983882536293823039 5839

698373949610171235821842177439742703646911258107515945266355464675143678868797822902295599477157643067126525971855615110357476610420964178412476830158403397936011211 8781

1200823174503714075700409271083743540108919934594983756706127409717699213954110952101250839813654956204515024366843113398973858736113065124745242321542571509170983114 140

086026489053937071277441244066907683167085424057300361178690524323205442682356865003270303065015077480473870024026729241448002051530677327019111454898739634929248920 6289

712904783044926853800283148753581059970561483806927373009686610988863789029557324737732183929682372659640999994766795576053071821618693945992778164452536969658122450 0935

645899443219172791686763985578925399969660988248722705869024920017849027121486353955328059468288469933578566896852991438034105283369383898081065416312507494946094474 08886

4690836521657106852902937901140010171041875820439261982433726156111356858417307628652063100312749714469978191038819926090166461797540691029725658474690450719252449465 833

527641746399517886169056732659316334124585451782580824490071885017851428871766720831159955592926726211782561190445950693489617572283250711475204412767765750508689367 0980

336668678679866809585451635645045670009882130467124423758025533584949678345502763107561618567611024200622908622178680112434591564775661362731079617325234681409070011 050

0979458634572441902662005773671790461232744208537976097262686877009464227258685007163645957236063816338474943499752065431904882687582535051569107427134586841794569558 271

7709802752816862020313195443597491646865492287630804666214313185341739865035226451902580545172423793197133547989443130184302401189808568284287635611511359254505175900 660

9029317534197237037316626763104552677057284108266195395396768023500467639232238198895390409926916785915668921976462053713869576979999908884131768915113743462702375576 1356

023595321292951339330688041066225809559753015227590711430726120998040611204959544702529022458298103260460366610790784614562477180721220945378224379608920847836988153 3998

578438358476233111145529449931635895654517074349410086456375682365245225528902797171999867299638162137834713253882980766128714625534783529714630139379788432298529584 5439

519676803864770811999621581708094439895803825070711358270798334780985661030080924836334010664378517170500867285605725665824900630381665925027603594014207243253350329 0715

3406437209910495544172192961728518931878766754091358771099753068394228329617084806583434971118140092278253026159348139475517360355840894266447493309584619986206812924 842

9990069049530956019916735927003422770580777725794298919248350750002625357538268748323634227672480871144139303257644516263630141577372991358552647618310615475055054350 039

7887915345327702159604456635703067706610019201793214017147969716738469733339707056058598922830925312952649427953618376079287994017708601760847530934791124788612396945 3

298233627503274176462432178205058631210032808102535309052281121335769067348278937719290836686403502827994706248624768867044020859538534724137046928259372275289596415 5974

991677578726870096143793390912193869911363019731897109456037301611109766624424400181780650555724623398592568655386116826127043340700951800886897139894921319480765456 1609

544651226431496693496974361698396168741124092692508791649510122518675248363561605712348638468927964566676498484647671650465612669908654854037052810528232541582319648 245

82861497900418034575969586571657893599120269240475446862562656707144162771143207057262320457058642548648538642871832592358827210050178192591032186210252542906106419 64932

19738482292467214508802767736310025106146589875281845672592050079006099263317935029302633914975478998055915983874072282012111603184744607313092642236057201406831 87407414

36847356693028138596844963678163554669045752531854826599116179188647506992213268388078880217815079875262709591652828076767314368740762605540277158332846667905622 52441503

16045686489418112599997950353070728399541880040621837690405286046378220683553744365548569477836150626359899363478702790903099746277218424110017648215901270567118 20876687

82275764676994285411305424284697967712936556371908112434752499188941044238998876558149098163133828344398678830561422069948656437056345681695102097143423812653705 29023114

89174162697598468906754938151368823125531785534937461150504580356678194431847685134829172679530465154959807564116899823793686265452254476823193821655598816897565 44989847

22336080402362152126378985700273204797098350733847558088526560046011136366410695357973449086862143285777692977138838668754917368355359148530578245719109983029831 3951375

70525256209580954017897695413946151720261764966079521063305486458118903323277535560804209288079548131044308314254117564693796449370088051788439064650598699529934 56228849

78136790056842469066898234803767228391414146393834419705052555274566152430703116893995864095218468006890113619130090893426827828837570563951953301251801823500492 93106972

72580570319664319775643414186491957095194411520225796015794217433299871249539848164321588401682318015676822058890334434057690623837262060541947083026988680883194 00151779

25067757175174593722384717722050820930704159311736220200038813060788840091117739664188836733320446529646445934419768596428126244512562577886323153831909565428679 23083451

24027616888359652510288292547017458850877854674323354131435813989005234227038800607714317834252668029966525159580526739682566297578541127324599996348271937105702 17727679

08830058490736336401316479936378780942754761608277906353980263547708948992177734451897291008461649056912644584004920708308526064556605541887941017689160277283372 5715292

63391248090560002823379201775264068935118044977919973237802038054845153464214411215740926117175317752534252312656527895654799495249996613418668561137172657535761 61246756

39363634658529021988359358313921924913934186424541359344281660384057943034058583059516125841208664179704045005570901510314279790145799585671974561364535372447573 25971762

41622166560981547765107924332845973035022141801910437784872406174681283719961462839166425348030966724024117848379051186988338391790267930764956491327966578197716 95645747

59333133134262607748971367197005879051641105756086803903926865826348706345405515763139861676381077414451225941285507544942159529485739898630568471553514877119332 27943103

86006628760697072692238839211010422054182314187838700284748888389056675063312220920514807870613610842837440600890446146679737158262720291116842293247482478917896 85877780

59776094181864431634002885026453644551355067121340118869078555749941020501202058436935943833843142118798496695796671231829694191171817815804943525795240601837585 09979343711

30880264021542881644344306720286303061244985371567180909678367412752020111345413499839171117253538517021424306732100031441372887105544078958247023764904752320309 70596062

07612027423317301765690320036774926942273303227577627517007941506913023523382295229380423742991955311001375700873574048960493014910013105148285638699842929417364 75578552

94153337949392024402317194271602902312715943693646130477801570469751026061543356023532272713255237816940552536518894649839034517885744396354358013434976027147384 3855119

84781089286682294577257535978429545434990952690776186980501260973242575667396516641860943350384149618387370350938070380101695366304616092407294362211337355372256 31799245

20868182716706419610045069000178351726815392178658474812406988299439446928475392120476967040089175169800447350134011378005521056304988254349320679964173418381132 08226066

190871983360217148256562393710727708004426030257864375469141406467379988133306274980434048844433937285851420929071406931327851505346968127345232046363566630091702633 5976

32388614244380188240491008510158252293256556727930099661715167571103702279090057724322645193485395815336762018043079066346938595228276348528073739266954153406812 86594346

99569118047243766089383156219243865641031313405891150807219328673916883238149447760719920810354753884234538967326548496844080635106715752371203070532887685369166 12388601

77353540400910880395892751210025497166937079787186642922014014558824562342414467031323035012810324420163216305765377138709905272596849408782981186152588849252721 86032895

22182402628298323270081763455612997145774658378547242967461824664384902978653007763159379191644254783948828268065501763316234010146327094725972018823335388133530 254565390

46348310517300849704261537676406203301379064787387371214645255533658355828253129869085396002636689072550568271410004173522898482171759994026807464914188870306381 4711532

96467895593180865122302663699720471131747863490527766941427342272327958087000259828052580137853878320031180872150984623227074316277615294128040427317387669976955 48191538

08427735709328137376056176693707288061211955806900815993982648764512003321785648698432090637602562999928897307612477412283832696161107516489152282506445462683064 172271

80338343173765819712463951447878320009351333318655222338955660251647081019002244674779398750108774616269889409502813107485697512863709006577191414377029675485623 14383251

59503852517313664426550285981057641837070482406073207871177055295453096981835197292941154182519583530833634749559978987631994251041743808773856423077173325405197 63611239

06352198949174390525500247755923939446107776141318676550924068922302864431715623756205532535947588834188411031832605661608570780124121329727491660000489927474140 2158343

70124815741912988770947116933111041130464645731987871069570113548766016847239555580872908971746912203692511897246805917112574573943911456518061060805637865308957 8477353

91188780952243969780018458236443954482446565562789229023722984608325547054940682218788034731789918341605402685996748721800813207788553382430527889525289709770802 60850788

17526844197474750300894424270027730324729381969579912766676269025359762295246233268788313894840039368170446872185101395713247675408223230437822817504981212184923 8110607

20044101737240292002257194762803149945154783344703033464531637313152749162692772871965807579765624902962412134273947494949960580458288719222182437360162808616446 82942178

44866433081941654909805035061939353734418444988481855950496502632226450862252087098394179516137292615631666411637908625996619584832695382064061026825170404948988 38579192

42168650982917047583952932601194431057299970094882000646282506414298088578461198438509317434993315875405684618443087248169038284969445491491211283899174426980355 44256672

05923759401508528758412056238186705431189016681809717891902212293518874279092195515518088868903134477084577771454928357845296172487465733320431666064130222407809 40992408

61486196616461791573178124113520858151699182455410544125676216433441070173512032368369224702394752231986468094665767008441477627112781820370673947730272527129920 31143845

13520199463470898105814668173287078775610244204807475308420410443016348726332678834504450244842310917551673670760528410517521452293493248028426483884846900209944 5286264

64184203472216570956864347798111036622042605284032034121534186240396136792653976305723917678082394197387937396993118579045446878731500279916476825885174078440005 61270342

80313124159400625600806031689679607752678055851584447737573204098686615089568326179598719347191082425622188268900233410539718917603630564713421885935991661004043 11956899

86685470952963076078686347518088766821399004676927370948474866326257852632299652649029047557265564262817106736122779326818851162138242414638514198006184825602021 62079647

746012249996967228685245060285138006176482634185967083763616017717878758478721135242571274834527649231762646257436709226621130773355520568338605430705415874024492900357

561116555571699857931310006819332853380018287774723250914115819850576954509512870720057652266342717714995753519942375852010832755725591013419830661178092084032701596 30241

973591496606810901929538788649629120915479131840927623462313114441025278015853644351302631195924384614550783343687132109051148727123095875779712220718360713602361416 2793

363027620066151301431755842436447252712844828608334767494112067999001847319319046961301417860432255267100830950296615161233914007232481274069433748481319458570184419 4851

954609071396254069592655653623192382949857221286126450946391949541107269221906175681772129328239509816329423697312472408434620676415165837242952236930174326841387410 2094

132231590431123090085591780898098639811472423431259772730725874965450798846085036494035563606421360246367250297582588214239709069638947515852196010056708757617434222 0068

818401867828964079713421198798942420054262322639109160808328221720623832181566009563766131150703525394313704384764067125707365986370470574729955770563329249287066767 1578

426397484164818987464420627326291809186345695140821112621118307784231880550483902301823855961986897258663753836854851880890027567239148796757475871447704496390596664 7638

840555139832085110518086946733442146964388936519874292945007963579336776306583547913440943749848749178110525952934946088696039617923756363527056828663233569387547828 4960

491495594355811233762962794911089627284563066959031292373894987390646454823452653011245970936916636819842940197570396110501809396370767769574613416736594918684071597 9940

977492129514475364355705104049071822680477534689219219223966559889264013833854256494287750824045655818033979875280750093213251659556264896843264720950845726942676196 2032

465242536113808128554108098613889932317527610005936827481891930572687927052706681650947384411241352252246216405896697737662669472230604792593765890405451089087271966 929

602615646056922347083077659724942225173449004958810325871173498512933407482889628228684514065870595822088546356622237969268457727080062428624708391000127132274669329 1077

595784116055523253942755096060607608370533448080696135747986610034569429050487428865415852057182474934302926645020129015228508576137455210972739110882540490195467902 25373

255943701093302223343533679247915548650890561031509201053297700331149099331914415825403937671038561512589708413151515283217981162509440783844635992698248514779982623 8367

154228185069666716266201761609705670948611250935942092576725027370408358338931609750567830354428400003970020286423333532380030836727576947167063207271563288145135402 6654

528053705616337565756556156045683973582112723349302665713737615780888148416954606951892450897761458277305644711560367234013737512391133422213509952010356177643077090 8044

730482689991779764687600344803644414864134634689995078455510203029888633384832818107269920008898285719368413891539811167635237013599604776794327352194624941833983034 1772

752871973216535239747836661598830001870135480072549967794815641222507319820777374949369340515926151214725240913124328535226609509917886242062147076181444365168205906 693

702697284844303506025219140497755126145044657256971231595797642958217968313207204976676286570471311714659716899414141941558552792132916553410803586453942360439763461 6953

352898463390144370977103171885263352978600986693086698435263934184369703188574390628654710685190800238247922659706695796912922827797760081119896191705187657704715480 4076

234201469014747012300723672006374794146524884880021869725445463705686004723267422019698082214227784721393055099639306665883512057342095362327068335190537432350964241 6928

024349246375291319682400703838920994682079708205085530265060841729064678943292489042666237149391487185204533360509281927220026678354509006721929344835718740048645258 436

1950094552025533853015019360827545508148367762943173466687018398966327368706671738897838170585143556166265753058934928379956883715661905117469340198535875250672746157717

5427929766541124306616885792146630482653762362917863634478732060481168564451331963377597452108914206426210773371687162905791090473783763999340317610132958656601786861508

4132025994391846880266009194120701567367052752104460254246476622255379685622112998222213661949692780512346135893880093787006445888955300930967808022056712229117324496219

4666336085015095949168091194431883188755727288072604920484225726950234777273625668174264307924072732430554923883140548639459295216734612059347331081226707443585443063905

1354861245846923527729555959508163391240344880846155748316511028056596912604738220096242987861895952287301938329285962799349844728509583416182154386610869136544064259089

1158969812989744271470860637344088950811207925632434391216048670980372692982232485582499570894311086376548403737521383697754037253509308684024083186592189475434254656819

9331920286973641687638078055942726490940543186086883587062399303809324893776063958808816927882172328401008007787446855284930095356168133698782188131987960915707800606587

5120413681055150138991407216054540732098424457112407869027529556371764996922732097345339032749384224697177560429121963794004392913939936963834312381057271231601654623818

6080341693779232360806484166851670088458007079905399019138986928678868501254679168252957297950907781400775837291959525923077898529009537496931446488560420183072483668166

4534888900616418487213140176197920673842538852960239842855401308933923814117570120808301554188871910117122035439604125881368880489293940966294766794056226564819392253 6

3716863010600798228698905371847071955248636144245298398605497226673813992712323269791352678194750815424751582595707182151747793330838053854225259358687900112017697150694

8468723239775696021905302772913441798953328457172256959513939845008081855050284617312093463238976743966867841162982536644122223504206323638997861301160460642764252472 1

1995387977652547098991387573337095875444606881810333079159233516940026805099690092016813695028758939377149493311215724159021236225154972698785804236244127487899865593145

9382697569314305549817950653144186683273288195130275199567217538237057152252291788997389839063017399172994785964732882451480573157662397274681272069283546859972049158200

6549321426373785365377765776310542564915637728001898109944412679472769211509560728023628284880601982030177055736035503431347690741276028424070278562450186760493680921796

6663050628579192569032160921550382981573843650495683338819914203756320762894376846024766103943520229448101014041168409635222224465266984042964025300170064070372331719352

2076178313500357425684523102015448618474112946240496328898364055154538023000657624314766841533398052677981987237595528463455943508759753081225407831152856459138266985406

1989075985920268537279230084147530346162184574848815554875180280750087011486020566300511079526262181650999704624609382003056246103553110403422874336687869698965895714605

2636013394740295502774288600482358997616032607498570471221845586671322714100697884695817126271499385898285859214625368691960786535613356845007576646743331086324711580071

0250863570422004275510279692222982886795051603903278422738873811183032977329682416042683236925334343948056151302774756556422435386312834139392655972966202638496945337587

2939469025962873887401488070946338065979969831651119290880251192560285817304991084121641799684325236402041202666033906135644140131389322120287306294453261319831333565412

5205821195292932149405364882300337137813069933752626752573427205471825985193439712284754974232282540407662617308559179774871262988700226674104747061146869802870273514819

8879330690780405185216982872231913175583775553832930641934332767058711857276687145649647500467923706677199070691387000871163039442974822989518150941074191583828498300089

05156337499709234458126841189055720390138093744926726325991473345522166651405838474159317937747013883109292072972574807268927486362545010226180365457994969418363051852 03

3485830613886089153747576814506890483050474323104175672544456786775405653532462442638450112540158811337628792279144576999324549020710057193890243323158180677444472667231

7664560768234838627921390923872430415110888337404826020756447675776852313457857843464984582933624074648014159794645214350928554441465662367260070571074429471480899305462

5246150539546672945745477522309885938349227321181401218199372897274783639130242229542283226581272993978869758174293364400546233397984792366191852240226254562010126296227

4256238175300517763286375539542768604195767584878682935658016484751646810850672530738693501865443186067821174807067102386061328973257124478239794432392685146855870717593

2678693358768547160344896167761171634129966733645058979211075460183012363336069114496418838793412184219493537874996652379751539176305915964610347152121465642747239185396

5021316325911230816528350294868195657135887476772396290191693199167786458730587552887475970901591888584134023149445351637158204611939295386500630901968781148725200 24632

5005859550973265669759408809248852766267444246672639849454940963335885449443855070827786604326376695588990942339213345324374586923695535840844157854104867882308876591499

2585877613493789393381723956612673264586058860998331302388177066929334834048048944814411828434656375280816656029472988769848951911289727995059763032773051776937678299759

9259573447528148628394441893629920990024135806002423326855203253434649358977527069034359880903887764835281141728985220615766577189745996931269049942745958577489007150 17

7102800505102185087463337480281826723866376110593672121511789100664460382809245724443141738580831441736587686372497575017241805055648156458882694424136256973792922559459

0205061321039231722794686286895619967419234007390131687938180418212831728509935919804705998908531525506061112902960745527654160555432346261016708812851520318516352330546

8450163097876868862516095539218201938430432208578808474078482365115168524579205376362858300472488949906232202771501038935424792067272180390323180854549352300215703224048

3035751683461419504518377046131294502206404923254337208233095668957898208001861335005068753785517389822960864579261346012293364278941293472373772411834710567199474988132

5566302417485511254461353073138250801775959002057552568829353541033294058247633459548911914800263059411228562902099198297089226752024511822200112557157924073365057461275

8838238123948024110404490718897939017921349177985750287517112124497071036628890526745050625093926019965399546707668561456593004672173104967569904556248663389377247350889

6872455308678381939797414983908087007124844406148488248961310269733073859124987903741870626001051421583904088761237036537135202929487876505488851828534282484100383985536

6231337645767092370447705443448028249088623220965866449859194336311920583162054533101445784822337399509556817273414508503565028059306482927185297472128532388175324000776 2

5654889901114846834623218046070717582378163621157379393336563514361265578824300183870597868274534862241033127098248694442254632051824297733063193782880524736277279010236

5751583877704457363802324310105913302522397460325442284253047639249936874122594125253442745719143579042619855980323554107555171796022257310079403792963224177855361778050

8804949631587383212867555429706886595155552168285084918184671511976087907235697860364620410919720649300412315018883777045831139762918700923805268182097875166508952119937

8270494551399204447989433788484231222487627259899299789548257466434857557926460375862242308540160622023123958379396790618608202073848623338500638603716795394641282798171 9

1131196147821670008308089507854458132669548181663862569509252316482191782213462414175913354379790283414237783538889205956684619798105231928257665293328280911966876104818

5889645442440220542748969112069333534552351730720139527954521612369056345572702560858266064853839180881791607076066138641953390753463104231631434614551495468392152545717

6523017627907134670298964119348104686248731091290588286930561548550109865716994317948320400204615028922428325398720403509193836454205680657981934923777973992361643392328

4412229124161310236812384123094008595559439212389026101480024118436732572185669286852117171438957735567136944036553418826630321967371220477846139672424239343820655757101

4652108390411058025499057430927093659718121325372945327612925571603381159932729512007956007811205331463611272220893115014719251795352428209704638353628132373305039579921

42571430996020339068800578351873981889314118130306073718915536987311697604985511407637751095130499113983932418267750678612930933343419446004710929787141898483801103110274

1884378329204828390622011459245419765221268914531787212145133126779878455604456159595249754529857535222154681714898414511138425081410550130854896234243628497910419440009

5354198478294098804365041833757195053081693362546548668506765948602108440320888369791785745623588814599483392712832581756647814893076301831884503175187702543381512172174

7380864741836917805118542713928305925515203696154426557927433409517408790292846894847675128182899369369638005209662580963095224478411416233470638867515420150413175353850

0081328771628081047001883468965555544325765680376567350202672652996826688446781066574956350999859617889432443075158310549771763453238063905820477318116208907400124014920

0801923869544920213679302425802464651769831056714850197856958739975556787502966173286707949082756613216383025793652727117235764937197729881838765016004696019807626071729

7338719586498709909968724717494761114101478259018801182453610215220422049819761816353003398859997872605603757975986803073314752247564007571518792201644858905540902200692

157100677635924985723995527632014441216849154357780055265954899985124797809019990908814255453142011174116939154833148205738280414543627180464686626291373582349664828752

2089699583629003659621294068548508879007970609715361504930307097144793566401490311205530808240950244384198758241647086058648990739406875441311017445035194648723317674626

0249006611578886109716268850323904256421665793019714233077714453553878628432514336021250189909452614454005268521377177876674943223919259355906514401227995323657387859633

6671886798210051122484817540292475013669500504759670844180126373458801066225304001850669818109719049088130229187287636601125981634073025813256626207879429393773088340581

6982294048034393246806755200085304321405383703681196744973642664329903378153525502252468254277644827260490649082697721388698375592442072377285780670194840754825024590313

6671777876794582809712007470914103345397923040702124704789597241630631679390426246365331538246802784629998492955829706777036736178091274486510961319478375542147347260977

852214220697744173339302375889154517446218414658881740633219251066399862636936880015612905397489178692275227186600000648688606943843341695689761910479315504073790699480

5414239314034277212056417610191846519831622755248470937887500888830317863573072829187673035632713538749271862784783688027045784857087073472818436550765235117062567559404

561475108174628886513889489019616918735866440178911396503312482862812773234149566257038143667039496496422017426965254125031386671549257792425824718393040622119315529395

2472956596108834910773118465066108537822372276507233007298643368167640663482712659661133551940546215024279598826033934377785136896658920777477038091230619879767544853888

4934886151790229895133551513178169447938984480551016044753872934602081056423999956531329438181181794916820664342298614174888862468891091040157838357890464703385062422545

0915261785817209601549590419019808172498633323653239962035299833581288184440031231668441282092157103650031779327671655485926788764291194438852182338965094495108299292607

756543518139988333500316594418749015738251515349117302529210911907594922425748307159707103394639477290177111179548387532454308219195710971163136558511133616779926745 0022

545674617270261926582280969301676526353679785621815044068552425020632774312078924592807308316095449322819905787604227141254698992203722558094193132907397952984690749 6688

497302828612768342445094025419733427838637133216298271886802877048859039816542665131221746650688390343880540662876132415147364980113225683524539644591778808774654570 2095

388905755986962043372578858288766430401486028813478566777522316770085281567031351708699368722839597977614270430443387536740461034096265981400795953733803766082066807 1019

298809834646112535217262647461480906245084356091832662332861476571788736707043264496423755831118108613532063045273135501338932584698039650758945995278709868677366166 5652

729409662709536189923808343523250645354065363667008123808431421291608489392606475750424360773370076978010486226866973272061313087712428838079513682325960493546269875 4785

069760870906213944467098705043539971706828557773245561833859672584152958438809541003637976907483263922322696763108630399810679689234160825946605411965782508337910051 7643

071960066914226778080310863589495491266601968216898487279809973651882678777310532850591912303608390383406391113657529384738732739647586951190390961292144657800965627 8230

854852576498319237213841985818804914914081220298366647625638953591530954244210238114009422010543587336021765607515378419898361830584080950957923288723131448256881159 0029

308704194884195222515577142536019696000711634744743935005684563443769003357417602825549108885215323925060878448597881672789669026317200106720647207404594248900508076 2823

397203513834604101168558951501894545922832586999099358318483888333349590571254321971399928556698368262705426927598788149893889555327752195716198846758409266923712272 4294

525243982523836417638668895023936487670293181515076725859767848492337458449430892319353106512504708445945982293233904494552069195194753050359746278833569846132022212 9938

974906934284023355296263560276993322927250894302020639727851055890584430078972050488044129234894746728867838915026228885291287429527504355650209629422473679346403525 5126

954479331493143201237484386524665786633810460290770251674067522547035447178168614852922458798618961723244658952819715140614101219273272569484710478372857694609245181 69168

799535580404984124110491975743929251100959314280976592191588701463865514029775960129535847007686116829226443748883090358961861126606849828180729560094573218650739506 3439

570125354825915035368957714669165095764450433626512511991682040885348821839921490374688320638921958396771807212607327885619513135576575243413453665708486593415984435 4645

376945359090314079655081066049468224721451838346650131789067411039280361890014419260045177269902946519122625302754503806032488518836047936463224990548258049577601974 0854

482383090288704159731204703163934836548926547230251529080407727077303618518251961201662424935513390875866910423295025219622742624256671592982893040431190067957041201 008

409478265978836503276031030284843132564211053477207583503146038617532938273761710329521381275669939737444185337163250635583905004764405043184655505073040899038926993 6288

620469402695844315063108897864338252987917006595528198783375541777917628876197775025249067296380621847945014014371102715759436540408833825817964419430833330859984775 98891

768522529027457842168969003684371389240369032563217458039652881882758186777509035040778567728215280374417828553924639303017935575994696195281418099816035856227245541 2759

745369637536195570980535234513266574644682623554839759342756845834543058091598382508947389287692102976887429987118954151732043580651915856132717351681114909835419221 1828

347036572886744881559980262163347287343430001461760387865900136880255812776602634600144920817095907337266610360736242308598843852524693473596609937969756917834925603 6933

8961564602465312090800867027278477415610360008789753367663944924562295092340663833583613251463923912202114104758373638153882739120159065947488481804948983636563622348 54

21468944007752508002307778637642480862819102818203471183174373034933442617797543695377194812821564298546505610964355938031147766430355133888829485169813786387513074991 07

404631767287595323655332912659476404873562844117507971390474005513323938961955512856978598020477697600525206068054835205471581242947362801219088311512081693681709227961 7

8050408945578359913093476322808015149969889135696881361110514023419878417059507765640482463118885901480833776111552376819383063571431940964257422671435498184739540779 481

350877889548854413001331746125100043073325401075314242536243267779801060578340073046225673594886885619126692356012329843557726500006474319751470222670528280388950057002 1

52053908057026851688179096653597758126368885914769419794829700113258576587857266967996800320896771899522050292034907180214016910200562865477296008692638027291114694014 7

1195506772843968772487327158127480428597810302450513289974410757541535707520610369102909431370692300027584895321251197846884642301068044903732891922605837738861309121071

1770587752870318746530455255581540388785652417324957369774887607670119504928941832459182180707700622495028789984059966097284353140332007498193707480320834245563618631659

974150015818925437235841475658435711934943348597563546491917525117456083081918043863357646830719241410750040948868501060599618501420586165368969551147587396066060761951

7162625250236781435059299927432365421787159533803175923627889878294913269018960179615887369958353631530305030483258619209352221459420820741303977987383790906406212477970

34395114163048800821786819413639283924963789204424567103999952897567048617902880627342384547397844728193551281160733812632791742199283209721893969611281497697228426437 0

275407503476407901345333133132456496302881157135538112815499523882630906800982570403252224821418954573119471050493392745559351320420656215368929493664686409056310958536 5

9861276550296528504908502547745982156692951273711332174559075315718029099432588340538699481640996723253122460832108993175248992344381194820664316769432057280142533520526

2187828721489083506561087419272164437457408832701675994309456501459953883255648240406611283333229346944738229814758202131797535505554577624296325512225636139644685499789

40344397493681610165919412356501732770121225749665266463117307115734836669975131418259115846143281085787581342173761329838207675195559755419571723966422267645474652002 11

98627822878202403645259902635461770596464704118323000965795490248713711791563189956210816754283526888091179891832700151008943093350226550988041726814890881229128708295 21

09766415444761833098113691278977477122980202133212746589852634671926624555388319277144999140834101059524872569030654767321855578141002838442058890414682109669054337555 20

6532913423204111494441007724598556415289012162258426352551147078595553213609880932895069644837618776325259427108701314679076750276325987325766791190434282627051475245364

9423116904215703592269758532102311719445897478681723215611444092844987448122559723161679444150344586844820617812142689172348256874778766278361683677432494087732449489886

910631126003804788658181342462107208455447364215184728768745937138298428035092082752117689938459491096246243754536774918274654184673474592704974484254733962575419899408 0

8113988809617887011452014324499095869529262140219339730031606517359189393323587541379546540821388091274730654354517356644909327584900617263889764584218459134590451977008

403452259990961887919339840288954323562790608249368941546323611677529403826834623555194286330995651263696800115648880489294626619364652318434098834586027305854133387446 2

705453538350203975715425221285252128218301302993326617904122692387061468284264574806844585557582486688721350254173093712604179691242764813199746114026534138005020247181

3955693977026651364840094792620145897591381816690481772316593805505270690768397746184553099126023123008399978410760478293851415988787247984223909143764543073187654518 90

953265023109477426368636396795651027995433045501962995102898414290991155800188855761777033734674454678487614755250966994450659033739301500313160838320599729844570358 0164

9539935756873406560221992121815670455920865803054460810365368880519721305776033697791282130976053685806300795151760564762996410624739834969589994916851870974989444093 459

835935232634681288025988542531965131635958944314542006815724356053980185968094371494286556694414900349111556979734615304236866068945047000453891921643594562912220420 4541

55483397818361155128983312621025076967436929855168320378541867742237633716951455530847260914293919256400680948773987875363904313284521095778702457216803676542536975566 93

3769168937215968264275753744949741800541699442708567746862794334574127819345975089292370144992534455143533966206168461628694631883411638168732635660966844266181685087 835

830220172682727951594919629889465471415030066994178790856944502732408037778427150341385373556050507742961538376582005594222095940727751019318804644245589129485936453 8378

4405015959209169180992093220851226729247492051440151722161097921221026915723023656934855124309230660305936577985029687173529133449972416296473482567376407878330294277 50

7365219596517539511732842538480999346666970118541311640518949526554950081439314826488144464640209407832353934236123495389862481501646262872514068441540323241030381292 12

02843427583407715871458381042016886254561952959225028937503794382926040209740029765913066860659527553275106987944541433996555161949869209406591523820117276786661628344 38

3124816662520388365762541266998454645663056269563005241453508164490795914799096477820024481302876707249446871007039715991137593470149710862440766515214719893306301860 213

0504808607034518380062795641871228536242328066541232425929709109770029297212199474442622854330868412733791162698475208548230041569057210122026554698035671548157735251 904

6914043093788933407978357663892407945831053809339861714613321315002679579725818962290576615119916240999193263215059982468159612035086196162185140955383507737097381655 129

65135202631895293924356239514539031754422363313065739019184123145894299056068429253759897791824573698694733042393692337352111541335847620381814269245862206816051705773 72

233329961192988887052130039696680004463367180969237088530503608758432040275576772967005007093445629219875590098517416117101049895879708446952195784811694719989150710 844

962635254670316057917742116938220772209323526005718234236239343941116792458409152606168641300255914305681111177970827773615183230788938566750909280970836933132742700 4104

0190294614437259108134835491707410473230297562216932801692770491932893191411149088697243890420492218758259799706756439265596940213811426712898167295742804076468583369 870

3542339906412061908925873490116640570308240076491226128501018729351002460461822256831532382619806998211838401071948228998960737201090757034014732667850030562583899804 980

71984977342319019174724659837227207421085576954325059034070334050482242261430876280888342016729941844951628869589213472341857679721081980546181583973331783147978139 2885

06482582508818965761697179187141501706091545003996639897044841309212546912320047584424537510041844647710493477384411257984961966549943416890264371295126805623028208527 08

7804399988117787752643841823140576165246619311886069040513242323953173313951449370956869460930746931416427616166093185970588956635259602377530748304313165254572812065 791

236728498218575347455538754768260270022662474815467105317825391142737172074582716509627640099158475043363431826546354402640405973284371165072254032651366553949818905 0993

4793798198580697821865502181327039315360049815369763298650553658774363415217958775612451858034830994833708860254890915258282377983668348739557098706627632980572519218 791

45360285571773767137829961577960518685217746247279774458945612549743742831904815350809536033451762018622460126046248802268144890302086995405340681415438518493965990384
5

9111745869040957358146683360303234564462806343052444122209949711134702414560239693118152077552892874878936396189283729487007933111713325796976022839831774557754586129931

10518107239694876579428203755884462077707173234133890415195554046475576327027653380107926590450199139991114863195571322212459932729859118497350132241737752767249833216 15

73711800569092997008018124471868966578772680054555115368197971988268480789452275424334870367231884210352035048530872411544593847549658620092222075697993203380369371970 11

862068678541949525556332466091158655042170617676506664877202306733521287122957286462859329268677061307692555141372493400192765224853943270059561964109070391620844032832 7

989866193987592464676893911495031828083479399860466583725069502132569217564902388765615240185217012211621274685898417898502864744778728805223523568228520359372244127993 6

63796742824483964378068084949230172094688777742518660795932453717290802613994109697676546634094312714343688065892069923889663399237781474789350280943736602759003646241 61

3009187349859596342804784736956735049651242784358854222294503380011326200397158461145519604154209120235447053382014780387410007225248623183888193780192165821037441670975

85735374113408393955228578747068218429437450448918247514388781233455586345077626270217716576308624209174705004017086640127559230703925990257096154954742574332757651801 89

688338285495571322884859164029072281974743908432916254992336031309377259116244716487414226815761994432461605729560600777557271214775550282195708859876103086165269974345 6

77368493453307531078550952036834215382026867636799048087551299768255483326900052195031300284825693688810115009007083340100905416054397050104488012412314251727926697809 42

466246160895311361541680530976880079062255312749586495569879991888536255300611324583354828575063694882557373040371925279780093240196575453942601949749977794696391673674

18736987987825059437117506136845125583580071465579918322786715428353719341954906222489359562248723500159655159582036027828741745205135745484097432753825755521956807347 7

78912724295252854753774163105437122739220305406631653946079295428011949722859868862620959705771365767457694642298585674040854993161468923354856786331441819221221368862 99

70654031711152797921389343628299637884132778293950989205495568478674730118373845505613147815705416301181460518125831815266099326685656447494864272361110063990201531934 12

882598321073694949529160862702977939042236362515914082470343247036819246398275189526910227950314933730899837799279245439411604715911973315412320997178133722888601536303 2

80053212834939228219427985955455416679258510853206403209848506278156616214881284667672704162016898436880694859910074525753019823764538420560472267979845632601505549341 6

18925533263566771752943034111901878705682235547120159829502893609563683611856083769270231506933402659423041956204596724407701136473169496910613049728321764084695335392 0

648150058350182558510308549034880383374818319442188130958477422037695226471444172459939593411907054416363079721417685593828641120947165620194101307656939546975599640082 9

76005354188661788231110201727155730225505952903350137402586928542771974669844636030298141183451832193293466211910497206119736836295670334194277916762383415463921665589 72

06710021103908596866839052817371965278139213450154756876629131833284334547580722012475467487682898189234680324971215786221858465238181791603387622054450261053527730169 1

7736647808824173908844525891385240418917053930136056205025322890909131465997522304654649626487270505042798563552499956894354846562905335283389052407476828234694425124220 5

2432146049691151599502745119211568291361987775744299750819945714978774047543664667121836718639985938647758480497379574310313556106922648893323794188621034303009967129575

6250603303359032217667494155356337239116889032366927392379605559478222318350257576956904849957835534199270800453710290967308048719319218734905095783279092933409790112305095355177878779742573025912661514733097195021283664439457963594272143484791826563859625320195043897241592694683925242421860380728712661664237383521768393446568942250005575813151840308505972561946962356965072584850934813782928372217068228979426416542578445420092847486454285138283727783813351358194082795357922839889739911377359314873156434126855773893828279744037394038580890547585376469202656818166100037622459230782536902831976059742661345582628952000747175198661393484522858606709892291398600737672164274502184189906875128210516407142066063895878028324319775810884437261383554629124606128103381311012814748306492500156551239064926376056315572415059444644757897193493099102668095708148423810488418579255005136768429135114112517579390222625874008845364163387127575753025819623722715907425554757340119363083529254662769457148113386654846390212303691690580495406784173105594659185983035030658313990683169764838761011995193662536138889762266165084667337594784663042039384465890384648834728228452379531706994116136410819446969967244407469283923641305679339230152919108360753578089544072265784231508405883419547823568799736621842186804968764307531395163670366263754439135185429397386393035047322416695286543843954922710470012092173801038357609531156944974648759568577239395946125495083017942186451249760690958553284814130398283321957441569941230839326637951758977091254917797311845939926153452213996141098349106184057736741482444413357246529150310050804462575025374527545985515392948604233021458280713117375319139408935171215518043017046220554479737669667478931730962714817804320524153739850169540905116883068125314868090439523232437602316225250438836689868570661653066181904568527474665587161772658516507341550659451756136678425907159518526492326611989419927697833028837779411117501447410422908089023660833919406521113932372676135978853892342828890089773054733631268033251710912983926148497000542686160299429609638488644406004910245503966529170234236346606504222296021098785880069442156985770536698448410868673105069630296090619342316066387489900188289495125615591102404462820531593770968426564803460337817314814053845044990759203550906445909909091445259725671495302827264521527396612218260413461475010286492244572657657545290324054714370601234431217752065777645527190011539533425054180753912191505983798526157337051622954411970307888743173998914579482534998715896943787734373767516156639546476243520818653159751036668970777583283865057523580501714023232620418208538777467983029508442338331103745806536419079447497062847706547679482962188669259068217600673139160066581607858328425138624683306260336679546791249223032388929568772894761215153639603094062765311976690109113804874054937787515860827573236354566858067490627165972159902992186708613896895137370683173156800919554827792046505577775066888521186692976521103907786318443369590672352028399996432396387686732036137916466867950311217912679120494226863196952264941526031838460165370450925925788004244059443494676228555085452556602069422687732189463339027671535820222547033122347985668270282524598801234683953776355174689122536788123818919954637844127033241873737633256130422332711844450749742422378266197046104011122653275889557238498505763648049409248040111162770871555279884079036125727883591241446081033368567697834958256973338399561692398900085029353534299965740696377950214085190300644954777473548411684055102877611086419856726622678722878593933583970287883595451263588076265414634051480829780026991158114186054762951288199430838665320484267405555510253322746134221993103619477720495243388211785032250462292214245695555003313775842857102650520168844871279318563741260727741141786989397154226727632286584861696775818739529467092943452537335065005696013460731802577670

90991551345034841583984675265057374239747147481254009357819535459734584707650074470973254939310359524408780242766698463084264446386736380162865119392598513341956303392655

95711671123044979923211736068489759175582631622390226224898462286579357279624910376403823104080481974942859270246321273900690507498877317321885468936424389420964928566444

24617526423270727279330815571558866484163195316141173690908132019027389528425154996722430348902376328054372916139822725299083690889249738852299592377545012678088732611 95

46163620900490104565727238121860734029185447799535289959211945518805213829562617740804127330379420903678075311256760080426049507406190564878080502343735589161553899 27465

63307620080099691290761006320537323126937279620073905433437462210258995721397889190970317550197355937869240463611160272063383384090322398941891672249406584603868835 99127

252142008927927430986616350795728379051438551387295271129836678961619181394864073685794226166711785990514038234729140090097922491302499787040038054145640642906637903 9225

23044921716095890520281781615990117043582891340913588943930495096599605113324294482509836510749851489015128230584721525802518501199967290923019534159104663497514809 76575

85419544419386603010239328940819094792784807830452404529657217508603472258594175772471071244270779869072099558068222519789986479329847328307793432374088524818024125 99360

76280749955223104632199106829980811820415770202403786723523630415869096428133036466253394949402426868625765918335222413541020048878505699747168524067286575194342570 62573

47366754736503125572083747217077731732426112803077590654692395022906758282061276734984710635197644664427167209384615607255682037913752094903953881686272030842192749 33011

96890650218554242991742579222620450885193664289287722877551300947736463000642410992990004593246677353137444054866848758777157886475428825132571024995835523050990406 81646

34131457314484939386983313477742735215011666976795032349208702667063334267712152648808795365929937351899242067225119480731326592334357720796452977775486743842476892 61606

81733916928124635913549460964501115658857292538244691171975572851078655298454278716583023412111155300074663525550482452345410485731877910034762330588407257833498442 54988

05845876271348453269204022568383164232223709764547140509123641344063619592205000038714900195783359807810085988860559728484791165320708745934356308021871621114895896 26800

89733082142060670637691930990581223716105016338135993918359318713030449261801897108928204261003161499659858596679590094337472123334438951788213921316896137345600884 72118

34055370417390886550221553897742480597895417234353658614901584283694749006043088952560611138755438584962726915091086239755480590136238040358262600530735745124871077 41426

34775125097739773093907252823363210026480288229727064109520105288380786769462037156395760476611450058080477176427287800131312774132443027234088512989686560383152465 747436

79823704601279075780608924227128322264403079900615323141020797123800533820530512191173870930198385367908173470015597347513283321602546122502724867125834953157390056 20693

53295402359717153946924644274388027596148649412924752773374134950714878033537786636311856230615348675776706325490186829580038258788485176455630870877219208318388986 2253

67791551404036491281933272346464104104208811093126098396184018558463702354359467248574915309080485941831260967299406674844194400842598321754803345986174002004313785 08286

18329975681776571031119631581865995049067237979178736721907515640653437764476306217057669856708749227075028087672462042253538944741424134363110166436639699880878573 350453

46724540669218104268811566954384513391960569769130745213394121929508769111217525555025022707891471738006854236503032373748677870255838905989363050210087166469785711 80516

61731567029479585884426271984103550946215100054672653266686174429351162104756739024197730966147242208704524111707433388517866626228389759505203128872018952354482173 94782

19424439975899983500607804123879219306567932572487287019768589246523250669611473973008800477104433865437226884395682500885716481309468446127095654313955754758050793514 26

75631381936217024229731878481227768318957407048462335383621659586843308579628537616031367647478472565866092667662925760874527312462223550441586597063699892768170434994 11

68127979282129692269276650708667244800880233097892820355930828885978535341197850211821400468425054674025497714173333668952304916831444376297734827205059961842796013824 38

26409005405005054389956220325769403114107296693855412386184534165135716337686204154882263119603753953515796511561579168752008167453622412708438608227154950444441606876 735

07152786736164990684079102050438446012308262194961869244003083142930447840925631186320777512643059419959491771706413089068677746385439177938197576164468111066389323583 61

84091595321163597880906332951732610281710148666948643961131017175536540332278904447010019763121234107646363624583097732689325932774191957149924498559646976173046729152 69

92474055576981593496630773754704071040347974975517710849312135392429329973675722029345975758802576403917756905121552802314521029313049965333937302320092050540968566150 40

58680341637934252197386129365897185693312537414578482782700860598921654039560817318742975871919106932757121799520873208269818635984600672764338764820681561136791228409 73

79440355613871405973903686799304742369235031065342146664191636140197023934232575269429902746933964408159301579187951544290533462614104915531568845012558555000913322061 11

27010184435006952174617306595986416881093816304249760877061545783954493791871917977821405022579049672212207781602380149238129400854632371404052972776933521421226846185 97

82117443585242059109702894198462468999523242239816555904667713228763619601838487431946537298122963352740628643961041759162471820635419743568571300716754368106285383585 87

36114148832792331046006134451413086470002471956472167985528236769001963214043564985333197688806488862002046177902558022709496834234610834925724753544521339268873693931 2

89850485958822328147510381188809460451439780369235292336510900341393458467850412291741350494370619605866619482665643182166566296456063733475093289553432517497273791362 24

27082393564558244327306600773155471301780840169319607522344183992270011897688984844635055266072691642378379008047941617684334383228458537684138118251167756383889258394 02

01206135674469196707575278047069344581827691577956860232495884116028959837840511525597978388184102901478476234508967028966728888964765797782306997988192708835134933044 58

98170864681797111283818200204419749911772154103420516542702703199984663491469520761598650684885220316328732259038150648355010008451778437545074361450592219424024187262 82

49035634126526431192655030406320257536605315091862942304009333102482986091924700441157769350573870042859048669736423070558612735640205404826776438809032580278850876909 9

48092573290173311716530546783923878398599208449494428235455111107556714179415670619711986419956408981974696503835985600015377631978659426847065320729130183465878244925 07

81987830112350426129913080412937786711714417436494170413095998427749615238523119521635415716204185542617257732897542748964252455836900805206971989783405327952496283264 15

69341561399857286730987422621402885181286001908460323175013330780748312361641782613805117793371447041481211899591909802237467877728128904530357995934005672043056440449 35

47802696346463971915660009263686421734741176193906414540231793906317382851890365239862900709983098625224892343712260609858103040957638851092438950976336494774858416747 66

44141210442699677685198498165067537953954532870805499052982680916273410272402610841956946273759163057030693604930495517602916771452567323503367130957078176961089025569 01

31338374511581734260872080919768962376158247232799696762659491769689035000181087114738574349836409909279336945471815629034528700893317786462600184845835027509394656574 35

1979662469396278967133275300103362167058028060203271745531773549327571281289722301694317565952958213137736407650325824120958394227661748987572818484763028751860746198159

530914517767537614228781538324691582039685564713187068248641469435543418642408986534022652979350425310530745564744585987765188270858920942199651867033066055402432358305

7412498897188396465826949814140327296804009856740944722029444144102681929134110942187415132828733107982329251450661078279210121801031119704277690769430354497528305709508

9336353333318510969023890698062935716343038815074851498982790171065936441270747611298518681777313704723445403745600286219167913221365782707024079509967914180334821962558

6266951974361124884659465827287656449897715944442118234927551749896827448384642155445225433577710157111531264690883868010392137927188857961098424544017603589273600925798

6197434102460936289896955972601919021738070097558023753808355062111992331474090884681435199033503297762255410262247665024469857171353572652882185686112393022122353021905

16277356974128634983540839032735481050807116389655729447668457921741120944840199655614355306898308612084948698645171066737076918157254460466820282052648523358699249 99 08

05955192782881458694136822989769294696665752303826034310488558076055117083009197101198492536480728361569767818774221507110373995369276615245244965350910034528895977702974

2778854153397858810660715689492621308277949265494362107345696787855068738556113335206511000951672238444435331586671798353595357645119668095745274000424340154400490626661

3796295747065274075783714609400233265567720783707945010252698047934975995363121472488812065999504473245405295694916113543765127592357126014280388914399713206284195468196

8858841752929226688644262143406338432354490358718369905053470947584495815394054529883914651863181509397362332140545096908945353116235023162645424287888032188619172536257

9809217969251069766315534866721289025535342766321179954188944080231117124107891063911791380111690841561471043264120324352019466878197773271630704885109519004363450847138

6726771537892816556212509319933084617355227009185624196261048152846404552681817698323603329448717990508744856244052584667057078202967384386993693546543645644472300445362

2737526653145299219622309881176456806767139785437916584732095046944304384883707718650184325692701151713494961736980782426940804064500922506823933795683717612153905708238

5854829100457156285405230944770923367992115848871886152442294590478039610523547752945184329182177647040126741902556257847710396875503365299622144881793526405867813332621

5920913490441894123125521539959779526570682574624076495486611392481467006572277196024957345080584087022130787994543507255603602188951099248370331300766185308525391274076

9203896980996769175607014847575779142527138022094159387748849404964683120257215335558064888199949024491248744387026964819045617856715182153214214586889657263425577 95756

2652700629297845964506174896425300021092247981808889462516757327350029677147050327998757989962912792874660757819647948475239203522264559406312628324841753975481745701065

7092602259317232087409327491400979781869702601437421478631793853130815571167180161841988500770608288114464967535146217195521385239251104198410816414048220442494178279538

0173574309636944364950635187482039605091070892098443141678406446191093677136150577214300475164292784627606125879326495558644718659029848182201602100499771983618037782118

9595651689225393418263501037136321157781214602370936691063219118839169527053631999425358449086023980831429872436326258410691536617915011710034582739884502300190025577477

4312143342156091021425630047292952168051282215647987913148361230334611027752033829796984764875107530862096284647678635120230686050987180875128226550528198916999445 05458

2528742745970634022960335685388694979938282801471099860087059568484121519647334347663376274351356005241363537190114791202992225315331886602515377068331015488909995369901

176

71736942984470666770482793224482341820656742587914678111518974774907255848915546647004332922203989506494527707553293229942626743329882814282556896573326678657561933719599

39080664291914750311132844675148739635772879342989756413360426581430509213481367985288829257251595246368667052647189839619635984921134156953198638045581484790717809078699

52927279184045229430102942836711083950634842506382085558653932327654528622726648834185074874139385493074176302795550310290813677069430289701202531770184941820914979238296

31656022298361980348204057952319381326654609721358839304485216628721435257511372368450592929824956114742268292183057942653346922593252263720317019199984476575715410489691

6082005560929388043856153939388610800038893141551425198807281346757093933154300525690378371445159973023442914505699864146759503031812715272778942502383466726331733374558899

099435443950036775670310919094071611411540557340884843428335359931921443710600685936602646993465720093603162344426382151424428391715961465813174247317051563253482379314744

9449200697954801646821898843327591528070953988212683989364231810654904480489276848921854710676978861350333524498113861637395731026802516423133941154981062670497502013611

28532716794506413100084685100521121363995839031402261910807289909816307020635413111542266607869548717738202141054740140240326319645639300962185517129143265244274594414411

8275602144551534404540028743249577705844940809840330272581809631796524917350701494846684157254659289949476236370059804890396576313824027694236662622690151478947116904980

2549490252587822206853726045733035973090775835393170308887085423495312230318022928795990640501773733442130627431431286152209051808992390909028791205122311868460241332786

09750024675079171260577404575292564364823259159993166445098901914614802860025482343871969033739870536003763110023850308855156273942623157390714593247386985029721347782266

071168099606040407499093418534005721719934091308208823412305076123526962124545508422240149861662157316029979040149669949172462378267301789409424959606884377389474315405

894732714701696198582509128215714053085784661918613228061036503420039013081403792861284302253788112844894688305371033031235700434623165907687630898083954159168520545880O9

4762371192788990746785753669928142352470441171665252891003605377559735588435425423746517694185144707314525476876928997005954613496803973823293623899801986205650562460668

13102252036805483248433496956896531003852375466758785679940094642537266615054675709129876236192670083598762678555030286306687993922650993773988460839066820820662301246122

99730378011574829644982605247059002013885624414633410581453363132825605320520059292648083031321203507877662958464544189169537342428139035162929142250483903626144425573588

10748683489251853965653149819630557124199950351323821951186967238151458339349082859755991966972170453430432731619357189676268639789788684438861190418094418582586957960266

3456303755247758311154311476203593869850737767074276510824819289768101338596997861434244300067684394207595549574076093250167823992960045740357758333106586041155550814690

850848399595861676625129053153032243940117244488740345224071805352218797147385313216889626028531253487585855672513707818686352116963311057203172237060467811950861142666666

5602658062908484881723155659726462424913746676244614371952845693522156184400208981301595486002990251885850735730715293231231813184991789363734145399134110220355654105101

07724359860556317482778472401379834746309724538260700785102526381660274367109349768677960974264539232419633798976824844220980191330548596452868142462908735370534116434766

9696472966752583007869309307984984271168790440833818632677597338032526824224332968073950641806715656027142682112303123429485516536532152694576106468247007076793291110119

6016546877992087415062186924352558605660814350070546365728251289908986529882762656823007437703610580622664547996570037627144257157725812686825466294687239956406509728322

4077614389187704249767623422145130917936877460629168518390246803136439897416961358829185973655389064577513081193455198291814912845765152299640726298338964187393802463224

7073639276859512913289625226756903128989820889247134957014294446868226555957527745822842997280977738321273253051416992465522506661465591357598655474894260157123808779 42

8472791143722367111530382747753341424240717948860202834057759385860474399509433265113648775432422147266313648590577194353727858104371833790095179817963113508959960557 05

6842175321440798955923184808524878083675500683459483934247832856563524764189493364922492977088581624231502049139804359044981004314932330105944902474718476398931631684443

4740475056811723746439216134318381268547094021092795718799369313860000384685143684582116680021828201299917480926398909895665394058295888600282744802699916980106424049242

6581075921272469063099245607421661169418389720402845531707331523151114205678509496575612039007683156747627210704573566171878402962624872115917692632706359138767846553619

9985983182304514554009697426850377501200832133206958860197305121870541843880610883814334713432566689792627702443419921807686310016485733840408240758539913655790811783 90

8542062933068122947293355971639367946203353915270906372069467342338151383426947141024398224388308667689896244157150644747905669323318085246429562923101600511186409568800

6805225939654913797813563666278658482540460687315167516856587130530521158705592738375347868908081484472246840335277000011912546712330388861562966960158721813302472824787

9162197849116433286067222446352883105320858360458051478498520944223549847131411318817352440769169838170078031706833636151053133551533095846623655392600802125833620966795

5544462962234869224431908476391935174960159592862044537991279244395476191279923041584432891273052169124081319234088661680918543382424029454461017088275160881042747219 24

9209453946970268602852630794031070690991309220431159645042981910857994426090015245142428334158154882276758200364152510502790358004805342809230236998208941459463316935938

9907001555983506446908224449861971704430762869320144595156484877012594850115140539797963393010043840531906790743647129958471833213887110909982356672212638054868387418 94

0101075400137881018393159148808747865878795001515539644076162577508186358217931011726144046668884818219456044051379248831971545712438419343896724755144423665893647185182

2607188363875070459294519811236411780972307434095521941405244642200902221666865483560810185144317155508745203070668135820388435215498794496039494729606547080639824562739

4706397440640513789295335728586689990085990869067039434348708008176071405043256903200881536947167076099967808378752608107329924794981733321395318387783282359167370125831

2428906473547421508878255149841414235013016726750821791110025579725174276007393731921421039260329708107698190589880389161813825799144434322105291569057878847458211103826

6054989466937643057282009457407529253361163657509360063673498502843887265817107259941078493030813011758648611583290660573274270603069996948240058608640332505107457077136

2035206886494623441190563168069892406954125655639819424718889220352243226097231754888531728266773910375723710106739337663808981515842676046852036203672407891490187927355

6180766175587818153830714572203881772971875938956629614975056785345081345976592828371823671924314009741054372978445101625754395876245790382089029349315014810483645135388

8928030438603690164265892316220721896665193559162749795550480092555159519046186067918808187226970297962020888002487787177912581021178942507276131979543109624664019772097

2322610026863745530018619088598256565626527936608540782281152515511826170482175695615416278513963778247995239435931246975243690592186779426383330733739209231890347396557

5739664909820526187772708461695587361455929521981268933164433823973142384726159453088525408871886577640223851051087973101722522845299769476173758249952632471411817301602

0183311191811222813508856982100481816114616760485187369995611471710489695145284969675812583453612978034773231329653596522390633574214202049904797794971403687215328070643

5798771212837235867186123284810142893866885703762072348447967592914421953864220121082077988765511025031152937169329199651538878042516256025746324085042288449977623894692

6963152865729374134301693600760353293697064178277877377930650105697394561162192403366764030907835414551383164348894499792375578158750144552064869347130864983300738980880

5689434606520734875952488738145285633118869663717726065392901299624512136216182124975855109246086067032927819860265252326322826483572273239183834925820212759389405715304

1127362818836102715025364815499370099624982612448826291798672824976494874375237721882327023286078786741446467607560018209489067346574075767645525563642243021209167963726

1321178552432544516726617845496108737903746066329417692646345850239275544630030638667795115643109766170095008053674662921490319228931546134479522051756730507822406654121

6430977472513044952288499254238165437964069910930372676864436987715967455838788178236999390729136295113565859212624706051237992070984453068067510219235976258038209179599

4036229322416000214494830694988847984025711575886642601488261724776092275527824931800612454680443270369494280519969324740567656213986896794541045778067938189867874389603

1549308087544850870841396937769738770431774286581696401871131875555139837434149480436738506910996892834084554865883150782980444292178135196816520654596288592489947217333

3292060737007254838521659865898890770988354659504458514681125195787272983135614348787393577890165063107586440694448397384650888544124079935717189272387604339267541421182

4046241390905429387061282390467449182904617638581776736597386933106850735700688392261440049675224410943938679474310473017573522143068707911389927519833743170803283017996

8893744273125215821356317063512194580243208618521125215636094841086274928276882562501686289940598158930419958816428678871076201109560759741287230522131712605071600536828

0281377723429999177300500943273641748740893193172289404803345998552264108929822154491181471363488518083295243711560960510757651097870387631281243959587830946173490398347

5833694660579154522058959144056918656096427650852458957058774828812041326123928589487352356033810431400035728912378562743250250463650074389009557274063694944618836528666

0543726152454172205829830850965936482841958696169115725932419827358324585383144016706816148734750861264875198978077618791749682700117124818918466138987978076114957261378

2398415346224573473913376690617788060546347492480215010001624112121713945435621200002729273788090686389119833516081162870910443237307015978376805284281909521539008995468

1456048093477272389992107712708831199940831181902204147888292396454286509798811164285982169556628812903107065000093202373656256754844374179931102181762310080201404616668

4899134498942412244970191755576492854239923972482065041525370392288105886341191972034439414870165759984553384786876567773477091766714680382367313401639828407880205349546

8329590014417804615539401292235930446998544589414974439097524012233423549654251547589721418489379143502778941401688528164981439552982955401128307290595547761225000882002

5122725613693737437692659490087374604434475687479171702479007356271678183376663649010405890049089190374645620178677383153699102840008228759819748535247250201437242104792

8487967599705577863311114044746980788275213095399080803117091873922137847363226332007664951635051082589788722053445147150516363923455862358747744662242736212811092579582

8761191662541536846573185406772626628180746456123652705702613999291616530551980465265030554548499179403755411083989139053949669149956241308657657440203839651663557873061

7890614207972461756714788535806936170019844579843509116551020022736519845348479824964461874462583047674648261736123338632176788831250491277920092668848423818185480897273

944618069227465490467321094749474591173096314265599310511814464984916465734889299095623605691807453782343016627431623595003450763718679986930718629506983311087556768767 5

20957403143485900228811237823624226101946741693084249295818822909711597530124904150583934405280223506510308621844597274035293927698964695385469623728428517188700622734 33

716976454998828682355302327643503763520442330312191859473589456042668354496126899092401308998337870135124523948795384596903934247060924693398335944514283381668768390601 4

541970792247310862821685927654723364212180787384299729325758725227419692110088899483495642276964023114275391244744324195895104686825138415711722663081780340834694446916 0

5476793293889420433966208647272271340585034694769837848111659256157715352284042764824972916255731478640882004014719734509654777030128475568141016328618137324409859068562

663716058890707199676569693695577718697000834782679846572398312806248566411958355946686463368241014706634969054555589265162962210076841661897345442324855875315765411575 2

350824419774405089409900945155868296661070967090242188567263513022076946164741961012483500488324399099685855418687667730961363628557499053167283427905552087494170066782

095830556996249477733433635071383166500916099445860307019912063620947261628162815430505579730019787551337883522835930191701502437749238293956379358962954953801061735070

050621153794116136924540177933110673136807906291585771153435290634174027337465251218836939159355205715254803914101411607607349325512865322420066828708349329470271123823 0

44819108063670170555516816605161708525775389709427072966869302260064896574146620726150470861373723724645325437825575900925675384226193084201930141705544579737519753004 70

41686118589733242245538937939958889725739655857866212042113770585010041536324733806348493087817055660615334112886073767434320714040687517684530102491894478822892467092 47

645453578114727268341198125157937479385901365363192228537725907459494891650089063531037329346048208127547665544803346897509941733772273008008835138661216396087695080332 7

74463275202112383366389075701736586883407143249573637613189023179572011933183468407571300355600358910191442671794848640975070348915715578801216721948131742438746984840 1

936835506425753888595363540645784595258026348940536848524566120713520282183609121072824486035709848598621499110720560382459194859829823543664747724431874436259385244319 3

49885976950847699830301144277916640235362440694889364336665191206330022706964314289141781565157624136606794086442773931508764532782000051431144169384502136024538595273 58

241280822840514800724995059504215819802203211911703806022726228531527782250798281601848480446388454238187782171045347616593762607297088842176998044060036095071630593792 3

53372760325679084296637714903184469807592422930615928875788093754556644160988810322733243261160955215205586557288084122820000875492308382839802493037018529206223877098 77

042781012562268286458258207489374506573468058957357127024469933040752354544863848261179497749065229880663969154916194567573417174146656667761558913022851724937498914231 5

440420360083760231932689109975546265851442441306916999881628089074859690483068772415009109248226072211953662783562688806669171455147204065469867244613418392754666639492

4222757872605386194266203793909264926951308081431676592825048764721048635839380560930913730692084838813775627306414915510911084412526884282244779976582804516708878961001

5596232545784169460313922987805452872155596749804085046227082625986099215606567760591964786619653501186103387301206818626988983175553856349333496203625458890583539704 99

641814143440022080602998861413468584239139340153564232472142530142859120565763856935627762470877172637820785953181433146684288614405778794843941854414553901189633756637 5

370590242787878298698149149790489982181020529068797424909942843227323929273792293589493756346044711741346181187637755675533171353332869446788663368525299779133888406252 053

5576727452268759094250269880262302902959198007187271543905075709204894742788081121617537057081777860657232739443689675971961324093497733807177485067680661100024970948317

7585048961316817104715760132811599797966971179648937493658018282931468853266706863564738459193816300614124738538845590045829273118304531949237507394953597139353273415735

5851621085639294721898095939341482083584956933469183118947980143332181463352839107766288844960868696378265107375503636164492166918087904910333994848244220747450873791935

4745696156441020723233200904265039496704594794672172322314104619778179232279689115160333870859699947512842805010007507936666864488336267250377590252110262588133484892873

1366640495934702802062967005570773406220297557733498918427780848503294924150006562102268729921694084302378690711214874594089050476760915559737701406543115136419939101532

2169319775060536264632456293786010408510259722532517309265096592347969499224833117681165948060880008261892370405949807704001739674582189242954895071687713310010397851533

5345314710685957094450605373359195365861719631497006018216015116781644587779937232745484370216719698324035215895168535513924782055912827961534916630151458051788841148097

1867069842722647325396278568856218601357551480258411912578616981540276091270507419511403965637928982882852274086014503953521126538953883867793995172958629270947841273911

524331129259464000726134649249972681810945151440336063005287128811763455747400496852423343398859029913064236650173092311725493223569364627082647365396147019361465648808

2366385687127431181462802429670841610218758998609649692350312982446820228624101581183380869587321577601822378961801576779989263352748408957310408461068537718696398441318

5846115041482873050231180669598634179338953174698177066866352511778784457585311149522528920915042815749224555913847113304271589135534111423495667324043853073572462286846

9222840837937344847060291693601399304095065459619003081838025272825630822093726975275298952076633889581100393467050195502721525407078422203158690015995130826384841974877

7684991692392219653723166544757440764101009965061712627959017809881925888477927081016545937332535284128020075751048259116525137578340100895184768524910186221173555348636

9698415740199556028651242236187405834090844768789715421259530355913270720760615089738197994811856383567450625729469513436916789182630665079524311558263643719976037744462

0187910161354244770062736476565491962676164446807721909229290808662281129598849247811486715342571553676031195245475615652369677038537047688660469802138524803056117638206

8757506758136412862878013776514061907267002330978458776691822833391651290787076086693751711689174162769080955175727969805201682102716262136069706041609156178870530540720

9797684930266376389686627360501139889883673359135139603154742889747138541696939293183086800212328963295993812716537221185295145016710843207469420490209766700742282746672

2215683386341302328609110460240156123049845591409523921046097565436483444615085503749970595533660221346260483996012273273641910694943992246524570005239136227333383328497

275573933564959431552309012332305634341523510569628228677882109308461256866114443870760958284067151875349808971433102743967023207286830105305585474061120710903593522245

3189737748928721223938874431654802090390081081088526692277246193143825290972586311246157795502759822070616687080487695005424198755620466992746639886393484868401422335287

2821444039547842289174547658271945439737217167460000608536862066597343051524833272893912510779922817171405917223602073852789264208037379292786435578021158357573583590602

0737813547108144950490459183211755301545019836835859869923267993632243604356246020284705006583453866926334314789829914678538149164332937337706472856119777547178021057206

5624167806204924960506810145055444000836915853863991964055830272863321210271592547804919890670890668715121943494416990158181216296916452965667288822617346469987438060666

181

3650125491935299515634267061483686297879194828582851812784137702694022167558068283567621527125100272253217550117539167618570824410005697863552015457177509609839035001282

5128730291388517597460971546374785158162912190028421493728409410165155706168620646222073668955493931454351822662501773714797996892061206532527541841391447611392215213461

4069286419355821654479670253694488544417123943334998602980270985027141815954167802543245054515169222688322276807058416857108824894008108059824363880693049401400360540588

2514727455882582328529765554366006097614188174543950683518994306440767709734158274975433460741103268881973632510295986891271391832732394065467765205701268560215837385465

8756251286737197305200835452057115115637967152422854375612319367176856700900637354250309455101918511962299514032597437920935437420911843155116834985132549158654381928399

5573831259453672482506711473896296729908788858488825997010742272525040354818049628522570520823634851226938315628067283671474350559839279732606684295072635138275104197468

8976767259136901730651236573171953243101286956107258052833631589009582669841735396384872016520049064906403549854492372294274183784110330807773869403934504663332200174222

4053633940752750598293574910869881854105227312991022233397661253324690730080270049025891197866095461412213341213891962638826661786556146344210813263307717607866380636342

4319091832541943227473241152377048775613809279604835764996953687485113088070624761273257503766591078316587857685506619883979458302690860270196411220225736605864073929143

8772259890740735987644733155270684661916708199094210860851403328060178785143014996799473634236622229182069032445301525015578576015228847367151823494492654828857179769926

3163527538455765576440677544890668248996613455952180097814220867268863056429154013829653468573489967197242017176041456299913025141444752488487885952192226843723645329643

3880788511069933979691277826916254630973235354514622641376521392975068099247765973472612690939787221392807455856470999815894296685155627040336605876230534031098345932290

0126983226920502201581434302808000223658015463288255290470682507363006307811895687198094076715851261033440193829627188086008611697072396703943181261309742471117179019857

4301239907540354716506288051264920095605513453336202926306649809154087581763402129960016194911103493751576558875248272142376797264256426977368198890882172269926555304987

2479047981000650468434265519305954864635159658341932560199539427254246373171515176352439514038074244238577247656596152456746699316286028264855040685912463407696639769942

7013669388291856076758200866943833756314593294171455199666977282218147062090620349763171208646603060567561493310916735968692629980336515985213008880240461816833186087792

9665002779221347601161748135200060259895150218963264153851311775210714641070725779349718225272679005993370186596157932780066176911183167614715796464874106952245376114568

8792311508863036238407176919599890035658075472701071991664477592277306588961119394275860246657291646477829595177925217284078607443295113156581166361862904473631555156558

2337341020541097055398105511858777484725531770169456197732740007627466291601557814989384991548006700158802872713964778855334050152578884671964802052467242444939551864372

5584409996006306300289818578112735319400344482753400429606559508922535186015542228917642946353097283905715525192200671822520448345695728302485815877783543719512685965426

7081696079823466167925926073561893989559028954229820069011350119992783370576175407125834805827806812389428561931111141325049331642605280233039814169006798898037376032099

5164384890586993696068307101231796152669245304649237463228462799082423347983303740344501786448039881275592225902046536608469102137948423688814526296968662135664767020142

7253205644350313117144413771351467367169726545226546510614403820274351610225460111162195775537690901298148136883286248160366899277390833940514714338487476920116426194819

2396449463200446384746884791115170448439193867653031100824238704981524520405573962058183162486945032953509198580564589890874425683122925434507473734151952742364764180281

9954873173772129012524214467540429458554024239064717022570044642487593638766367067747962024427680943778548257665128437769009314209866607101870293385933104377328530343716

8895184802547799127333139633667250697624004443999842871548273540021362280036387835780861930328130999490596589188937507538344269255703320890375832462846794994772016161849

2448067072606642394323219232071760037252313600260079381178885240250457731492535024973994254405139909972427789185118923895556724908833225321079886270815640003225278031510

9766866228334677738349417461225894542098000029093296907437726013869091027914062598515250597766818440107816706693400075149348285405556143047553911053314375746422948620974

4679084475876468363178927730859885111355095794744954218629866774969961594744440927311320066978611380859054531762649459016781969714986139949226972233845270514380938951350

4644375512254861117089139668089366777973099438488086193319091043578607208496730738259379239635993284751004072727938626271670447540757192025039341919725451894831172790246

1404966895517907998691025376489096484598167090171393512067828395799612263119733149133778183433841931852367538662901900463343328532834341824921071916092767308013235369472

6484892764429798706811256546280611726046073321917630474121506403020178932057956876051027752505304361222709604675845312738661652142419408683408375891400951141339295457700

9473174811688536409418352861099710341672355817860282349820203149986794017404171914028393621051948060481901824518008170137104219745921279840402426896300533553685494164008

6392177456769534458650883286298216584601645078000035989852129012395900704202660744988785062717607762225506063907453194771889250990581009367298919540995766335386583780980

4479353779918771214297746197640947272121402353265517772361634196607437639153866019450177691400624068412731629586080635067629377512529343675960734271996751352015800873739

5483897343990123825656829078591144258828521366169201402990041314621368063235265086541121808474998758441118362909790891983411875376649604809362648932910439059772563295581

3887767446954618157978704191597558335000377286724607351853508854954642029309715856369497857640483741505558099188402093433335128195514555185733348881649167694427782405554

3369773119201437225415419274505976013798414437359942028760525557189378041148926125798303618380107711005004239239264415692277900570279474013694422918479336925354324840487

8023177744518417795835582575476184254957539587854945834308480441737087984926741952893230314895899160068542287957892948159000687353733333338638139084209948725511726171087

2944088868447850811634558526921415402996720988693715886232220331485581742682635950689362344813422473786939460923256600754721256741410342184901311896396765473291674247189

9342433028029092698507522949079709443009238672877439362551113136287615919738324134916722682612282631277938117508074089589439719002542641266494798017644201645415881317601

8972659624614413913192067850352463830657764016522973204708920114691713372287863037038453134194264414249664984474404861773851352937310809985947281511173650719139881269307

4826906140392864227662784943295588285938372148475070778355119896224911955882370450645820056101702053248444902150229456176556185921457439900955482294137515441361329150430

4691417994122460933816513626279028278842138122077589424038840622943317279892594183668293679259604224184594177065301954864394817033552410287470447311715070347750834323733

3116325876727636834685844486562539187239460827324713500448957086803150325038676401735315079400345572855037001845602769961506156829141611706104616174082484626703351891525

5188248212726007259576567889912567870201497866790847106631074876467309890914719787925985906257336497349823503126098309873468616273935805790735490824683049728409773238111

6708249151637346808705105219191762054169882625476054458177111799377677869654216992578557720426344244304207449548970339043450720611900769736351740195265632213928258312052

8240037466954590928045259687984608140870719534254761388363353511911214414314854855201235813806264923135383387675809188923797585515732036588317624241169167523814588592075I

640352366843726791759063705392197981126597708113947351671962999970520901709075898569163890664270423570074450277512010490394814832945744380974626035105789819540763987060707

87581774072491184506978842994138342060812428390148814872599854181480292949227878324356105549156410917488770670678201198591095890983868839511718380134914825499274914269857

5259517765361262421572662448896096148297970308420210516027967147856159406461363824775890201102519923421532100601752302574212237542107495918728675189552155329945322689425

1884094228267577442222755282076156072771039471825668024710660677383120630314662847443386204743259568505689287162653290832787843996507167242206138294532916604638087256307

595241532881209370099229896006981396279686279567635198741824018562931984922623336431899309017048819952588273380795326588514493938124115432732058916457864229452684117518

8081840509569046913813443077848902114879712116283977344305484614980793562435496714129473332666422483050043644545470274917232078399565086801876173270331906565795475395206

9292520153070643504713068812890320538545563987992101095667298287304794661005431635846234487441555407133814323441311788424196019037028686962422762465024004208137135046401

5993472053275567950353390671272161779060302162399780418576334810054483887031730716255256479959999869353196813010156297097871167306964940274916657267494343208132314613886

9255334949294118316387788990325940109341115727429801331814578281687913514243955787342059156145877368988080791707114551154762661682081777574724842787971298154823852036895

916524054373052466728731303019258295249876393209857312091610561357220176392716995986861088676061338436496169518654848604616848482407438381180741734222448394789399827801

73422230578654102177292648209140948503501322415155999901630714781485270516144322582547801434401095336602393626424931385229405473564168364670415743005583729847040181455717

002969623469850599778855736998021822303549238318797629894156977212363844088917892752775284714477621892057425453151037479977706862362421899063030962822117732197082303477A

5455060238585911403898969166622077876832601239617419997623655063632660169906611689088687741333780919516149770549214171181911014954347869382062098082787895731327899060020

86339112784715368430438149815097030477088688163597419253413412219139628745781099342483271410917827631765420963125071366926365882135751199875401157902802850780669833384187

338266739455368319656571703194629336410927266727424499274732291380098357445091340799308568360484677866535421058668005211542648674972395379364215665358720818554125981945A

74275161184169366255632419359158569508890535314121833662399878022115042456600150236469108159974192025443787361022672129412282988880533679439503294084049616771107287881031

6146773037150218352890241464817543095490594488754104250168040699998196719379820827733464855815859107761788284975054685679139904011384562436040357440246191237694402200760

42143357338726039253724664089664914714654730072195292737417679613565233067804268200024230741218667486694628058788787382995232780501070661013854028801191385573091138057277

55893681940309380718863793758215444351626559912030878482052393749351895597158155182804814807831310535096823567161235509631615286658578590195287177546425064129849451057307

353324334890979972094572917009587840832214044050147411438170956577125267157798170976763847864206487613679394578442385860356496644631401310486670561346276901749283229004387

271120819229564158948152826558422192618902530513815911803423108851491226653917944817433904322700506874288819967061960688500673328353614680498596081402879119631127482546610

40437038478288400798157913445232101668673741121294496813451012237984294021947946916648150312431873969119602156712114227943068201040876756809373898205468421478560142776 88

20188071717414127493674678175132277109046730015309467773069223290953495694491601169046818182723477816197512572785153556986162922029188474530269830156638749177021947605 09

41748668411972766044408548329750498907825806153902003010352180271792099508178391212339700700278329061937133682183336961540826520392716151802279152864076831503299740609 280

48310506255157226287131327681824780116915493786248322401723837072716088641929421181718546474811856940087052996113141059458293870330528629791882649324742017955124423294 76

29034883495272292830804012226372276110275203800120937520885624064531125037264015399643712337079037439512032028535025474827798093030202196632509329094488190574510298521 72

83069962242195471016519393297693706545763334324332564331363912210835291355577008126505397931646814945134120751218835054739685955538780919473746566279031868161713641615 21

55407745354380414798454584063447474508568028062324126969399311229640560327310927938491691879333596148535170018149698261648003440278228139858790148628521286746175384850 3

48068386520168074767440211476655696438739675796644295886409577559154192615071966537341081706497482220419135035223932036927792390736955880577997552601903081435508492475 98

18577979802981264125192698083107565746594693511285627975905780034193234760013696114472901311372932653188773146721410741275221010515155865713493912768855730064565566635 935

25354509457378968835800277072108075468519790215677663559608589952449272204977529981545862535929556885865521908620348857429443548834017661683055326602458359944543309529 73

62056428294745396641017593878078757904014319567264450256538964052007148706853706562711746676157191810642280464382686780641781701710145579987859492981028674877471679017 43

55039931696835285921164814878612862891163759927684710887436616177623930982949823406413782807112174123783422214406998486328823924065637743898084631704339814190170417704 5

85646785993494113767101851048288345847591129859911326180631166520977109326025457776640478113979812027396279052178079656593348303196386372636054817261754402626843572645 14

74111436197474885328135254323132104065537475356091405133576448451265031599499398475912338676258998862826742421410656317028268420274277535478527849806255477967309562135 01

07513570108143767120973610656933898112877006439508682263524195798998752445315132684357712526955116584260624979530056413423679403268694320221277155284522945590601316630 0

23329786184272936970542564089572247314796084227996064312115572289888413406901570607916985539706270694986922613155059545816240665427626536098988246922842402217051092088 45

22641938004050647235141065446293973629500626870691857435034689078702526688990310590307332020997707019556267080078421417500402487764369382704966754069144895634000929852 35

49912969960465402832129951288171829317338025893433608886395618814530549363321022467023712349079740681256872488335280182087968683767489328659515635001584373454271486153 42

49712522615178086471835019904993217819915535387378562345508714313905038432869651387690329053401523842824396391325686729272248708477969680927365604215169040230075017366 19

49935854471440415221410647497242161523037246612553017965334543540889800869016307189158808455522198431157422312945196703758626058776263336684850752752847903442501653764 91

63317928320735720436314962933262171984120414969390021578987584055068836313189782803270681820620114709707824647760276296351324513522519224278604476601784554635866895641 834

40825955325433408871016151070367468245292204678084853185295603026039336998791252960230732785846937162703749144838891633750052621697332332152400708758098485646289779180 98

48253427212024812725655515050305700418119895186092167878092709372414804711203209275216382196930053667750846835893865445281522353475575907638023785215194601827245682363 9

5490568125932747289545698934471141809813153649597485154106854609786330324864298777018635531367450811076763860473346402957688211528898135664493094931043589036451413175524

1478099253505914813095314712262341354064609059803930072954096921369456449857878131484983471061858858025833538008663482207458754757769206741006757177870214802490680993540

3404978086974752318213687886719206942034418621529900468668535749778710147528458215215350982709863527703575629384931393160423887168990739823956189887221884206119027709429

9032275736687186079770707721559232641813601697172777407028841345717732062530298816172640649801780192696415503872932862487451156694400901469103714665895280769075063099165

8019040282491239634567983926956557159476136694387393330068373830818123094204415929595171871924926797929712950294915811852694465448881606448909108574738497225266256135983

3739178319495991506543998379398350900517433878817182955161844732570694229375365255153471178576184977477973316771607149942145263817047522727295570500173374401356029781216

8718861307295720086827047281509944936557613530068382030072487238918604555823405676318762623625890434079602395529144081868873503916372993471628157658628534654447671897796

6400943420249016434108486305364084940912887711996506980457880713797795159120425380257091986045596738860190163441482900997297139228277863871263080176692137049349769043808

1956561880129790914650754406845318051694049624845495251863363242222261458699500929790456268865970393098396956023564388968632992332376590811424860358035451697820676948892

4866383306879176594263820092637789031658270502911378572975097400493867535805163232169847363385841973520487633945539464273928852497451674899046514594738849778743657589470

7717794356270743257228553610149292967565026585100600321814615361292859106934321487868637434297702059825451095805417445562508116488195634432335550915407022873450617663 48

9800492811148486033667358805224298027624506580641661646093309658322774120168700461864708735460759460023553734363668584630008298911593217465432173511755315755510186309658

8560274473288174361764536559262461764387774037997688608018414576447643035712983995556915655942311745306594836250206036438221735440965535395090457436938651043129342716658

10521464773628560410427152372939781248710605985932017807577384367065631135068593058886083620234778064987975853178401672905978765702813875802603233337988320303896853999 14

5202912912522642348966197633971826810602370959759662980918534819012504031589962520471707676639986835315319660868752912542311537503475500515367406868469956767733057751079

2350490345778778263396473876429056401644333168224565090395700668568009825182733058190424322611289485184918822142539686131003371372243018683955964417346708417832040075715

1241680837554131482479240045243081253677296897044406796431797426815935920313582151463967892385331469168480192378561366570636094221329625671822214085376580706834616858349

6375655648152132788002082516961807984840474867542864812471892324572784880995178492002381321495055976860978043922721365667914325583872939628855575318985133904712378084559

4071844832836115336157802570583643858593371292347805105809160715792371735883271261645606567945034809907997205084227305474190778110649402160166666295094593246968082788587 38

2689608487744561183798448657304543205368926840349142345088153510167857570666874956376066630729363649798409105284008454623326178181497976141218187286126865346071166509551

3547783230627411811482071716466526542709914847880777689282345795486110671504947364739973648066529953736779993402353274964801034460008563976976336737322294714697302776409 40

7753962883901258246629396938987701586208937991855169330846381599653959133926292450056440676097423169030780271559700744418327838639200591504192666472875278979549277034150

8263915478969268833081660479564209112379312512452382628913239114259259248872353135341403716733799306171199648037407664440361082778150937198926678995883750525547609294200

1843074443896493496814253542442287548263467259939978795979725472244117287600771260884093950481192159674887395558626710654903021324022477079265249776944955205108056410482

2068021729276356152237738510636020972031378948389087608773044443391070371387133163749092821549212966094355995765425050659702392398001339719254398835333581932604235533464O

7090728580486521415339035222970604712201304941234553577895471248630004705562889455044608622266485238473107331722303856418653118021793512078836789316324319258064485005114

5282276974702881297556723041938969669115168403425219126668605613003845205389336911210081205025241087952203462366013906294095868068310486156703491895288995944342180753633

1295702241260765662837857763105841332497008021096395134916086420519405513798037382651472598999021247513597282764870480302409576924624369255111392235317869482928421855572

4374736182657795600351890158768759814426657626582583777353798801168581116118854587199032823772576597479255550151703522520432959456689504597798593133546030580462871210509

9200265391999022032802985513499055895296203508118519116239331768415451061065516247882198069414695314021638542948465099305759018425477258768576414740917080050765864163376

9358146607818722445058379424375762870854540335011053709267102160168897425190172664771091187734019658761319238244016778669447316036433235910530463926925486473405762211664

8426993833030504817439731193500536761645050844763047325637642166613981817216715311128910245585162073197007003112540505208238290893128757564218754167861569179700970093805

0142984666852120015915113272059232133736245100927185727349413528948123231159923461589690978967445413062762687256330468233531371596513000406153320056264213843223148270354

7358635091988446533835684004841684653835147529872668521047590091051673159246264853570708194303523598755049297733307531722035573470236517079315932787444549816445932473938

3779436527800310927753543734929701120308885081138511036298562059488346961563623208283108425028737613641284672645036795749821634469459924402320386636095796527312504891201

2359996284808751700271355908821282445921236597263623626193171749930602780067270385204438694420239626594307173387544843964935739401654469425438918315939596750099329847553

1832019709586811147403411464307749078732347323831843169775800695104759625347803929053122789720283171030770441520409620275360661687750953874913564865489935285108137546623

0230104416440607843379938865943431973972762809887255359642183688625252868429209407669353101746760614361504022410610065751838578184594309738166652742282235952326563156OO

4363460606617992954342681278246980231107831922175488282484434894215590509624968095516816141784081058068173009002631527217960747234579744753439008266238408070488767662012

5669754226324768414436029961270632036424767255662395841363346963420183396378254225447430632805351464492080284011693507171925134344917591634464815473816005896456760840S7

980165902557489767118772253952700017838346180768335742506196377939729071866416577565549223989511605044361810205941685054187658471590296259403076771869386054624379389529l

77382974148463926922111853547905990950972997607473872962647276052104595112827435631270960595023755731806195452598765319888191583707806094607301046153140796722525575778609

2730919024691929884517398716315076857663518649379762973806603165031986181720651187867335919765239567443188879533652290886520080140499811495657664501357148699739221946031

665890119202261370639549215000080968080493651561790444139936547321194227076271680662326799317221812574576091524551670509765492798757928496616737619368534354686O426007927

8210505029351060269911077292583322331389423980871654050463146586178672199499749697290597915791846480370351873886438202706098049340045567860930765382619265263181120218638

59251656925865948327448316434614594036961723580597179588829058381523697885331682301331452292999154059230227405855037406285359045976578596457380652160026315917742888l2480

18307666463770771998546947143246041543087106778821876907079694131367846057701838283928309479160915776228351293308159945269760684026594233928858711054408683063663858104204

73632094496534988776275786269905353915532791188119263351069193127339766674091486139745614525452756883805182285266087134963283884183684848013626458289540073065079156074110

28890834546596253808696255262637115923594826384545608725836860376198170672633744812876372682998877722944540407706789940164381250027554201698302306954511629983134310718890

90986254101600293552441548625913761597478709185153402568888130198921884456611164910632688832423448309628880496579742710340822373667382825577250767715305186061648306335590

80193407109897163098840874672591044598887595960530329248899363852119394909682837108645713523456015804386429710353846683059689879680835873497857626785565888383649516272350

45207087795334676586720491617208798131491416989137566464545697421666338802887093115965529089096858324662094236872595520958765867639645258106276095395529149176518173767290

18987667798425119216807552581600469267289283769650286492087987822244210714046467307485197437139760530743223451552532121235870851616285214538008368612462951548039453329680

37736424862964570243581892316862264994992186349832468121494673751199481950450744965970432717604013835522236081534000479598193685223092588145754597956696947562389495724700

32524848660706542882562876735928839761419619059132908399469163841005321889186943256572066776523238247316438079769604445729694465028343061665221698708026175757057051241641110

09207357726203970411688859403761058433130293144482906595568429648729637716132195073219971641441702565468946101270308245972384182836040311660348915371607623286396661656180

56000946715549151974673084232558676741097334029846169676542107632231536789799885806831989228308597633750929464903879107437077400846349362362400830084950073558164966916750

97778658081118230477074246964360473279351820384719889611703590083004556851252805095510667699602373878235337663727822677405205106153201899838061149198595287500266294554227

35260581489488099794079523817886224433013323645543257410388370423239809214161496327553959663617686752438176556625101337430464808207552851564911671828345413047238795984888

99915627591908215219594535894398739198788544787855883930195317034712402500720728681966631889154092375629824773736358962930373027486492513691978890676358246153690757238110

88900038683408930397937519930653817228736677538511417161821464063005399340793672109509412332835057142652831949674694850212250586274045481109395874053728880966522799491310

35041148573758189949152273413520402201771742697561032405390025938558957849154940687591085109004456698779029635899957858043995947222843355440391384551575459051012236770940

90042907251632725064389114149403143929338160492141148298696151260756811930048851604289256533437306862399621812008375911431730894419899802933772305065225027098497205959630

53606069301554054235809356222199901761551336947568289739010093518988689102305620334560674781595563737372431505131046388436846166022210580766055016338443951339275390504710

98111267789378425274293071714287376055437904153578146194855984706040401016336531189306836965939759500845851332728473088004848774039317183964923212125759155986396836831005030

34405605278988766114724308920739067717133448999590499556528668704313897455638641950652562633820072560532502544979125893728660693665274756954585549379365494669059562648220

16601109072227166433627147859946459999267490495496151264230852976404043695030783887567665223227853593755677217005441761522757890068849222464725810068548316761687053704710

40293293794295759712945613452455275456568249439714331385804950973720607539412574156291011023260518521610876296113466237967436112342517799340059601494432947964164600310880

37766887059200488620180196337697614505829217689137668547554193811607470643646183550950427836763844727462071613819711917704444879751377889228995884738200829650704622312720

80621051269910394458936078904429066128912164886732932771245059557649995316456888885393740496576571688091303924234856937644501999120287840027354636310802804883903986446 28

66315940784010291779687774189882662222702853291914550355495243566744611953366897039280307633613422784813008059024523229864536534759379247009771237251485597896223350550371

40823738855989999635725768152824985732302910209663365529809751869164292807619274983229652194481696043715488475908552723586440480201758142575400529343879461964319673490293

47026961869732830560212662858359414291417643270438839971403798486296507440722652641634634289829191540694582238400532287197813012034284651150002141596938756059895846743 37

88321510445817334910293140849261962544301069475586326736133960131549360027028822653755016913194817216772727748879790713086459125176896227023434826717374447625403604436 12

38342663685290979347302622177434409361004734383259470556179624247014795775993839612828218670404659473671346740038028868906734350560654196291543982323319166341577275777 26

25566712728756776474035878251863240142935123099265241861053652623906898057679590093782672125122055111315834782392514171893266684696741049338628805893314125836487308559 6

85236487095735227355156417316000570435683375128586486798777626878911970174198926297503739016966811455310563963891334919478491126002887272022436295541133021032888312521 79

0386091534116785776233880138563643471230253133127583492149626816843461805655064376864844321691600993297935886971326345948047658016730287623626375404641773711712333755 529

07616845760984120314814906712654478813087496692652495072837633958248312795542416998444914156090821234201446635611514398786928367644038199996073613035658764068334110290 87

82368523045177162181480494326246784203403756912181002048571334683860316409189049331870282565113342259651395183628517923265340031625303117768583043905855303143470009499 54

04289931062006909384298598494625764243642747550200929598219970537138567540242399582249361468818357839052925627662578147628254905215311884516726792585106299641121890474 29

03268982903001953919556490734818124668433899387652291244231198145746620093780807965110820809202500372217656466226546285167806975616512298469040287638196536023135636136 49

76638902219792361723388549181980957352297014232045491394760020309831182651347083195790827421779732291100149811042092919972707939131054168843056476680628814326393122924 8

47528035155632827952732656083410881616979610239630665231004370268230915339990110185178408534796664576963995643862325907256630756039470551358975851270259376353252345454 60

71157876567775171323140719849308707274172599083479908377522266238501618900013696653310225295738651936909936258963337910411775860518781920708777656099941098051551760524 3

38381498578844158021483003773807294222057611422187341912017380974191603908969457081286919618687718234413339778059759399170408525740245952795358955168367104612241414848 82

35070420989494640533461029814181498384962874854695004074324843010042370237702693594977809093900955516428125416837951222634081034024006017318562283671178791286350576512 2

10523464875470892286547579799198663491343367457317808000101592280324580415062040588149733690546883830568311188564269274466682562826056931827114971107908022216742737252 0

56085239809705192876172818294917079681084956342092215680052226576590502708101059646896004191594139713443207948752006197036214365310768194923146657468333019425617725802 9

65628710574635667936196429335090417591547605643722848231520544883264267630871114746063603473283432300941904208674812321963355980892981390137432172319029440338021383503 78

46658247849282371602845709018509399863087689746361213922874464372822230689814427108087673984745198685973565683247585023790229043874338656501635430920494975139465171204 2

39963804851524043827140049987366092215951679436655209716246719028784972343068794146220933065403515146533353438176212140974995290813343668874219608546300562941871857198 97

189

965490389226099400645813457698989029375252603159479352051728738371667007215479619563757407012855926075633358043611785695703271510860973326600211008827123199163759340 9971

23032026138109889964222097317552741425536341306159729484372048567305363345356618321960256972367743766404198553908458665477313344243061721260086297605296720785932556 22923

55846262846333189292831983610002592187113703997150557749320487597831976713628632018442823365393471636379457127715884098417687777055114469465240422088538202824989296 39508

46704657019347597460108210900223798138939388917366199587023532221962941499139604842992793411658670487785711133726534928566536892887508962608405860497308118077965006 01006

81097962304559548011897685661967624238931275652416559831945661471675822057205150733187438649955920772921715411527070251573742932740603038899701817639249284449614498 921199

60087414559201197110619178135600831179394370021856788080126118115799464147173035479067670391626448036915230016380975095498647857715301933207587076728073824162729979 85657

21981668444519311512147215394626062415474788449253147452223428981444393566896520817283520184910446850123635346659443910177128605907892320369817781441406576579531416 93427

56627442756479249572256177112244150820590572608481471404592395307959706042717245927521655060051371539677242631825934316495999408488134896294439801928050446990307433 2958

69642023477994406085552980260168762745948165247001320475095699315825105659565545112468121674104418920149712917059113204648692995218936078725719895433268702764211207 97111

25806343717907028536568681881306493152903586491231366257190281160804499253317309331886197389137211839634887920380921481773975151215550607142123784782495146494932858 50008

99517323133417944919571954937381022516204335585096404429995065494821817491829820980285016135764697993067913222478498512958741989762340204826428706202051130484543207 45643

63054682985899790358840556461006745899723002864895895241453303792218607675721960300019984053461042369673948465981989033226447136417782885831861564108998106333256695 88014

90875489118552944475487777070332542816540487011452902591625351980123874140104360363092052813996081857821750591992044407167990153384458640154205422766039495098525317 14647

65665886681458842246276815471774897706020161939694773855297795179206393425938196355238038842483958103927567056400517745841545364779201707342682078953784241931871811 25134

00191354907522301885366424561518864030265041520622587526974644830596060008664808467397587469403100443050711008032857561852600337068332178341006973073703640321298250 39545

91566451625180163585488925018864904176713086234741645120009780526716591409458666952971943266261362755466627639747988313242076600247082565551686346541573227303798463 34974

35417605339118055495848517100306504714657091740008220220333420842716210255069982951669623227920520670901249991059790172386421758536906421361877237109133881499560018 522736

29653609206373168397847371498411623140479545716310285083096492946561428907773603095689141375737242200843195367673720751672340211743272646707775707205664465381708610 43049

13019933047479859635164887964619449295289058257309790409506277206668356944279880549619837512732746507601821215205334922080851917626617083613572333425100696831041782 07971

86065027973257761157165072402736511990861376703898866503275577438942275815661730769721836436801942195847681141757576578022594122936937628548376150113712094447772678 83907

26376506189327449474109264058693198959058559969600972046356710955864259422250713422810831610787854562083865526862249687755895674284009651707897439983645219402878969 92967

68855225099245481131679811960958449994512830174395658645542534924673319276992922114986442884274282189623437770714914385776080580848564273037171353764245306793786945 79057

52335076438815218106566079270751572528839918515710526005618833910853622127568842615368679981807360170875673642170133324263530619346354543601726038855255677458062131 46382

05488997794379948659252807324771277015335667243051287455111305227076922621510652614798139613012054825955398498290382216585400027871828179452759345759816062682566935359 10

91970253175878958507864251238169022868440885550462338617680937438712550030627677911446042717502369048260534968220034680627979397528237235197310708566623247325548156691 20

18466156539344860475403574057353264973565944918990736160811763133524918574724029491422191055346694230286637390630837957952652229165823026099765074091873340013305234310 103

0377785078752063516583462778448353683665590270558349479612983505612394339024757396930505229513081104670913249882048315378942261322355663430981141830986684768102825715 502

8682748383397571925873474011536646716826293701552682244212771105239091796931923912773797717688015422468834962014009973227575932939177240864634892836464443538761811383 414

3988823534353062371783603772709222067901491315749461235248649803680609448488571989384119196847152483689960814722241875431733533374236949008176264368931636788534545428 571

3963603275099228957125803241463056342872931869971175945658363103616835157351924115291745856865483974709206998158899997133603670512517832216587046941565206271294188303 672

8616973277595474898327009337617091038055159386077975183164791131510918328012752515936515048860793988485860993081193493801277896709108478408470943154894557705504550361 821

8115460265023335986626211907545728433334065465592365267796790097240225590376662770157178745301067836845974624241194146092702551816535744964080370763592182048070317078 004

9503051527919804684148029262375914410938821175845422300258710509897434053801960809670600172944613159435393190215524986391770306032510939596603820623564441379417884276 524

0643899787215289854873908254194811268497400341380587702933160023497523314378325068346832390294862961289177107044987706230858455830072538581936999562749473165520895617 94

4627647768827751115761470499019482284323868027347074673862814752675333720121010716466861961600633771352137821138857212155477397216933044019471665348302682850980553100 367

6041798533591083991213560094604211825482838260805945116333374599933098540090956755312822504067646007341266354406248261660563576912527998130161592594268648699500472477 533

2770649999384644411392955896914652204299081224390628600046503573812695217022563921682177336732020915867504500478981687036460961518460048557414398817260973844721274644 195

4501983355932317688295578306458020898295276315490735500546541111137194763043626173297859028465131916146665516766350641193345758667137818109841564659296236576908362716 072

2338147885085506388631774907872349191987679896746362545471260440954168197799220710427640251548433196477390558104961000164705989515785294399191798519780608939567864719 736

3890098524223400054422376311640315186427938021795265684299781428838321389598243939588770155799078394489206709170655203496769986054155902105765147715787378675137125473 044

6603349942396378779627117683623689624459677644019969786930278957043516744315098022898650128774540393407397242671605002558785498888940993842091001681738749835845628935 27

5170791170549000543621962170642940432780627987378666767302228520183704159213524224275962837503729973495914312511292396853190129753561162967230848617423217638929881754 0247

0391522806254684791030979628293430421339402288412931047594043460835668343232495613647542579862544554498896351671364445937935735007027731767348845174887099668250810438 991

5567709984887401788299074948442331678830180605973715188996429503241991398806477821402001041117774789688395551080742111648042275079877931031511511084338317728872525585 366

8013231927523396907869301592580852419159088376804536260897508342110444191172855632146461237202911674061466035739643190133267351138318086854427530611568220064205077627 624

3163090763395746802731529177990810125354943612591471988752386663993748525397978887930776928132087457027296121993908125462554625410900834305204038710838385722642176756 302

90493017692951152885790854789541342729679084551526624977107434221537656995492169311900356755887971969593608944405602798532282974592300572929022501918690499605151 96346658

73845766286165583709262229549052558715098878019153598544171672135730700635697462366450699943859586530458169557668141518863751672958285519470256814867521612134585393262425

21744695255070076927008328053505369516383702482620563452119086436094550878637455568841651474777446606868027320905274568175395157132037546217069826261484759071 06925055688

718403630005305138681095753887480633463299948769553743481082454131098108373690707510452638146157137362035056374120611543698395643113673185097174963434058322907 2358340161

8290608716587101601304215919171390236458970590273481568107228986719530257968376456398971876922517593315586797334077373304321675565391075575423891616070212271737592070056

307136843747821213528898979866821776283257253334518323039274761865403919208796565258928608193070403573964208098932108943220085199416761171752735397888202661874 4562282022

053152831427848303386602799651751797558198833310256254382304160189423391703863378357921032037786682180618908632648674196620083453772672138030048353496960106622 41386427

252925300053729177370611623486543873603335731512271393004752577030921799026910568336986814700700466807131900900638769068942035418676510749733938247585986386289 1901184985

160567672720635681836412579224498280289965361861777668677814102903988473476271790234863835367198841815377642843579068363587905670058581213427848601734299273114 98525829426

381306584979321719602718883988881367210302529373730687284284252676818495032793295747070453752303208640889188204692025781143740677421043931464527865896380704085 307448240

78053943147968235436205409865242544613096095689729925320004693722765192645252347104386806162844150263272906125719272577676314022420185623514890590306131224408194 38963471

259810672223080736248740388234464280483917599711890693049486869311881573358946333731378306463075822060360727221651203940831736077127229108945530997277130413106 1415677302

487550303033119745936782356969206929008684065517843550699097961301208591342082523653383197216400257843865315526851804559506527621829780729270016262380753984531 78749321457

176744284224049806304873363455955714192165559906931601168545680475785694458258146525410569094213041518607874243640505075700116978451599285143636331419198562203 9283636376

980736964248023175052955013733589796035987444259036369071724627520840312123803089261318802320710301404611955903967347832214727621039428519154515034699239286860 6087894827

204013155178548892421185875032601207662275879486610780619943166946023124667003660469456988033789419169279093056380077125569611138551110682307186263508781301659 1596657199

561663926952401328193712268673839971023130237186061442402816750002785237013297428005731412337710767305263002918854321849520377262517251966548985863027520198580 5552865567

017412424781398441147945614036133723592910024028800169542825276370017260558174084453430805114569010709200685375107498005622995679393703609421655852401268260862 0139896639

979818266838146426887629454986869695601835872133246870517519561716040850360702192659349072702510994757252219108424455920783085143427489795831409061138136873218 65624780

519973309894011004757071818985229378443814143454127508270984979196457929208023550364363535096012517177168056554998278836687203057953323354892273581439095605122 2929425451

105961566598998015880640054229431876949270762221110284761808261596446602704309729054929180957757759026962478243427196842521066737089539128792695710391317015524 1989566593

796288840942869051952349190754939683374338510867886831129748407742561428880242025456470750857403395398767464470647241224405084157459877316927428065793845108082 9334713697

457317170780121504655607730879875787050244201825130662513284579379669342674659176754473212872979925532293956541582868358625639296227016169581436104796463370168 1690383700

2557364944013958190229025904302917933014131985196006053939593015118034855063021486381739005927865937796283974600501660024561924250559351652393899385785832922591711647601

83315867395892269179867719926426770722804451657455450128217130748500736779344454702487148837188476688243781859855683230039182925072221247233954381450812495920572738582163

8671741214554440720007746256677999930338858143839522468418606099465055117494931262474754541149008992988802447511714394147171562048431661614834901904600119092962556851628

776043685120922176537252030663261027926045712112346384308976910057607603820506269489451831336812975700849465036278304450342429885580336356198454450541852394839908251480

866797159530872371873295246175261964989059205469690840452251974665477634065536705083961295269437798297198225507400872467579607375592988511922700401408992230976925072908

2437252930253655458496293343701951694483160099816953821753975089393183088183490256194526897726401106041349080453131450107139376971054634766542938733278849650778891157044

3389987596862067766941170235325721969700633559801276796321735405701737387345600278846155575539188888190157937805544173152124711048525279597666087261897929914561575520979

7040480867560544694283122745402563321911571044554296315225230436082636308442210335683371003474828646197343120322024724439322933802978839273166096657341966481397117329057

6318607759419141898287478329113668433025352925249647921101964649065217824280216584448011941483756087062646822170527885558666093973084921172488988070552950041388669076368

3994300818877934803775541180454695193374369307401500438691562902746936145881457045663729762794944061609311993111741804520449283545614699712876823501751445373892838376800

7204168069639536492505786930809325864323795839791850836678526870939433960987834815131423664526254152492755877990653256163320812718635364405049101813148648797280648360849

6665024892073569753621719047572055407926966384435426213099435243251883097135308234078731415494809665748791962070429813217624378108179660000362780596168645858331102483433

1251029352663857765397147351005221181616567263513538413677555892138833561375461669015061175377649637297641275729796185460065305881543374664637502467661873682860135938997

9661048870373212998988077189303455828424798632586508329716478813687836934043158699441584056073105170080740906242322983421483645937540666898879902452967750380630886424623

5400067920501694025766842267333776234700738508610419471106994358045344557048916857268425464900093712562476105936696388997312784658624437991441399515694668108392632522318

19117286400745587634104562885751255056808152522939279259781486174745452694776380447699595505060703040572307480234705742344672410312296534950506517011654313242852197592232

5039634918024654316346612918022256977140890212339328788604173941013526311520369111453920038754109001217400800370764068065824627805087557204105425815704573154793095847220

8291904617945375395547055895534904623810081663731945502358481019922271929120161775202444668605900966401426557247684365331867035220655801458913652142148842795655866270593

3887491878639391231094985612126299412921957055098215904145913861125266562187945691785864140841846629181231277170185607642984140347592485979536413929570029539960004476524

1741180636090891070329935760123567450289489667243683113234827356381370078481809220454448786971363944071406838108537502379067925491990743545391118754416978797744588070612

794679102610597267850687549686191026642898726604155400035593706209614687774802119559012974434756190369015032298765074421165940794984642116383607742142679643983102756815

5546121057167915310722177793025457322874537292989194954644443021474349443399152619976461756050044687614144519713784817621178977241435546735467024250734783642218978843024

0648952846314388163435075296533334782074398744784424394834378621580005295941201695844997595662195314594638517065744940644541377088385322747539546226747217816410785971506

263948291174373316912308779475584016245243467316076527923038336108403393227859230437069614233618515120201630037106473392360159409348761413931578014737552099930571439690

361417808686920081272999501268940633815459426180425248707055293695120120933850061826700896112557402629284289397798199536585334676890135251331654478486211389757939533763

477043789600054374631526792261265540350013294479593320389950440369123410089565126418333271896351165130651176712022579372906101484235246743785474046961258221106399832604

158125475467792932216010039689183516686733683039313629853299887287731336514009920077181159495852996478186606788956873041297968193338686403654299825024897608389872787996

322479948612906215381195278117506535007681270346380699853213452396070408503822031138373124673544085400354980559889762848218255029841751924213819952951835534403125784310

666981982362890498595699397614720195040741534182884971636795557299811576928990294536517544152653286097253112470609774310064281021572319914392872471828409399674138326975

137019206039345242446281820941620536688358917458615199461028929758611633262539363485791650895497475561088704308265783353395487206043619313816874211841752981800450246240

321388921541354063163302382161565464533991219188665880282830367309947128978081654878897206258768190147486576432647455435864796605054419314007833058157665753933917660149

901725110135193032764125437908172466899981078707432988286063210542872923068461896542162578575560161255234030058019732943125553442148865206998097004301429834507905380923

445824367943874916281483401529428782991908462113223908762350246378918770122675476308791945324149114109125348773470452902228567697375702167027235152036832281449865303019

362473582464002265301781786230613182673475745562719183138593703869423224060783415856173750148103203795100191326222685916207583999928414258360755023703675143734725006108

565212317820690269323271707170180717500766710702438300882134956211420966662792576294753536561223119630348729799239612956100144117684309914435980655668473318377111148982

166860528824315829668157736742930175053096196635158767553864128836864640168032901020983641453901361820123106000007757966076771445323744903666453819033035659640537748486

770568919521436067964328702651141982058714099621884954127590552896627174371637870154631182349958080516726387898690116970574332525233066881141023090637160965393382917542

740572281318248528656220140842948358903054331768669388627252990137430129882853320414925964724870842887086812156215565905552011876699209953492885869512934925620588529975

459814735055255033151457349813024963566600571562129412882016229998481452915802727452943367134613398954992519726142954671341106550874868735678990505261841864946700362615

556517859007474346830220335306212633198282054810388329437728480154352985423406909915122071140819165328421628682826366356108343235666218458349658424351140825271341844670

438854605640891533111831123911860539878116762767613703658428481807313920056249047162328232991080660606647062640354285919007969734730188705856542992129141065083105024921

811525610063742292432939047324285859043691099780873687016127156576086252725630413794618685327159089484682802764387819688303469013986309935348904399396141313859646288438

448545178614698948363593025023338886138660578825349382042975420534819660539437264347797913552987090440975167142565491734186574028331056378689930376561815662351820475549

419434971521548912142520647079290437621284540363565400426443885879915446526504588441502208348189980825237626378349391929908774186331195023335350795095501985044294646003

792077032647231653837167797632255104615083447057278649843104069195521190290898190955066437540036659891571296690373826058688622094315856762141694627700344614655883709614

418383678165326467179887047516242002769841056810499347974174742755566238344821872394925598724808179028869791780821130782667388227175699536815810068349441770141103531411

7098279679604574215573467737293131835742900988172510707001027056816105909256197738725262954967064382697487631527099633231538978021419584628370235855228029775802140619125

0393900095497206896922935154870995796216818651729414300236510071410275558943126234999452621742518340731495182266541367071120543595047705298405516414408231604009414859322

7671833561041031952334848460795236466106998696317658850281933170909277543809750296912928290287182718686765605656010388362597476899331918150868463510202044431914159077236

4683025539168088913774267233213994712884977597953977347922967893621930992511200663011566173906575736837676365937189211422368499120214755719722305574100572545257245385556

5056864605371122682267950192963103945844174558065212238893467377781174877110853585636510480400702351384140839434491495873363170337452474420769031046589402807065204049010

2610180630714409508901183652829100104263112612172301607303914278299826054825937487197078945590863342170565376911559913371696415252912586550719274834593711307562682233144

1930750599400673536365037293550076798042151214351972532560562264394541411316747298778368714099675896563070199695608246280145049119912407121857480537069396403892123464697

8712255066960696516150681300606294107400804707570130916173507733475564877254736912252150981350331929338334382107885543833236187356907160805455809843670057435085075319994

0976596533687210434933222806188349920048278738112527503049208888174846428319016539600630466502915620890532294731999667891042999177453412876891030909976718814803093271626

9812365720604331596496493420535930974995546353415998438238620494250693932014266637836894812021997614179860830589838293900673951455175735499971707538924803152941090118514

9261408759447372115939328926370506958329918100862084041028648992346326034564237042482425706390912938080170465965363072414492818375533047948623350633663758865610589 0660

3353162911176597900236530121728575366201890101649815977724729054022670786268338777301117009431851403577621945481676206437531968784148518425893448140529583996384970253959

7621263597859456691106370601332279533419105303795404259783041376675167497687874652996491783423364270007454754819494713598668069156666458532914903082320598902812058781268

1576743093072101761358226318999288797323093201014923212637326715179649554896892477661184315456716175749393840161652290784408942315087668705466527579323880554912666169377

7598953925561080406938118224800051350809372046686020039055009141165394481995941732187190340971882559072629584283719770085817941850701023924759988289613573778756435885496

3425611467807833405187109513125759999318051909219022665615695039282619479016017023122433057544312606544646250086866227480438667441944201539426038115578275400004040207121

8190131574640109342733574833610146940368545125644320347075284483893175456176463157220928781027991202902189223828247035054633830941944779746913088292457205029220624985552

3551056541643045762988641768062017411348089234227288347454320119807266068956925882293270245447753472176552839806174300254282815982876747135983139248677545318468446 08380

4815081566354194625658132963595505948290763301068156449665178803283777234426495734762093975759557930638671004777439340064983405507236218119894844821271727778513899568490

4476270086131269781577572295464760335927355634301851525659582920138209150226200880031749728538998215039065355933128282835302291042484991010409600808129227371345158145829

5962514381717546634167630658001246761930273939749682827160964943723113557562907810196124240298111767982618671404839546685238558072759405609329731240555537304479929325649

3279811483183320213731115623240409254441936217129357321551955582693070362086939103095626257928849446571817529163361050844013855636792071157188857988513496298042756138550

0828166953039086989620802903035538006255632140366817790818021164843877948278512937817115195312920602534993978761754640800188549112650947737794033808138516582177365474212

231932820826980804014905348395503964389278202924723734827169643746781354156230792934930724033675169925218892240566961469968147387885186031428976353797624324269819101596

5618186293922048809589153523367722568736438466816976741773574492402773244271852172458249037944402883067384456400185365607465417560537108254681792569364696441802072207679

50152974675083841352311356162549952917635141688384581887938427692077003633081667441340571715043076445505570862019621875090213884932155884931463611851668192193333310687104

157219222584513623221990026708380222348732157579711911396826803848840402814592695923283959641886626796149305159820371186283109450197691335579388158951141532725042249535 8

88645052952518856976647677540641989646125632782259701702333755852088622218260028535928144630861090706741716126123002253062294326943978032683573008816248684496380618338 12

90631720666982539273397400494758569587351393454769511348187322641752631190577688592198023732093522988172495982321802051416465423317460267710479573856951746726028070968 15

250437335898205504740080301330175581522716953096751972016120092056623087754287106964586347137428066751678319373513256522148383317367230811981653422398026247437655894767 5

692163436877669156494799894490895333548260863798232539491667268994167499834770469902146408996837582495058290814533142200622637026588908567589263050621772504590274990999 3

279197623786666529191863955876879356638777647427669516078960839316235252927832503641558467802461815988051429269144069896524719481933963231368546341865090928413827172521 6

9538362006323009210996206249425060811181486751298160865486378491683891420244074612537349911807444468004565780723476210113068446077979422132044175184816160101908431185778

37369230285339399275611116062550093838015931151113590785216256048538691432381224590429977294696432227371518952580297336604535600707534380412669670586791436932809218304 11

3925179378607725904330105369386056453122825753943173722335852116815430443635849742772083634227879617830153625028018585784844259719867132834248512769481082148289899874 54

31722097792403608326198732536158596560141938413651762588232316649713661871498820913042815510162243911604524963384256540786205403968475984137295093158148777371240187179 778

38047898994943654297767325701570538126378522274697246787424130079364232849781847838418709500092027232765464981769718563115946830120997154727353173557025264097429486253 81

48140078594209375626383946867371632446500669475670531599472687811256170460064074445580742901999701262105369442807140922166151821307934869897283700119529308107366672854 87

985787148223531687967478735752661938542362400071391130567555033885290142377185416489229415567163845886141110633338312041108527645828402610255558454722593759618723463643 0

9939638012344489255652989827202990036784391510418687495824429546262126155525198674509444652902219649632955400007038521056321967658248422650733125362054626026868452266096

88804383804743726232331616660159127936881995940502571999932917633931027100125953609469437966385968083266431931649658496339772919441415183731667573688536451820230238148 26

37706530113911289136913462491327912603253534591991632345277581424766602795479540704330509573058571051219829120944316533594324084681380748875387297085375441682806609884 93

50666996372897944316567926876367066059226649550352104308343571403351838775617420963086662250481525113784858825700250405158386183616450763591975664564712150619898520627 46

107207788534090089741333788845290560643246783723784423811624539607897311433337060960525973019965093743954562466866281243275277881785764509786865491389233961453938720160 9

95472773168875997332167711844199589484861426103891518757536353139008447216715931361056590552306882270163900604646542371934043098508250773501548519931747183573730449150 69

49757248007908426935982914383093138985548754942322744949162792191174416817622585153292309062702886262732717271373674728853638621232215215659814014346174420864122382487 01

84217621137984800128918461150291340723510400996828163551558849625934702284245296564463145812208779641481397405213189828785428426965782243489218621245340214229187382487 98

31283655208388110223504587132896511975297239395600114252964021960676967957581479359799931418498120411115428221651323925590709874816233710191611397919813113343628756085 1

91368235626377581119520159283010980245617011022257367211180753783939970561978347858998010395864273294684313310109468796463429787722682194448056999245536854601675335399 4

08618117622592942776471534124000027690102741176192399224871721293219882821321458155833081032767665682157622328610235788886649557570655197671423095042057620106160092703 64

20023404509720835648577181668241511926944850681518629351905611712803659267821369850750877498220731933486197900725897758266414280631975955986314537097065477664768007258 78

50063099401087894554709720638038948369303697002697458292541889356827697675923286776710169803509732769360272883121039870404729507335731757223272713912886823089697758812 5791

45493793429853594473006165593150559024506929018029493965319372764130759794350301861951828494006827945565927733810621490164498834284750153040835721432245892652000275214 84

68861352026832020123565045190191187232355916703723297903459787550694692772157114718237996680746390722945122990853465326174679733821683940916461117621210756826473382614 61

48780221208854610941607249501461647180175574523157572564916552016964627970618347370461961194707193166112537770751989446868916112464067799340622219650686591370531036233

91875928196574161078398298795138647707406451422449302925240762368664335949866113933096820043150156117174402845705892934250488997230280359024388557971513154588328461904 66

31488813005446165010619163392539632499566891885581207683296137502319540263309630469405124798684956587264528769709915657361774733919053124868449462587226912095502070721 7

07162187967918776070558048101519229507432633782002791831155368264376788757991198116796887396317781452995442566577309275671432912484702302278647522453354025140790949182 54

62535291014393522896664824298455993425575591777585758242352338973747484463340047259313572926244839848777017654813188066970615022889309610656882030057928247768456505759 16

40181294558696880326458117051128812605060822201102948137886762347888324825282503946688656047914724349593624085753393149412855950658025762800089063568814923209519216400 31

88097299919802546703286321856744877047667974008052444941230192577061686079333377261667952372310225603045607349457206356031317070375962721464901942508795425966836567349 67

84779640178627750094170839259651792113718885026960580859287154656418538911511244828095751369332743829787840259129242527015238018394282776423077865998007149900112718626 72

56977974935585858827620784419210115012275574177976386270585270654598937833296495187700152644142164487520863294241085419318611140968276285129014378480234391248047246668

28152772743115489646255670802912294104703665441279619458724516005911000089599347648268573468246049695113680191131432130230724663416373425090192136199058793486103011684 91

36703125159321122314328916323155148639038920490730467537906033384814622379047633637202231768335411432973333114189394247379319875513336595192369206055641815362927457172 49

53820445747470974338111998252069390559257390707774360691544834954645439455506517513046593883516308482474634109905199496236797103589927135620109785056162499523189050215 58

14684467724636155469783168324827569631357175583147078970917287783371804851989545984382859669977686925059438280995311638096443817997760350113087073944851928516254905890 31

10872332963148825592074274014344023881471254559590700119366547096738912587270273568527307775193968862038706430537341996367859408521578977244356095538320973712223499363 62

34092989901253196033994721429578747514768554321732167127462276803323300056382327072275452949421725751941251053916721864335859445349268769220823873314300392613546630057 29

636733155649090795980489924726693864564374620224968338000050557898268566769671142801876726217786557664915770035246110093279468256367716153962437927712948381379723447 29096

852205312331820157895844795748404966542625020395674682354733532589667464821188622077302441641396400574395899344512407081002310766672834298600575549363747498649085556 3048

804573498089745784926319775144735958577735630772546785298233325468702009571974896190023249744687725350453317273092423088905788306207272855498567467904848611804926653 6573

811130031825472998778742263224505235417218301325634179543964983867938294564119675227721807970790505644450838043578920044101599100871056208654575358619881337544255212 7302

894241653075803033080796360397960666424281579324486052872495607487409036181364062430201765226239558683682892096274924609118894291902212698768197461064344788994633549 0493

675843048514349980646095449673288773001522154402956812345344894693259234979855786079219347904248738644206792284192513273004963905388381579374791162995933734710442586 5737

219183591342311892468121510056417553783567312752772033942084577330932293933774147462912041433642375845322780010418199175484164690798896663803490314042085797512767023 4369

730901782041202312016332370681609809619376238353166428146780856607220894937814058508265825612064156690803913301450378737470086191634207868965841313273363314363382372 5986

924785670108198872061494310116009326430205345194366867309888936587361852746283046848650865893166284417428158134399120583431847933143825136125768653093775747737196608 0824

138797890956863192427878213653414535171512012415632145577261257171154237575230435611185589196631444230869366705119913815340322621062159439742712076546523176519896642 4475

262047151909698917445511843743312560411260000481062834177449518999061022134126445340065348576158006336404668827392202261921447571115941454757056524353886708217499556 2889

089016770823973102051948388718354983432880888607110361769033231078137963055727389811291277038682799321659040531468963259863919429152076441218374053356689581994265209 1520

607223478701115078494599263794258273810045609373403738470052604324004765151033974426291589785942161902765461243710073141533133956067012099235705589355558643733246623 9368

273381376660985613860817560855257518818229823656205930398480268468925648315727203819634275024490538138712272836538138174118903818629370668796552740183516011067772154 427

487631866938516652689190966932412414576525175477138616790707468769050286308364149318965595623542786245515875993374048688880593635094764046404226690237394346938131980 9455

398305963795278500402818801715189687315831068254147354750483209376687988786820162497720796546229659986970928634442711878404326348465824167244160379004862906335227396 2632

643689695218637458554773792770810832099800656037854978617928168238018443737773965967582913220601255392869706131183075749736498210147313999513780119583630465784586218 8194

414978493700984027560068549802633573502769035013349159994106340615470787332906037072311612403418705507883790008981769470259740812636634023320523475949462864772027962 5211

368644837784212775470890607510255301454648072545962761137431679308322714444951825155320253068899305853194831806190976281801667723036121660678136951717648759790641767 2078

274900447456671143903026184811709882218889166496393868513193469112598949862477341922210392931856537346282447189895787586796112815110996593701003783460366990836605499 5393

142099942349260195172005600349859404096353991054727739469799041357870226320692025454984165726774279463961297439136852179485034061807728509874281227690114136816402846 7528

875427664834717285451238985915716062110621038611504145324978791301213806889378088785741135694023601086117274749076863458679625973610019911702986963967536056330358598 5059

024044857052314430822798558580421066760697846685856139920532524870355545853701010773500886495196354119145399547557065526277584357657458461258415631074749536770190450 8846

0451789610356300964498325679803052637110090348336832738009258557839639769788757455040977616660990279542918858930891795098681231156305389211681423379581310829419130185387

6649838432666048909688018607869899694639519523554122354444951091490427932659316891653845246476121029773238049491026972063197314457687223018886454723103520336088032734518

9251460428378272357373813799044513964614587724157800991829452769114892564495544578208112585497462991227232646490397507964230600261945260981238603894172852569764474274292

8937816441022597827808080938118848982866564507504699035503682364045702878619264007846083106256689672105151078759980626053310210012235808990878645255277392745448389494246

0977309768103288181048377558490535754369854782487209796239602856463370367224447256026531392813243226027749446017801180064353804656172418288441537932797330316564262549 64

4575247852251514033329218029943636689200450280879301458362655927453036932085819923685824058244928679704700489293410367432496807871678052871557152697953873176161393561325

9300309851874585260746042807145302953344424836352786309151458111677033843498109034713808687989512890927047103603336771077811408348446246860614725498836547671435078971 6501

4452769938600227922887312461611888686514548757124587583194866819607135129780973428900934829960916137217184080568891284832030737804399029801197764459526087245017878848 27

5822031416079664684533840137300363393522391326174642283971469201272934108032240148629789074135636204355195830151240608782357990545993705598339963964272542888444321935083

7396044412680349885498678014241201398479946947426725134575043284161152831893433625782555576248604469892081082951581213080747248488317379714406575523709296204687229229750

5739043255293584811980266329093403989497358092902735650265209686282787669265416618779365146499963351891859198281123717705808512908891479895022969802277536978183264365803

0659460809240080176890723258414412971928242472038004865907624832984326661253026851033499026576578579963305728578158044815065341472910340118651075275759146213859575557984

6533174652421247970356296576848971699687784598414328136415062658803237353726720510020391872086548894049163339238448057175638872509301213137906973034464028369489145321 8747

9689036890800919107694875226793196883973083136328551644284854543895511923516267055241661011619193956232121540447458844931776033264586483326999953276131156081986730317 98

7770968854732069736111069687352300866613257328235545132889270735840381580895934095400477668633738822098999477588872525139289471025761158113058137307925695089111060363374

7143004818074544707123585696701876163044024859281029412448165330430019767812518488732429146546537476530840785629909045373784763916341069713494831641494317725925735989877

4141042585256506469922575333866702293799929902483384497963616582601770376024028543525146892938266773788401005040778149801065156552974765023025304818482902235166349070 89

4498768111961215106508070884528940298303191343694788607197230910879065454559290442696210241265089620800237754863792190058221375417994513677375502934421394783438454088249

6830055969807227427147523461863844963300372011058488444262822847183566951057836418070908711811976341346792304827999192942334443433745265711851958275817242955697536554765

3558571537158788673155823731791203581943368590246116895493584548438102182098056456671152278111152866314879791224104671503454122635917402310507267578156516907949753545695

4661023435063517892628200738767157841844853233126474816964104500932952639391072551402638097234507421452663467912487992195042495063983505756316700242220978888450142354933

2644770854207329564200647999045672882730897363424016355981512716558570876026319347603574797113673228442544954614124103144112136532959073184466267811106274019900800 5740

8513360217113291910191481723858110359763233621594459068660688581745827210663607832472110553988228531116230830772249320718760314283554972399990954386747911904120640958 1635

0306921536539525593982910836444057789476865590642695907855588927801481153129605282973963783052239656879793291341582956231550105619750057478825843506838948027201316800544

9241765541341553672296781867226297521935729676215974572929982784576999581857012470410652755234709316760210887462018949830999056268035473239191743880352852394519685590226

2991234055623668168420261406594601661426898163648963226705671454527615520840319977552181122228306946402827545830907880095255382670011748089440085824422344840901443930995

0760458749929609291944867872842446552942603550401536630485728457506789033420637548565180260586465453504923154636606721495597923199612838925660686245382196621379095614 18

09481962680234837370486445390985796041171310701243314722770694381624261607791747358930604032216117319023629010812414634682390910147478453481264736939378034508669020117 85

40922691890721139084273626400840225609527953662226878036310749929518963349372478078424620773854564629746985678187794641332775595949591985874191668447830186688758259850 63

3580953047305926279587882662813566957418566351836629677963353661625690006588345284789326123079425333212093431309861700139402245159398630153300275885744826155520116321655

8305401109024799131744318609478889442471917656585739383361529389164631578777520692609899223778427210762267547136296772652833826045803154881842918345790562055852313846748

6880987475741829951436089527571148969698465913305157182490172640634694738316224845702956882134783218600502197579493279575371401946919871033919215034601174895493500576 48

3118480038321070455100031972994460626690817032846523263802422485817403542298510813693111624820378156824026705402650609272957413003945906744527919799664729933275003238 785

3445521820309695277254118331319492983745429967434508349456920794558330895721941669742554468253386456106651416194740273230689180554887144239701482488141302433322116028 2

2892880315913275460406393144756504510247078478270031008306239833790350592085387823674384874690715910716613835270975585158434913210350122255865928574296051076914689252 902

2359382791271100717798787378979020601040537927126554260278572353850665444666703866631450870991655715371206920784426287258032345585653143785432590390181683860053275496579

25726069037995759888134730688880747447547111932240148507132580064528148510649450621787971736653675812418556178181865092565915089678588385373460069351497669402008541424 1

19586157738199026180199575558106225483616404106207630901758560745738847326450713338575654604173253371130280137894079385793643399533627809231210839700233028364694233862 02

9910331376974507251928193044810605361488981272372755345364515179843869973757297234479617541226385432501523178083972557268785022568712168135342539028347810191915420343242

5934609133113642510731939943370384323886137588056827156164854936538023785090393483362482273182074099684532966420066268037268782026486705782985119284503022158150530356 41

7568937754210631387943008169086192587428698754675509234944739615670759250191683691129303774804955199097039823363826364891350584303996481957236211574077257682336990702374

46393542927020534604480383102314448114591379538474923915729431278074120440608030975872966973107181984410669334082604969841515771659295519486558164867577942339368988277 59

00357894212761604704734211672187920488769423319638405449947511159325479617623938498538240196477081852094864855924228773249668730030102462610446676908525456097605252767 54

2609722175804231532157673532907534815364111375640334493764477315701074417762358231865170035472053759404317824599133598339181781909632989815139125754319138983372544052815

08154570310228968029957722750833121165977042219982689658324694122451128329498987297002528869278200885612715994977513562152414462120118572601525246972290915140464075319 21

9928173428032761859964570258021156017462090635890751861757530000424919672009214856764015969761597040573694259035682705762172698082612584580596167061886633284026338564828

6426103859307490073261464768345904157879793049982050590432130023886393302978738736196275533218595801266221632584664077546476579633808544964957096554919468161205115569443

9108079681353938460597500671795063102561224795271996046570862538431272191945200310509319123694795458561223792958467448322080383851926631532382845070622235541635575201604

8458595787727964356640792897731517789254324601637419516533248548662125998117972463576280561784916758392325772236684392313904636250636402302831723001378553839784405960356

8332684279926654616067211095960638834242950430023363651087339459142466082468786791422410972599551213534391932339520885700159038044698266311069545853638464299547406804067

9609633400896596476123350118154718221565202593800562590621502729012224260404147261584034278162605459873385724563714777921604748274433441112967312376761198629786698630802

4179500473099054868078683929009975257836056842325273451249388464642611743784590176544695409651594708729976507112762666117578696019032897428626383480646025835183651959147

2871025416090341710762667758150465169462544957993828996178666609857273572606855066958690375723054995572090626217139470396850395236312944870874152476422506065216697079552

8304128108686049739229536592431541725993932707486079566741459387068412327500416690860257614875887987623593673475462930459683470578124339237473999249931821377630383752999

0385151049213862685594862828980290986040984011665073222899953224056223484778426908827773333002520957071638789137572836111316150828240586774806128378099765881224500097507

0984791439925240141622405328598517840602677533819228267849527886672456893337594516073195686216856063512035024630992837418507807604204773843053254838763592412557531636452

2931635117865222823289644483191026924741636339992805353093665926179864424954948846174813289956813731394830959499581124735554883738711252190337005154680402367783261544240

6566906224043635791995891916187319853048280061974222320988366936470840181232141747627673223947330600405172628593548541188246139912254599360462896971413416520309350257355

9361261145139647646649021062754457374977144715508058882686031359661575978888082513408417512408423142188759570263961666676842178850151166502959559786184159605479201785548

1164654185831131412091278284596904448081819980631438903805220749709996445926804268473445415578103344932059501563201963053976214180739908627084806430321780024760901439771

5642312007226354349327379915731551859110065247277485199710198477976655689449196716486168670790375717066483562808032596486767340458420586650181937024269525364816955814590

9093659078076868124386399342565535330465056785065548655571282118381739659983633576780308871040572972063848194704823330793522090608439862801838670895297949455903398039750

5682344935367837444146988538880452810081360625583295193112119347517762845206651922252738669692630025665605713798746774722632191103954463938461218455857756926921127446966

5654071571414181979492296144640390215239365217131970168237910162539056307962867690370366999872010105519727588390490266619397164834811497423911738704227142950498039184409

2035353505646482647090883162717270439141214238422921600927129123600670102952490482688952858136084943203538041356964515930924338273010698290740036378191072142201091900692

4187216027755580459326490152150594142335313528627788269126850570087709441767082111178033916132310778845476958923562860626769063681154808619448865059856349824207852248210

2735571711682860437427930019053242147326980760359336634855924681912023947148092123328418605006258537910855539500693514325721418217342416965828916871047738311059705099813

4940969399553351724534645472530194525002197042367256339419925959139030043726038976533972333101892731998036678696654613980753800150074994058963965696044446341048889101877

2540175813648299270189893187435147571951190164326245651420062310747654521955306220940907622656396770318223285959360902816252306279768529106713880771274641902570560244612

3783078902648548600649008785877722458281102434449387066147744535679373207145334080832496103686003164871160455763852964000928899056590388078720153816868605784536026239 5376

15843658764131568954201845166804253112509061280575860518512612612637270190642362190166991290755289148425500020316639433680320405711411604237579803585659563124749469 5698

49513586761717161486134211876492199857402360452581553140087609299196474462881338290732168822691315735551490184217278167785213038366037635612900768544531563881090587 18069

40817781907895745143359593839035139959232215815478747867262033422730377952548101343348870112698825270694572970442925473858639589403162524181768010338837389259782856 37951

33490722566439821360408060769512299054794220774558697799330344439044199973576079750280203525292863656227166375301043481541454332706533916509959467358711349333821669 89485

67055738078220021731209391530078265261286845142429072659990570958254860430354523299996515557814617689528577805366548948592631729117062620382979801608412159318818869 62902

82871579787179368849040257669196947891197016642307944711630047969166029633665411656714129266583864167655842583466265066904169951079819904156389709616578360678034580 38588

75490439836681107946259228094843989003473574576572212780205076636124247453027283277919753687094602692110264128505204647715113195909759784752009540677372174118439959 01350

72729305996362535527518365125842604722808130501701636458298879529638747352394422898404127423076692657538919129379270227358715890678007404859216483963090839234039884 00951

28732229830618530714944412455289744543189585611874000180952752096285137117256845726219543878759256273722400118928592109735917744530909913760185710513365526899609798 2796

69130166471364566970732370814846534938788898100994832982220454610020172043612751203153811658299101511869430491144759374415119921482776988466723909198092150851582445 61052

00887185460698730137255363463299564455463872644523269557824381689553110896531340698420412186385006905379902901129655602973049646054821960184977149958716016018632187 92857

76555148260986889146641786743554863988576721640798099307846447004158925298121200807754694044486902285347709395086821323407338738152117640444760483455529305299958930 20716

50213184557608516378241629071597940866179526322650908706250000978529459880341211082557760720141887727490710128389363243734475532766109946298202924784572795959958669 06046

00010182553848674565658275507158587296867716751114872652122065893051470298169911459357597062405238204272016760908971936440310471427018901122919727832816021135597563 2554

86999922433885045198582826291038182261226559925915970829197862331931668810629752541068411710862598703071440860838890816173401839452178676169102178400070572151153318 18346

39810904855059841908065379840393196765261825490144962826381313686830190537277629022140822825320503157581449110521496934008005917184288674742328188920822109870751009 60942

22717960611868752310865130548794073032475855515857271295685166711502658575372152497020735665542874804818838168943809294719392497078342691198162104713083095702791440 44887

52944652228809164269322120891437353263434701894141865498758085443897837542761116048219449857339919464163451058827596415074970868687065333087674685517086367220041335 50690

89822703363808724832477362384276712729982790004723750336569135964850589487117971773589537523517270453298809768706037171689381311492611799076386817572277825030379948 1053

09971413321067911176939082497857595017347343274863356684315499742555434287981387288858840828665550630311692303426240065192461820851294051384109438338309943373235611 84083

41561095254744542667874415394527438085478157481718055429230172577082542155145523116898159841576303310410248678593445139563372056885889069057119083999697995213344839 52287

59228397786569316970545027552498603759381380065895257056548715399259424999717317167976251830780718006517305152956338263279321271806269594806341649691021111602364643 235890

4772434776651725043701882621158437644226666204751278252351887021789802726728027711822773152103035872642082452898843316831579166649925682161144990342466951532463913570478

0768526898998489320525337210124674485594511686982516508631506559023350608946133259647585740474307164510219889646668576509308330045681872757241680767059216043554681009292

5593956489533295027971567218920273257081506277090717387103132384249600991967666572621248871349920569723586933240738111770557323460902157984722345213827715795803472205402

5896644153905341213744700508903387153834939078365114580792027210147906297989923573034032499697624060788973218431088244930826619080262204999011285974929556277217464611468

9939294010594942097338175943828780250444099687534019038860177170124591227676884675715236965521220057724245036539737696902851785827201084335983398454456817840625749043191

7087743824104031867141420417203681142989503677256546071050128601831884330590267041359144689996884717833470408291468124154743093009981538425850563276047390876890492793224

9240953499190341621154460821389836454362734525542137157891521320603765643461287733464232652794907883819238644475770595009119444142257604748727334642460795924629817046410

5346065133084095714764871552128608657847073529551768127582320953381637314429047417984026893595113836233619036522193686948953929298610560654350579702715511242608643085927

2820573203824528633753600405371597766322099134912226229821552464413415294565515770151918986924072986258052706984831289548079625435319741712858425766114028200953496918021

6777875090640963889389104845046640865750372940522104112800095473196600465004394421684859422090144524292722255849059366382440277332836264503030533539930969877703430907123

2576249414960161079242873166852638845275126706030057076410952614298966488181203960638734084985550094173219877966753274996201186977256904144690206247765655350656642375914

6917333072748915434058300807072214809831677481930217368453700212286491519944808643918607136506738526628029015686800738658482695676254210529864116593386924825383378787521

3492229698895497703342047861740980716210184415161772214819110043733371169367430877326697308599010371973977949610284291618684127575699139864921142478781435310883070128710

3774766452424063043283708377315256119447305575523791098461773531290644241044708945166904880695091460731826894492787888493093950009950702290353435392133255894765250328071

0528542412429468783066310827995140399264853819433461802956014311243285648469653171701739125387897466609714539422772741011442393879589598789105456641042681478911626857280

3599678303987097866634444047449502378053749408916979173853720974707345273979460572492759082492782583350682569083780835456936366817395591500548911712294589342501939702089

6398720423360813109299527818850402771287385744354705819714490571304159192507557151185687124513861708461376219948138815550829388483756914525902356397006311126012211800437

0384777070672157882914472504703857119508588093475506374838293531777538088558941174192632937495430008507222483922287543944226262695981911896956305252790993462946153609361

6354944787917503777137415907062317596724558368532599842155881031625624442765549902532750176313629431277960119164305097169595321129920452717744965463360265153656582884193

6387933775355221000752330624993891708860305513498245420332860149882251892444978971365438878760732583968694123983880820433462661172204379981330742351367849908458648166662

8620111016787728549322725897317962900838938093783688622430928160362691807321356465371535050488691647787791858427737908188613147543573704383964161086684544778605185698836

4354553941467448836225690758229765668051777774848742973152174071722240899625202713446380289384331452516983638073117992146783653334217478880597728426160203584308289244514

3907954874192059946703109376776927346001157263547865330378705088255862616916954752736042627652426280382477111290356531092061375501041535163500294773602803042651560687044

50345658653273748883138871363601280833913126850405697305826236750606618539761570417421742759409404777662958560993046444536969357131235331390184473101670285366894180350 23

61632619326787257731214972498245087303034204762585225658713844171927986481349699003368236635925882060107514213059935405118239822139359584242128868015569496013163426588 39

22903517029638549314312026857517209243416771675953673854595891563041515434088850041606705334665408281307008328977614882496962892401604211416151036179777932102971347626 57

99794021546997803877847940159881608521421353530414979434853189195283011575006555852642361877164423608307897345825052329324364026437592459180056329087230507629703648435 50

86967621331082877017716349377640015740770619290997731183271570537209205365402970776786726653277933085996580010922173623890413584279519335088822145436519113434014342809 42

89321478884830738725770497425332464660948353516657817670744653783648028470969261013186450293122961659943735558202928820234450728277138137353108738681566684680584165539 52

40631511573679215491126324775012652980708686532078718046128679242880775557065292782594194300758842780120959101663775041401431445651609392042139955072749878724457109413 55

55045015722897094705928715478578423754955553068827716278606818246476471999729800367332877207475226535910397549640886425476524143137886190662271392033748303555538284344 476

44598243634065327813631957808343899184020729563312274240406329113697622685654000106238986047816686297777840312794899917073967230675221629509652878963150378284676791610 96

74558488735934754910866919617224603794332653671753266796427575943996049976680661204001653302850494997286070427542509372077385863700415441026059524493707585003637841381 86

07795676066806461723429500752397765145943294896719001839450853507115250826284545345273757796912248333719234276158203080146280176756282502407146025097160563093176724593 07

60486882487900538471580520750744820305264966898199940251579494232115902784104325425835337177788899239517463665743261372851811005292757346641848185181569944340408818037 68

75351974066363047833728440558181929943124928206995745614238266530509813446643009698835798899933364824964722667678660948382182006917644779951541006483149509234023132985 55

02076012976597654825143221427726580666575578245211435118357337237730492437142595702958551587115638880779956709924041678394507854184745979089815804130829675468152034660 39

69878439383081839237477162789771382844434051340951168427734092176907252476310892244466478911687943251322200736515345623118550138742152335445030893885398610146110690748 91

09566359662948150468546756229289415108923729431931956149182318256112906025836428781721949291753453882352799362858896363679518066994355394737960678838639432992182785568 41

25586779954979546133028786214646515713991659140284587348702705261069817062568063586601281644979724666816416196987988677368474709635196349675767724117193889891161841016 75

72490652812301639681693150030825201489745797388900152682402377798309724004631085064997410591209501570775841414891805283246730618351409517491370788512597534824769265004 23

68546235843601597072961828044309168026370057859215113722012165857223365413744447659413954964395531325161609658018099208779083525790375772675062232251858830607569323189 56

33747331807153668699884986386914370393791379690217509625869284257169129054200208155469777726908846915524811744999365707260485043950809189432853044749398963006031851877 27

45821052465577410812254858746139841024552315469728605615990049197301338745304316023246449926512501704259895981825565238875877954291090967219088355357772498686136615280 49

58762154417326193041179249699714236480394563630993072083516922495396202008388151191837244452275263668892099295766770956673835188966195961605375350086023233648287212672 79

45718259271787177324843095828273573981941079466564284153735209789228890021037317202758664290118383502401038262269418554517462360055390533684447990378023149350091043351 55

5983611865012604956960259134576936587622265577063207518025910297097449295178885420211929716912663104142091401078642523861320813476345472977531744085673221032254607393016

7741895602027299203162479753768627807845971052132685726453736242255319250951861718760134511243376299053555461959175254804303890441737617913169627434487853115501952536997

8077934471263842488993435819171956729612216520534622308553410454617535160285303531706014088121363733345863747449007128683413621307466431499196264587662383247945685549526

4936646501737302526991190873488938380822049031399905062754439889044634722325658585987911886688740806970103202682039916459653939827631397675758675630661191248528924485699

4955452747582595838059266980566909315279118773020597897063137870893388322070950897673353300855615294578493640663912452296012669596009626249236235435006717356575093246211

0097876387859400037811122554818628659191302654661216074980353256023443563676312411815896469738561508295193180065715412630622899429863180944708829996977823600837319327593

1245373029308031970839167921123822037759386912820570401257724448615797828970323720982431419952191356933990701376486572379409018973102677314647704915431246335314923116498

2846119122320443297987988654550488550281722412719641296214255244081433833184835548531648915655594728454941595183299317885179785723333498219716745220939822794703894830938

7006888095009500176018651075682619018212259879110673576574102371886627246993510757585531487585060688926530110183871898352111721543535127593826432969022233949380459763785

5809073171634956875032708986570450916388205248079423522275356870825602603182812925772700379918530126352379901061942595703521979786946037934090907681690393742403294222448

5649062429240690413555579781762809939784678087508737889272173505415721686106252435312971752870170021713892652686261966868712382220018081998105567112872178866928665140868

57307314203026029417580061475433396996144541957801734713074532518466502487391866312265912601566023637862665201617416091315672128370430238318580118360359563761144021963419

8851777358715380279808392396230695928944502881270771132659792205383647524900638650538467352654712796013927057425335723992843925668211900759207612282727238608562138276534

9933921722883132892944445541855911852332513183003512791379252076038822932907683177572843476710568432799744762007968785459609023037060154744372426114673976376141025089181

93427878230418831155679580454124311921034413514902690955017999960402072322609942749361962125944584701579942673840826916807034200373342817358824220480345776191805538441867

0610927864754468711090427656490961336789669320618650990481167965986055034049205961741584031837366244498788910350470616500092549942339456621656246048636275236757195846212

709710103586783737628594551903569480784184420461164274326860787480844054251871227322200262143479895429528267493240381500981848046570078416191838634925747769264605691767

5531836754822789203318027266531093127116436793381962280095954133047870684275861578398159667863531221264910741628340363758489867323878266137690359301362324296156031166345

7950334806960414680959998514167331564166366106381359333702910558095186100259659693358538133702667209303857425742167864290636425462777371245161425008963865385952758013222

8918653790607932844115312395719443330597111674626649023894326671796113079458231044119190079783881715223000670094400240063875922694362285108398908252287558339675045354327

4241265645394801066480187933702647796786435630154195928502562439369802444070041489810128503847612557665321967998949975116516655256803543642604731499806941576711495717908

706191447997252271479529327963073535876112075829068056459270752877176117312662945676036633571335348362257492316090884997339500082398022708340184421860239715622689469759

0112111641763528196178800201355089860699803555039550769601752355717349136765805036634809891766237472777944388920986795169389361595325081576104264980771878858319166602587

5455855802071249602035690143602411606765094417511639955251176046355458345558683733327378353855866576517556532384665889863983862957934979865807839120288373221654180551015

0451839104689209505644292618713072718897597198522462131429519367378454725128419139172808431950208148721448681809122621419507962485836787251200849590524925366352557539713

0128252266631931267472711708740168740198182053009569012107994936932404638634100522606283359784777509936197411021609915334124174563222729142898453154604467932463165850820

5990084443044529235683561305958572769849765100357637380948890437898149713389441121553404782853834102518662734522092981378958015152526795905016883811079779373785231407545

6423832677537259113972316858792687904119567525558325589432624746316959593237932807739721523413165102332695551004256713478165687894432590102030761299318706117131114724468

4738098761661329821589523733673235671671545614175551623880797120001523453111994541783387246086759196531569694459754343526241471737702735162352186908458816943326495436096

3069032786367827217343337872342122012779145307361528442917645937575452766551173616624937762583412326694703866180109952641614144694368666554121465355725108823132604443205

9478295706469642084050663972063931736507351317300388044526225960365484863596942695148372191940537488664822469551380014166175344131094397669630489259497232083175309962615

4307845299331801650845208032363582644210162743101366129855870542291846918585358000259657093610688867074083687084907612579947164130996388859772806057254383677775945949487

9039592655876319211032276505838730439987043541540772170815337122816972534708480460678481530339827425354606736133106255364807823517193731976382923197606879318639973939709

3292358465259252548784756066958736726025074964686593205347603330267022587893997155938467034638222108716543115987110832363590150691704625433514220561500233993163622122693

1614164516346722487008908575597842867209087050350593911028556125649913517596449387846853627188016450215023465660857544006212932808424563547806671370456568519259333592247

4643444968138939640957288030077415667409023826142450163971489926715058806214047363887717765209161257743011347519545220374908961631221049043534651801900090993521519640471

4588647735602146512247941780590042351820766796507858002638510826166678848558950749158645126619309838305834826296907131706731071719728080776848480579658754441379929449107

0474718297280178725901703403330009230392538061041675484669889573135068038686174263992443690005077832346982406924270331223442533816869554568542413174368857672121852339245

5145195436629288774904004394559466550740842702513881542258012779936249482761255301095759410701557524570174019788509769995259561720868943409159725896349283592949176418770

7351188755733523909753386139046117441975491085823789453594881734097886394506170092273558356900405856227237806491162179925752635876771733782545081928110810215902388311849

5295970302890454635523927760376697180464550093663893952712937566183649877213557432275835211796176974028937456777117102096019708609988891974822781834064514963240191341985

9546667518630438217822980536476808596413487848827651335906070316166008073856450236740863265104795655555239056932413659061767319157743940613838008162286942147836605812279

6601378688103642239344928990155827756047672145288752696518498409757253114126967999683583131116362301667967637406820847352207648198751988229863052801828873740084635839

8398351191797700578244819958671237440728854255425882067330386593512348849979702623134280932584612061873144905739334295261855683158647234650823356311168990392980746237305

1858961751274620110241350253016504735987391299668773165788793314681146209300688222308872061583330968583647938117175872322118804407563604562665101608344367723134148461871

6693943558094472992220580954003417477116924948572249761131666871509490601304242787861170021809547239131794441841677024576301402297501831938044884481604777014040813741893

9545475965527353961460351911339301223133299133292565079896523080905985857406951069192032793347339726614955417056166160815429521787924208501892615890842770899908154102051

6351796459815754559688069426018001303557855470271598760068073314518161940783672212238517331773178365856999962215938487588392248992911068712777416931374725695651334300630

6522737924551056266921880807811832586731369601120907714665794436628833001440815863098631726419063330440445080582281128154889161983229327318187999588512737838729385096990

5006909581505599686237771294231761972940408139418136027240514820820737951363148433542279393342776536861035254077258399965486187460681108544159937826344218383938448584852

9035304569715109912504423138252492529540089600277653873866357415638827431411023599533922274661428089503023576301847389996569687857964761481338527462600017284007661035335

9970011074295886179093448568823861402212026281009878419477855669576706020829729728493217283594788507478562688354139604520503273390011597689975923488237896003704743509440

1262181010948157151330500526967347505159579302414192467718837799078970967419856206206033657762606620039347514589512123138896768052158583214858756210049398274337792910008

5634596084787274433460556184821630338722910556433093467219264786656314689814420038187211093438973898630381721295790929911274307671977729157837674854628928521808203967143

7850793363736950077276642667558541999588569725888941718854732457041265453872792938355542304128413701321311631686167649503182078123352807867036057531692611111413192742877

9044645590745310830304595611994778892976495070646397168145146294531242943635554886863612488265612400644093270462720991938024993538960597443351284906593630428301040581746

0612320439445967861994375881821078870552909724031508851804469767207049321932642732747494746573527063140084600227060695596769687699358161220808761428908243828225082716265

4510150765071122548776783127104898581258903148080115323482235485756250163613432445755489708401852562487855841161559062915706400186193826787191671454229784270131855635300

7592339174371504225236871430802869738972306594087472615583200288560564305436754730126975992391214093078205163473056839443576608050549539129145864564189519842126675570326

1404016526464718191682556178050000436626398681263260515677037375763228255723722403373265634399261010727774913476271761697800242024174542196335423683421089982050983996309

0096683351729155605180032912375027974138336346557054527834853747615770729786028264231698956616215364150611916911869585076300887739841331443323522161086480753762756279535

7326192768075059179197538220805055360859214370642558820407949945680532099845216195290348747172915878804337898027199175755284124951645538973669036679774585106049237316474

9334851389131076704781689315092800565250807400336026255879746587338958865045549105944187115107898947642772391413200859027313482305147114710862568442443055612645731317975

9466908304134034969841895264128081303923795354233955145164368583297176727033879966070502753808844666613249520847059562360351328876356928310404235142307368983022576189398

2210451471164975086495793133102305164007027404187229255253125550507519626309014190428815837860801872688533033426032418874199474309487098090077968962215212782040363871184

9869115568928526788514369208111989669919917593944222765985852960731611027277193039228750373527674603136498728601485908780108902769648101784192505137683639443277985897834

9967075106459160807997814986746212959429292949510345561239528518996648389315134166342562538722904207857003100505766723878249942229304331876293310759921633410348818530789145

7862130438446990225971510953784248396405195613952649940910541934802783338141050549750386325213441665253257834624105431372420513436561007097502647497506018793298970049452

6351080384880012217081134423115923301825065523290008053290609505296747617827764990338046492756424607844006224775102298409044645760882840084383965375725163309086936595013

1668059031760653994130467831860450062980954641179322276064394545097464031190004046101363637566525147633922056408539488115656651564144995469997989615517085192389641735048

616440196028228370231053195494455715777919567588597530366098822024968864962402373769741490896227826267171647090545458964342513581072155099312787001810947933780019428811

6971384559781303437591174498305039727200508838510020486452766917597291317151669041458699094829963978857884712415821518191513119364659265502246995225206586123710861707775

2972480096285573811172643504397723991591353337279437121493974863979262479245746309268806016126833281080137307606789638853035247624795772962330363206079753131170032156315

77396114141726709684579198896768233174980250090409276983988948694658406197088371187171626789505980379928186879083596767443867394110895371969861829767384589543924411217443

2607281434415683337783965800933241626531197588638631457596991628226060779033373953780686154995433439666253404137239336058376135713663148496022846567534731883339597694060

2425850333822826967160251488264837670313936615030465569440359099795880865306477527207618121415394106818553151735880056317978950810699820712326357146714531162758252500806

20098353029985987653456158772655590577677705235068007347211604498064883912344314969945662968838109923179375961515198326501205987814416179328900586332095639044967327409677

5615657921695125695757343065011177279446769975484639636764740951978746985740500256663043049693037828646730889074067621720816291000927084108802198036646903200660489829526

5441973616508087089991399726806525622796418049464541564487670128005839444300387771355707804656432991321870606322081514275062240463573963323033417129205307955186159422189

134174393464269860039630400508002042870415729593735309957716677615251624524092900687484094452448708296372347946634083467678892429045159840508607169534168927276857026100

7189617528775251605114575792348758289688272500582964348325725019488704916732932741441469386161100821207301489450942209156113531966015060095153004215187968005009256702774

616975269173663358847895310487516279254152291573647418488012937060082642006574185024170407054317172875849453496755969256058106800091735532339246553976580561150626685714

5888623136877825027075077938929852212898376757339444061238960194625480149686695371667569202794500414657284567068530994311959103987237024927090104353255259805011571212263

284707450745903405964385528208729191184983955166804404872154753233912561994593028105975250334168546567651849453998852432775783503720834777412746381874054586392925447115

0555432279981820868187969864701046561341186800539651313467232381249441404741459647131100723335520281233852214050723861536931660894605930756758626679075053701611040703409

0926282674887750285975237510238545267796205530602752777436781940577953384076593719123390416122993896587754050420933165780695965341258929276900210583534723405610651455542

9320091707450439470199566983662770224702288304266506904551219438370568467083823573011816322854340018430494979858262176841561991791993512415309491580100793747740403927636

8882693841499375805717517134475581593103274577418757637118775713224689227224249337777113957558484346931432056193522817599764563384463523667932694166365293829299473155416

0262278104340459712825822426701976096925423356800253834894656311249517014689940046400911880740870773299643030944043581834084141758018478995026247389569075478192665221993

4967405417288191105033252790021585459431757345887327695863977413383026542251574881272858188927328497769691108525938565217239184042582582314470310647604648145020222102865

3330865637440225958804713218694501443668175385307315892005190657548015823449154844317770817671252046119649394501886256763741453768645600601421512612924246459673064284648

8510497611191884672049866454759999268017756643379003400568470375447562878249348984087140077023629650417075010454030069683112353172934030263172213170861762029604344548209

208

543486763695249394153847667538095773551214637499080503829564670902052304673985564355823807120013405179535817759266117929746014869213202630600393179956484944447316119372

536942062347563745047826759397321270337280578134312662111738722249280633363983451930368038925930968899520925203527857198099238057623612284960652327697097076479039139423l

4021333686903948086876847433080122688699462034708237163097247047250281060331416470144741205421382403081283471910093086202697490915360722527512865977495195070144683595703

4266146025302815338482727764497900776889807445830108058076258028265253984671218499044394258918802188062975995631428163848511209471349952422729096625621310572002786025213

0271652435730201321995053171120419333856243218115968535314364280988660109585436852601085293446378411885371826272165074541415429092412576342816834642480358183339986607357

7309409498606570661584407016735064684551083448304103407143306886135064816123133500844233624141744252203847162068581577800344074422408997737952507722722425221632527079828

6483423936031993600770157814548549973279149571524204777183259862557471172176000475918686134657668007497138823888252605950369614563755509483645524943318493757958344708256

511202977130548905985889360490419843661556802806391016931507941239844261414631452720749063421674666562208207338740544102755277326425726662603254341416451806464620135910

746369048140913948817349150190905887658865756080546331911539579519922691510175261135535365480449153805235243092095683913392366418602184865740565313865858918038882901004

9544105395042819085217095689709852226604914977034425634152011335435956016208048334570567858284751828869432645728470941550706639004202184742841481507293211425552538487682

3984059640369467558558706507780528881184370543707668539402043667908794727576997674271743908241142251517023195532601492920532485233164430336141058134808121187844540168004

9841863719537750791225098168839594398437434905009276907430497219732166826907868189429886554326039347996027915286663167424520499357683219759829614090609600277500754116118

2341365951954364768265974353872503433557560760309426459371768443525658456189413304362658549333964488136678141899403537812518636180986143109380468608423894176571709198375

3027352946053182177484980676410072780689353948778562080766076462886434975781384916754969480133361645855359234218744439048728220232748279000438097437222409055665798279254

0792018271917640731568687889908698234433324298461713480779415792607894378693787932181927623832088562198256262053706157053365099866604373352780043029785385825777258208143

4930689509974690284212321480543625214410992658451320869835695037934121927877015483766846258848514803532821005762259531238439549984559771836617746969487713373602949024320

1400893695435246428467039062829612925333175609054580152415336269704633415609477147038783311408045532720266985169165505458088526234722700918568381152913251607557514732983

104817451169011854473890500270796628135737332891519473513171507142118196223950640360662507937661011410909757198751950627907671992084156183831262327870536779495872093480

8531033700370796769814433398653194730055395503613737169035404412274418482508972554341132914140992101050279406078134345274545748910397870395925576771973032662920585002184

1243056110147117266588588758581109801362242719178360150563008450612609580351146221868970392656006767752271278115086918240931263983238895143307309208061818367053327222199

6712213824392494133845030855374291395100606121406401527729920472162474690601996875361612930959443531967021387026224274713425295198346411502152585217190733435287605089694

8985966187372464714844575400426321116913066108000759019927685277231229433845375373445561927073184207700352518881985100009105886970246361413394637364036291676485024313959

2226108914312458431080249185337663654737954281019800639467549165673882209372067955022453974952793604321876041673712529418689041678756050715819184628397499517761898470120

1417284922531659976424913961553463045596737003669827139244746172240537756388452506057383117220633099522461382751543842624841830545461614239767958075757928929553582315837

91059988100013635583580253819304960874848440623244213019673128572797569638089158911415106268293671325339044322044393512256271342583571937580755554709952805865092748914856

06150149058672218630181453955271543377441157484301460454210472371065909374589455601850747065551250496207976349526296285857419106111986843374359599027675135124284216278738

2704073617901822642116190565452355982134650098443468474989342551941906125685394712155093835778399733985359799208044091401401301969205884816241657792074385061659298842954

2875926765325764612786581365386930230643249514872291568341948933977325772383110186072859921382743995531670717977869463102412096562992515636907070979765746223022481118369

9337954003728904003552813838791669645146305017446678302677339156584851313373322134555912016942819944635586911990104670257248863039143131892690272342788076559567143550810

0852332873676188833143306258402854461381790916491511389868861020424410969493039946188156434692259238227154237325561865763731911138393508298447375787630908881906697487503

4626160773620105614764919585888952617573029866000284492088330986935638966693573658315432198051146302918030353282391251226514579604511291362048141607426348368904872534774

1460497928968665471804311096370206936618003815649264601763228235467525772510063481083324604963974581894856250715927985140068823456587664328616964994470287617278607401627

8793760031340305365372001633610673119735402165742745526613772464146797016342322109483927352053992161846684431657805148578748738995611735615742292710799789378304117463423

31723123687062988799395637969323438140930773031667385587583365894879219229299170292532195313106313751699564895562792077326334439956069913401234566186600272386440912957

7681745674556204269696639791492485463176155558347884911203132980810938202001667082142619538393913519494332455874157725836948116349214048473354670485252762669909591444857

090831604238663433552347995241933217431270826427571501537811447046489614779234323918836597604173276069183964760678663226749291933231611318776739191327131564491305693705

8551533950582292262979366928009889012737440110729913075848391948372516387525152681209356155068966128156527850437743856737065968657129040745040213967864098050162871632426

6422673376138215295623652204021188430916924496201670398377240229007751911990173388725654945167024884231446673301697956493138953858612398111668308257202233372469857377872

5176730167468852701156427758200593935709812258690125889277275347751245969545250389826116680212877573805636856315644421994581874028106568017531855645652955822886169552862

7420028196403061439105900321537979539693021851942326880794685279140725877194846904117641691522747211068243909658349368174072439125672601413920553750443877850971869061283

089542144509045453485238152261213609136327962561871364143164942213935544220600533827345153079867230668801356293013165017655377047163009150624792123739193705754138726037

7844094217302591125049238615493819007039732269827084459330938371631480611283411794863130846199594307896849641116870843379334405700795264780255324299882702122395890715726

2154491883119891179649225568725795187376437265210419618862359708076370825191601974822344820994433323650040151033080334509873420821279214120042048018050107979856172321647

3504401369811485541058811972608739539425490862873783903746809883208172214796830743130269381543636845378457615575201994791177412336708593571769209592840288800006291720871

627379774153512958050297099442419138076287230850635785570200290134327092777298737515761662474840473928515508631642153028208352650155756311959083682307340304392715181050

0275265003370868949842881323568496500724939884740105694734338056373384023824360725388733091113738840764500037734478470910018648045541171100256140540878369928869527413926

193790851385229297895810628398059044992413872737639857194829128448347659740140414325918850245391067059176832462269483973611571814898528770650237921321789269536116379304

4646771025399165684431444320202981168593661665925520655979568482691676299777340317373780308748208117848767254586837334246334524199414107801536921241598778339974669973954

295186818200906496976103678215280989529876069957103569096028371008599789898946334872095708021792336657487071467736736097246399221575321940236988042560150000758404117573

1957498416740333035260394809737885653902886659099013584031964803732938986045942039868966162227548943655076000234581410708961509394692645277394235114694011127719149136925

4378585394835272648575521602087204812491691018527179951837560732844266771847679540711162180153539879182477498161858356464522803086323997713849928359209270555307866515564

5276895916218692034701564555734028766926384132577036016059645969040322551231020295599709891641809071291457489862443029770711389414628364848487157678567794878143125124329

3550247687268468946933878930968619372172452421687392718595032641171831981357089270756010792991851456275028662192436979841781141404540319609164247843921439027833868372641

7115549439630419405888067523818774291085517880020788117679147787731136388027728566969401182727554750200093592740494837691964174174743764309935882811490241585671721186545

181954032744630850849001867213652717887423627513368543825843572609765763398593536487906202036888810270158500580830280052905200592040995400955993382815605515780805692775

0376405932422853821694589247308372693348034515544089081803000097135903138760369506462952558119323319104201739769685110785451020588411747429566742640862762260667721607633

3832609244311716326610442814595860625069986860747754290412444646328607846279208041631617188773654072051469312121432516923429453543324728241729604332479029063540574426435

5101657618868875755289249058307322668857940636107263471314512803909679409368497933681487571329869781321192473059949601402778186483195060983999490292484822645196907669493

6680918529246540254800772199089177291960931566054867802714844783766043906003147146452296723373821018725039173155245869955423886136497929893405730911868982296875692219878

3526788848165620121799323557181193394789972941131625038237716074760122507890913913600736981616449615507711562475184867186427418752220983699262511079458767442710260549101

8377141493953846017308993344936047697330192877913802703135070732109818820990421774795824467599020083571168178448830478739147534493519381401175208813705984398465495570550

101894746331785135447806050024532228986959561098127445372003405068433161951983630318170374789862663270724406537943927775061138437739100370156045085424417182946223230987

4159926137923103063979751962906214954936751493482955342557326734057366254563878208772478017012519108061380763294191048376862061550399017105775493794371981498602274574327

6873586654721193724736483688847386369550472304586168757789926241799192428880686466327656215489453775846426936601705490410696791396585650472314269792668892085692962878423

3165154008797079484044606266020591420715726414911842725772544518517922522992289989255241793132955616293338761041608491996663174423087794058873508394787307283099169773983

4979436847746344801579069104208387495354926175917185412293597518990089222621721914459306788619760356221881278063881322456355560680694152552978974969052102555871166688039

4625316581559026470171799000399024833401918474186111779142508795783220037431338999438878790174932178328570913622593452398799211941358296486042387182406741709862281219018

5733368156857032359430919908413100327424163085607040183967645421975659821525834228867964906175332880383375925188021355788935119163336359125653076196344688595555679031931

3426753383306631643406076972503374612045657857542749445605773180947349044636421152998533043536983846609846137669343084617000549083610123916548740215520180743832046374678

3904999935185678107922897258746903677495791338535035133607550732132560029191031551310728317711625077255153125730496232722086022521464840220284279859329283668879907714081

9239883785035622052945336156802820375313954033176431589405632531123586823947289210428147770904529570491287593524120513586876032845142582734694488545643193344560610392771

9582129219413108746666766592457491385391579446355239886906767996953556594034008392663094891563705155183329394000503226417088488994601744966590766868472286681018342347335

8074816786082569926361457941434157737969727619536251481604750464457139352852592175783952785644120276963185951519925370647385437507348419804585231099524566402639452567114

8305359686783111380040816192340692340326098970410456803793483601054492920692731323910928948941612528772541725880715769800290232769299265396722223125954237897818796172297

36923126986290413294069304779269034079608593696955308287063349885358980631703790655102345550359811047226307843243788750268086652866619701235843107548719344699613511024655

3823076326385989468574353065605827530017359129830695156394194337812121118044070153615314384579873166672036184046659229140007286157047092318388883712590113775110154672811

56831261731358454449555693404016062515912214452026304007316786242123473984156636061570022957579512596067890949180714872199201151338603342012490334326901903152471111117705

37674903792507047710800078072437972199974867805127054386580977383660845571415922313112507699438505749572084470616176464289701673424853130726531131411329692244997325898699

37933620538231043046881167270296781793531514792526227715087317841347201175082919918140024385165187293321922367139330218393718018284319234620686122487712481614464372623844

6672273871692704343226676282133595720865759252993357046930167137706293723161840281281466540339392997647994508355635293134044814997460821573143678277060209036200023656147

9481796166953389820330364315197605893882403131045864624539039457563768870592794191446351316565638530341380551160738301152349650160262898331175266540895142764548739436429

0003104649756341864670633839370264043271999344450176536211941839809140543543120466110710229491479572822300488815098928800520298222195603137155531918072367580873515739494

1586386346592426492982712468742873788710721177960934285282218507517612031632878650509567859784696943970668574363944173341905638350440729438621752610060673994465714452565

0821851108112347349770923533060731560343833666812802897283552949783468612079306536662298488684458973023686950573180717092710488020936988600525942346855824214663237148360

3347088790508346751416318599787318213771136093379679226713048084006357127404476190169902128050362495514649375628572555347220155291627360048420717450788079507618459136069

5778824861874794960279482146832553636964805579475195804226596181757545537257578006385007882894288015404514728664507693636429261656146409237609919442302530717760664764609

74890130071283206700958344414867959642884426565334806269850089001435859793432049547007397901976240218318645022518735576819538466957130700628882926375616032787642159655933

1252924924221613295745040653822016726239861866164858143429888315192031590579600733663059264478017682428057937751925499116138740816551961800806191448772948511662569977244

5150821371463281903651808773347375322139017956945546985589460995099448197662316058273897392430853104944407870847269906403178359352904613848224060814013067392119403592658

9229119690168304343622797153048445980731590371353990852305554135850303305994263075300844749731213292281390041493729257315810390367157343929813723660967707037239975647113

1383650360610405488716098619039618499361053903230560359089527165216882441207600029377720542005493450599593284765791445138948041792035850852529938744500649077050320426246

0392913479186052964181514073543083458770362016261363564753671667605974780085903142936133379293319175008485509604389718110409887752004921222370695585812496257133144760665

0693367882768887310324454243982890773510726773326678439323322019262193656446632205485521963628598869821245905948414538983552431719342491299006181463959235638285940616902

9756476311262291466746536626582357583297221661709968921215263224954097253009276090968909298153985477810475460397048056409701943861124378321613301721735201695386677893805

1847299996113301753095363920200797294384334427132419629801071959511894080207607854224753470765972679570860437380133715614849302325710198132874943896149485487837329709427

8163362544363330826939905579688104948920375875078545376405053657575719555719402413921082687609088769052263686540301870822701883547584723999228940340790795136796379635072

6344224554191280957705903158772555892207789691111868653692807238302403892362710498072888312875535071751133424809692876972015866751891421714342557639048959469858075006092

7981004390924945240817662509527417274156016145431594518522171855749552684752772716738913902901472023505989549615741731698904285539944028302783837762365010859093691920154

0276948681440882683921593425318596862968867707355681783603419705184191167919396618337297000193822980150272483083508010971773091110587089489467230542892715118246415440910

6645329532393545051930527899093172811499649272530603470215987196134565002063107133613651786165441649703573682097625837494386303922487734357175596671101166931279192046208

8982875388710715363136881554667959015531454623653372771094192927484041691314541970017754019166194192795232601888253730337752336012196171161255538897835617670837738785260

7563134205865681970192980420505034509501358373027020583244706964639692236906385933110811220139677100067477245536145173982465787334364145518521246858040728801878105976487

1776307978933254645800932156135484194842177917559330935785967194699195056191293167993441194597294243460001028579758994049694365666190975793295668054246701383744999094886

5697217529786249999444038632564887331676004076776907176911021713324616671369335776775179668748237981197416413893296651098321913188912306128832130617534730645943263123690 2

1942487650442656800640372373565200124319482373191116415861199401623456107055588036662603162216687847134896499772571443635703187500753298600633308289970512791458197779206

0780694205054949268204440463425629715753409492743557972516015720797972660701996918700986122418896922478331223069883519302680013335158234809746911378369744867632078154 66

4881263924082841865160249149549798644272098660518203571764568949935602043071048929581586506395191730563855938221751430131434807598669332650487479903805646835095656163984

1602315510479659885931684745229784532711563022567963824580570873833598486159426999235303174720562022037261527089760668460260887447119518675285250056178297288871967524660

4895413107910513002573103926269088847423715618309916354673745723994383500092516391918101842370641827844631996490552885699393252826562628242869076046959129338888 26367

8905258896297992600436583512928591660168162711585038595099200450238288052578716079991485779511741071458789278928594455422975748926639149060612255901846742420489830696032

6099241605373198039930958031845874187561196551694103025884274613267715286041675625997166890091974462057087606193824714430682007869995158218691523480946205994733672841 6

1874380183762446843114622719971835404199085721267006752205570817790802076422387700830231459632437697248430228137474801499459678297652451111647562844948823911265802232072

3393967248735348379683470903721727807182199304579617087484668322607548311946463631629550461428918187034402551606610996043968227976008510510903936291091994211938826655136

4311045937739823375470223489709893823833496062224144588181571744869858007680176809831354488908073098010459298840671012861381855597791311265857946279763440209325404642 56

5232144878549983704521267864862959677235993867028758906282699279492148880889025975297177478906720299367123678763451986595708011874771798648451089825195533914045264040275

752862201560983909743678839234336869029379490238806297699255692023125042770510894350978320237026090787722102888386917306520297074268705923543037688984749133115084 5727240

8927276852932025683038229026985498309266427981696215482643789646128368380420732092446348106248237628678481958108547337889172165033709317162300827835446095355000157087325

853718296069755081704358349922043478239737270858232696362370176097674845003041090604011687748131236415722492541367350659699997435146831130004790437633894638115075044798

362249775689187063312159825736569343060981010581075128320284644444466922583587764541076542454461602377782788423014324148376077527228666458631768687572843620346387 2646470

337810455803840869900184700132949067201629106732086656015272003570377008772363933708461915283204882311403505825354128671849769189874018370111971124740816615401897 0145776

023023745038112311099710264661414041857261089563696058312446625103344217698695531230691835330056188518487114675653747128425378375360270025404781526769638680028149 0670826

847143662704493834208868297956055915309143195923053793897090912350168317523156938922081723667794771797136271924145558888086019038040749460151095181573219926316308 1536727

786395339826503769350931962174930710360546827463851923840108589380482153796057037554136341945317910214402772440022859505105250788530056362548796203963051416750538 9028154

839893826056184605969254309230501184482024440573533399486462324698616427152552041418394545883390645070021120281626903764372367863270923848083570492855665422565700 4171664

346794270566931694659035590025009811020461599706922269604303931341501123852020843307834927685212802322911525197379137517432840567171957865482009683483949354987063 3410451

091155802399789655537286716719980635621882290898597135945659956583900719908411352900703212798486817378763769197501596503476210492679280029798828161442624704549931 8735873

796114461220750417737680384233088995124892738217191170599517713453429945729725215214038343046073402932129297183599023367167551902048367988934857854281307409171121 2749135

188966503880595036488660019305177479737720060385947944314665010410771923517224472581698761357315539473606361022608171954947748733465753076976238292997066659381130 35566698

283830832766954761096688653181122041132550888898206279798064803011217227923341819607984129997207200884183938722113903472473108515327748366983782486796544853960546 7332451

782821837371290684884322609750319063553017941264823895351147387863995054860224600033576913603183825695223930439416276778650226367159054260821794162606278653413781 7872842

038156593007446364067398966754926876493957184903132121136562023902615229840628730956481281693035018696850371310954723293767472224785729641708198589405216965981052 5337889

233503198725849488937286407683296645338400341397336846566429981296207453255651661075482537381673969227716956365366825813785392959280486393462406800456128973677765 6499264

445275561622239943174891109978681413003408786160964068909193446601068578173996496691929407151979770619673556327808303748676242818953293799350174377033041267846394 107423

900040751986059181564659477586096999941255968962286988157890242657789777920452894572685978800151203918786449716049993622646124958687721434771817782721540331579087 4383331

809450293538191572814541832682112481972325972143223494026289546957476210108078742584765714780160883340496255465063241474442371159678992370589827766568589771945792 5992692

904083389327233247930254203274120827944369363547913979579720396366395782104958416314403965424093860475283332592434350078130659491794990770388167056714485647590147 08193628

846063438708308730139992537252983668910410331359439434771325452511125521110927987277859513827089163665909520741709272521299026204554103080348103227000462081993134 9774103

396993570520081349069420803787220379046438289024990240122138046339798024218210683934685088493016791828966213042549333879965487438610932085149105332758773226902672 4129296

6256736459068755914896312050174243945263893979024423237032649035968525782137809651504544989318660685590975663239442261822295196564215725418090320998044966138139531209853

4796711469934269493824514966747515985290047518052766122600719775722496708151505803914346182401252180929356344269769077593665452090820250602780467149336326778720581681597

0250481840076454284365495003371942335655317061100284423030042053052952933086763760286456546110382753741547484910093128725956442056976109993020881040353186694892623960995

3565722335747574315878618459048319664295632227261052716481983854633794370836572964039488820397486960694333315145791639720748072343344270697628656850889473094815971906

8311061200102867523095201110639978597041881942784387319179548374603671903556930383994015483738186262489261815817546723094662836205565121691747032788728457031534544485852

2369606979868892249345328209693435676167926010847238262589759905263792325916415032763625594607746274170433254493457444889477216876271827727072947967992940703725682106129

5089993702462171998898944678766862734579413522640334981775308339939067029665213380346983727289053247600661939254582065928970141526129507274262922793246579170438371626932

0753650111960155752594694061918181848737771342683444424052930660057334458886905888039317748451177410239735877528228438595867202823987437435295921155624322438928963002729

1056872887268161611773035695272316977436959291424844621189894575011690312957425148284517441987133717486576746353974745761595416087815219493803821906317197854636480687724

8861810391894489750730538558049092079632148308935231848037909066681345271782335322466125219499267652914275908909226211751081746705000595680931351952840080439007572617866

5775125745288433053553174841429175337424877509489933543735835955457882706037397391292269370301243965689712377394516731859670419317393074231020539449279372556695143497880

5545703305908340113045524208837745301823471483685405703839803008349014666157562782972454384494736538947399853428754328227478538137311632469929383670295832152967629316901

5770163764597073115455682766349019482504203271623554376160628961010317789208130071331349683386548654072526999424381887465482728672722781054698999243638538389171009159271

7082304906672765961623781686404410858757454793666754385969867095547499965912023647186302513423428660312083288725426148465049133391408455713489742121332627956375141 58859

3834370232883676361427210910916326438119930711805813705320521818716890340408422833149561397141000910171509937355016250498698021280775524186200458708968444383063445989495

5514400987651922004434826887012701983039406922428539144344376925256906037856331636359169597563628551168556316245250277545376196294289043661945896802391851580679143860655

4576306665385508979167196722773975207635829190576647967180388564538818866035064543168335512453205238283278272109259629794743650827334190694841174771358669416235 51515418

9766502736524377829273750109051093082586789846241868494722187945092867305656472937265721055693249737530172030034298462939903576040553269480197521260030669803843503995362

2458649675717353644866229608676650221147619719066836700078786195257277249607749530158270401915630348962763155359129352170209298150999571189777124690854467614485083505413

3339784826134395314953719150241321492525701145762701032683591978855164103614737647596250976223028811188954047134804246179115385416323284005437108464269095036218683372877

5644555584451321260709365088896899626104066097271490158259265168477635062365733527671953832978920009067298565323545222748165410194800740749839188230279326391449423695735

2889952770410995339475528270108194335733697184319508165751817731361789203722200462320225025710195997579524022444777321462083760083485538773062739020918868352510196397670

7841850327839303456164014187654569380004166672187859301833249780430668413708789977806097087513122457331792104776532196322692920624441186024038239395826938487694 3864799

21582750767801607506536356019247816328848950673931704750819664627195118968792595048559814225373534918820225232232545270640311004505864934019626832439967502709419772579999

62112631549818066293540715583610274971906518427406565937254512574742135652740612551420873683195358915340183005643761475560005901887559432489987342354418536298897724641 42

9112985184953106090530703685290951747470466217592782127028442720276324221886500362829327344812778190824737197178733122826245293390331056612313694376721597019056278625102

31496508385051784954746257928633548467647505619389560487127247631535427130602573246197073058891449957628661080519401608773839935947596879342063061649761016289384743 78762

70839809365286890936241353974223097404401233773452835062258300768194953505737271247291463024293420118205594285875409672998247743329952532893889102882623850029186860 66223

06007695414534401415437802743546527798112480595010881568865390958105517925178926168594761989018512885485330019719136580509343086513733915671442531069334558535936906 80573

11121352209014898432261639643263077611402495957275755180179589419401319774573422892233099739196245423781531637399205324766455348061014367306832579576051667436473662 02346

21205488325796206777946589615346666284962251255998837366356154573809942398223413977857318118526694509219333400278395660522190434390795218769528629536258345114288337 41880

13897668334834519923543727595099724884754998534821287541602121420007167425273228186584713024374038012472127577155173543806869321781709846930477213869343623935185177 20943

80919024767919123501634197498300194349251439227328399895275284543098006139755700791417081678257933982580345050330350435599716301845528168292642279637951739982625697 2139310

34888695236503388767235345917921388311578797662440444585686266118761866077854423457825562175139151512175069970282671214823537616753390299724794386940098439803372392 60825

75914971225249699909162516822418830277064831538112236871275612260858402325217728238991975461696687100468066683951394054683014706632437280971730852617500405405846357 99643

87130602504665324509851371135047840666967408120622808495247082736778489675066868066569520461593590640327826022810236552083797774909998813393057249306686665438786938 362894

31253517516138530476569608483426892163795317644541891627305175221678972080411022372283886209656630432693750538126058074357155644252030153606598273724463194200272636 84000

72903913523216097806820898002503971154135638074184333838437755945688993432757328763589953934333013215225900120838600512520109318668826735672604998799535122658646076 87845

44884183383413662542219697146325189211728250097412198388943799647742461118287564927400801095806810716319090555440663768419924830303824453861204763918078777478409553 29367

73126665062304634915420945503013186992838587040497769498762308681601198225060378407782933137148693195576904124809600289428590147156300359952118751134960928464643882 77636

68164442908723542365626241849131109707115881107599568488241862765942931155326435533657810786249366068097352567283282438804714953365163044632204199356237763659235469 49824

86122240436933050644547086698381945613271731670628721129223308827822876856611293670404310973668158215665253095319273576065755366338130811412615041827425917915846860 9756

61711115535926504724528901397973074836568456676376607503000803886827448092560195250182287786775116835185139009239907351059703270669619154073287289116846607505209092 006071

45646383935659156554266871106258607999663404577588827698230347449917712874165892377979611704433066549908819497037199281218530920424550101018728097074432904339482702 8863

20072929682300716061300967267295626979183862541923927466039000712109973496105323355847256759415833536503896957888361222771610209908180784994235623092109652020074968 19097

02336820464796210938523210558762150886567676848354321163469821573876550837832037338143199007420263497842810116848957540102189754507098326654276214693339080399046753 11152

02415024832005665615806359792361819323007628827294668878379078853974048969533931470031131324493093227705326130028850543429077890063403191002285599374199590545341982948838

865764191083821664299146114191054104072171837577155061513515392771400877552001228409781872581662708931273964546477259808894904638744412033834039847054748264843421660150 6

040978703871976045332968159456039741792770386611691754030605604275947492737317580608753112966079880717230218830918163103554699678999676814113223840584872150451109777541 3

09273348085613994313891956459179137461296122743264902894505828696018397663266768184863672978442961082457532735323785581012799169576075366163284457154796775700222592039 47

90124564718859527332353801320498670716155091587828895672746134396154959024810526757899163956156292280024734147290929456542144238427975134894573060583395554662070210 0

141027670794584352116489088168624369796568234197708223313013015802821876841167102851911373496255504415650032201321878072083632156727583289411942930094201762773431074932221

6301696903711021196817814596112985080356782471755722595233764640402399244994117133227064814092208903934067741659079335822479617612719575790623216075333480442592524721663

76532812491737879135545453182838865387075647639734081624444987933614312318569653401386442209305743912872763381581387125506737122428830099858186321015635349402278310563 11

7033107671249904005132001293489702721309952154923915078590421402689313004609865615236140525303927254313140978670372367159813508704144415568474093424285806826691887058701

3314646320508191505624847600443520708075408782114949462115092792335641676736833501642284278652933928327928453321515289204094301200081708618584107504415762168102606083356

8283697384319713651082936212468002579767991155399907648403804992817180375653459518384595099340093392603110508797537641335490529395708765991342899729770181614294760801328

3728437159059062879686640047061491784659514338089797901747228822130531415145267504796951734362334726153303000930497426565394579474740788563667819470875812034604862122 11

9732683985031983988067512355607212314224839768206933579702545411426785688286857621814616682467550295237752661408949262179410234215163541177570269072944307695709089606441

4965881671742121668318114963709144779139340867791720363604718337590738200996909450123084402978243198983074299124747509659505402432113462983344156393868466638451130417168

804680082835017990969544557428581327744030140033636837871236116275322391852009315108695478540406038514296753370514492285816823175467578599332489704331947481163136568762 2

44209211961639847807493990632550658610472649946278570911848293076400523023957169404530229774843375344969347910427880464975509169928481102733559380944046934895784831966 19

16191568767866474420917676961460159614300711876371159818435709487634197399138085628617818195168335660513197809453225854265525165340525641898360419180987756475470093335 45

6463863745881837071930899277474777651940071210016021292429042884377518855696937984137461959487866404952885179702994403417092257126983643477923420004450897240195642776 84

35740004691806885788296382555685769552433481059235369632377665412136136594165896489360126908303912189079669346387826994625689894323842694790019544917649079925967283332 01

50204055056395822883229865421520127390385712551158338946014788267961307059368446271407317663585074877351536087854710599450815737503768721757586689747637142040858593475 5

2037159289414490384555188824778224888605678179484488505424827117656020412562510816987302947899169290417780732082029453912387288505780471150279434067819720679870667734 6

899175968570170964221498843862131723330140364840906229633661397312051267854801975140106876149786782238295155301447754388009420919581119084559317284191284475424593024043 4

41560468496036522322070391819795473902374792889430628755879895504346332729229426581981899384964339039017485919007454985194324377468897151635061784044765817263836980897 97

50933160686709027362906796736528277031546320116423755537799298474640332327398535516109777781075212622968949860513517165602410287103772412940827807555891992535075849671547

70214909146833655432231086547487771386228875760810079271785897902598759188635196306045666633363192174079445303345927730124049043232891698863107254908590395013066665592730

11702603766298106832918880154007740068222930213859576454235684172436497530339103424759546679776970800027375943580064715248683506681994620785001781035428128258352865340395

21232796603536324082231808982544771052047503704252264797228699159145224300070833200074295977322725795037652993767687202659189314667887983961876650840897212071621470802505

329655306838233758647809970173621775251826622594488975554791079002943280737776954120378881938575336245355575553862151372157904856451955247827238390439225555586085459833783

24204224899605866221584236888878281887503287720578409778708991012397962235928130415428146206709469072943044276373570795194638240638535397538932514553204039865813187666650

67180128552920929028138846494499131489621511096573538273671105194612560704832112062881259687496905332546516609855153284705020721844897915130385996182707552535083094177881

53307371333348324728777479051810609940065062184695791416090258633376537026950336325159012400610772655185040857437205040286964190345060154341482587482135948966680516971622

04129218909013651942661633491015177709354187823411594434257301845846047967497734112396736746076937584906354299939745453007174309149674014521858837580790841009395282512239

93941887800098000852983250117971552469662980523935942605334256683484171065964689960240675931818730076077165696460027491847845392837597739561010543622972283307967124272595

81913381790783409621408277302609459830241168133924254021024790829584271922720912310498777743600822820404793982383576317324431719483156971330010828525340178091758465222941

74735919734937221873348650377566376945451734805841274192964806237884746960032363296456071875008561994006290196361814327696107910102484774499481774730369751899228135557693

57501446845470381796435604192748202966411484264755388461609284317366732609511714145514664208775937211606640057131871831940378249303566026421145554655096381561717598833263

06606354150291012138175740705460347589437776573433474433136145706954995855206815968719207955264670235032289325886926552115837404571767926978698309365844160521753983979699

14164692130528871247348215268404863354416036667164545205728789206539039689657009883303927822831246398832259368184889730076202950191392217469639190081298224475781030178400

71241371181333474206915380637319634203722700135312823156128209273078733306573188224330352677536168514401284814216046927928006261490423726475295528980672386898011246352600

170892236094195142983185054938776422055983978423549606838430804446309187389828110323261749424902459296870542909598932771886727881814902205185942496497864372201915084587200

524151377330865901634373899036909618394184660480476412857748573396024328884856159481653913099507843261527612427437304198132191310971382332353652196256565684132100997793400

65867125309809163123694545655240867099025795737378690735707957623330415204577601513883455847419623747926673163943170811046161492806358838918930129277650436642892222952480

65961964742501563936513045554221841369811550560226456922468844270921908249138797460468842621535222232159695297204600356284480180514309235146490648315581470733739099094000

33063516284763645243070399022890691032269063357760364860551940902782680315937808826592838678858928333981443121074324210574440779725530487580754382718089738160582946051040

83029383863211204406323798531018120096800478401312104193172311588019894128999509449051823520285501747845472762059863697070009621505367367800710401866180814138596278076915

30330859735979222742977968064432368930843822161613445029092444424134286820459892391441005864948555982060284922716247787026995589742281427014367258362020191046924111432400

8113656782388531661678230591013029577237394942182206285322925329662810562789429374661505175320710232540395606954202499821431539771325543297586855252724801325259204962363

9186428240229505652917173498207387727486453474499266638334680804728431021137809271950366939837088898079287353281533984742606450074080844329450210486602352792585313312965

3132245368773308954167066148363110682779419010528654438552547588213894308783875546974389267645496621380722884257239345052083454215664457739359032723197581717659160914992

2300536477728381273341666225338414722462994241192462240097854472979829127844039263998169698249831998810282024020189496060671263656610074693970892064689403357049238092700

1070510535093856117942730216979882535416280152727203897968351604236902381883598872104029201907105608751001679037111105179391713754662368383254144717859386530297056462626

0948159605973111282072557182811133240676104217747759645483911179713618873474878682539845866897492106177035032173667065708216985559866053152727023642929621060332762951293

4921752142974883617498973053872979523177137651695656076009410257209652641367228047189019484536577305232471845768564345313341809126025754013903941163886109277630735614771

0429820371485588099882807701182076886043581805556974289513493920850270360998529136566724200040934068156266480004775036926701006718756549830267840394977902849840804112864

9042737318787323574923351577792665464058755270197331741525534369343335878177439476676986534109034241828856004682442717355891956299625079790827374560677653849024226242434

1394441051476683286260981169869629957329189480313093836578772030544063568880873921733643168563019749588240778565646911005844850632217485602786987082544923435009462428781

1424792557092875881603713337049894447935541357861767774259300350199487881893546457069989802388428594340235293957359036387799554858018444355982316908424883535500656784002

4862288539779059190210100832664391404237858347141539211252119664412719679850014037773781611395469455626393439637229169839703443161802263801853507727479038358387760381375

7838647012136315206550285043820926854716048049446724877051115295639919846196607048019909225438759193604899417764243237878245127509762457528554900919496634300563250421582

4785039438656573470302650698002722496538162275541258123830022466850559270192809527663208055323913607488598549525769899507927614076643457646429097181815270408434116864875

1952915242506986869720912197272764166399894803529381557206103652854299804227933990984630926287867918884474582281838491541379025757617305573721909891733580870609521181391

3922837017330476881808509917109050444013024907362722374529981247942216588118589638088129678928602717350247906136422666697096556330601017905052265542575044259197988430962

9281030657817347163705682133316954675385257041275570407558625868324966663999607717507142454243476380734993509255726525019287640864927618497173045897625148716488591598955

1255375228222955357809275515771773494254494064636532864388734334217530702797210788843935784080519477769825417393212928035281904328304226608115277615033728093222216146277

2802255172759402589140495678040276685368356006483751211956556037929089017497056882892495376800055117044514709040276477095266126178911643527084947663336330375047626681813

4938316846983860367178618075709038409994416681888508857156733750859452381605828850594971914115773519469163826632910009363196937626265633997143885908306240500221068562483

9375451664538977952645501454384899174219683121931401372995118410097509794199237468840954250132123647670753289568716490593510289684443170331513881484844754291156551495493

2398604734701430994530959296573299640696179056571891557139592252223779966163349292409269002121723513543080937580132161231377545123487290335614609148164275810499520023608

7135985496480127596933372611487822048271416542861097287385314981069732753696855913268893124658863973778605279647314577443870375512914383631014808801025549750185056343837

423264942884038079814325557800163924992969085284589836391329251793706254015022668173596465759859416332244151575736726621923042569556601862213826190880192581203723881546

00776003859044903886176642120015712766244087630529283978600293770835310900439186588120008743195074102899806565937988870123097310069701795579926044738696361160786515 98376

48650167943285933431962812865551608452919059142677920493990411896788778786674303929975761676465546813424735632581831341226626900374833698503187200117460321357251155 48751

0227692201433444170414759365038920954549976990021428049303569545892006008908812289808480549761464239812417065357454304376304063168249955435897677978729067940476948112679

09513557437820175241646753461329758000981295779099727071190393638094497139562962587268238358947185416563847329242758695098427673777219233217527653968660617728237159 65708

21130355810363815182791325694461193091588133531942734066542884074159970819492402918388497625748315339375254667365082843965540113896630167639046301783547536947865239 36554

79832094346814763378984065784724561420759944989774128751744945814534657383789979602559586269280739502699277569907760849202215589743467392637791753036664387053071485 25194

70220957089282863695049358558489320909558686670233457554319254514425181288007849915141850033013801828358291916795101513506325891573568794876741918382330615913768592 23498

33225083199955278486255059084867455046951602420521525235667638266643206244011346246184835538231919030789790489156492881675694951297606022618515957509091442772246331 35683

57223509941125528091622824240694113232325689953200039030202196968217386130987969800721311638620390327297191995557705918946517770931086733435970190867546375077749418 164213

66245871228362571309690425221424092741444428629467641111573339026068992839959142073445918210129878938271304750373833166795787273900628348817212343279316790078017927 78205

76279424741963951177750945739527443895863353347367966070550527012142544299479080491346447357381092368930786035662366461150744159419230799122635805375263596249358838 85499

45357860888234906479016662455720947882310387009991721086516270017412776478931430636070317268396116681796735997405243721801206234819404677351511011501357530393553613 50398

38765404636303172924040044393454222886650437559521679638559904741481073663576222693266433064224662251361964755994793944751674283039388495387860866631365660876086740 16868

25424385950930983900950824741461151827971514792538782363231160391015979083705326761335653050108925597369478256693522289815420801446620485501976532321021816166219530 34657

18412880165026448317775303785757210757216727037351922240314148705332812556025237909652855171060447447911806731970710020686033490952336928994351734601699107184507049 0952

86957774131794120530566393158259808009454159456847401673376199344464418595248458578067918467182672145793302716286484233254970208681474069158570524830142625134913013 17931

89738382452549317175403435106305944151858517993322891846398568766128678082982141129066850039256207747696005665324348516251485428270485614233397106339613269314052482 11840

22802087649328246007079295186711877074641645736456342222618471812428438354882660565417559049879536936295956649725411837193356979846993998266970823283120991093412559 94808

19873220386864574976150073150130803594050406734056012325709787469629188299464967009553229932888316237622770234460841617862958418100330595177229060066089581303058313 13955

88588048276225962517551839426498063120045127181001922219497057669748844596592692997691620797266423414339698096085014545299116867845278772258015085742859764318050407 16225

49465121526895079761409835692430941746765451817195667472044498428603269680371841082593827335584974386855805135952284552875963138602758981945310017076089442733247468 7242

95891164782188536209845829468303040751100830546676612131694944656386623369731490536304878907883287404207267338339692583482813533324626119663972767295769874440365471 36001

65916747714238681819916453062722898155773566229226610897177977150083462794644093605843157320637834761951700001658106021009287840443568206520194527028568264322176071358160

17522219734672233277802743989435971155978038127627806526046703857508556025560810516667781838263691621120275954763575092735610337565179769946577949596114491162131167 91600

46072342568134822109174704161025408424839924042350962196912636812091643490379266934922254635101740341015465437528316207061053908212266935339714144670163871339195724562013

5139204050918614732321829361951891236454920882390497224882879142572991339778222478186521013391414371601077808100071612966209368046726337703019405918478585965573088936450

78579360008786286866338079763689028078061985701002322447725140393303221957205696718642388023218633477611259943548649924474751661178360316695263754160436006356 6323871059

27935792175687119998228111108914246461015465535659621270420999440319758735153439833319560989417309386547454651409993893976935539224586430358876231576156258615874 62872818

78133711235134587883785580421697764398525992789049962429065388962158212221897887811625838296590736324849697798761200713068188337251900370377403487225042973496099 34766072

8486400016319929606695437071427118319214099244239469582566542054191451204553614236488414816026053774866149864110283759516290999622091232981921678339239642561390727753657

07736393725119822973696786924689151371652648697596513443576712282685837531440126280418404622869358897358246373787848447481641210733816758775922822307915498221080 47959043

4831933876343307339949925194243372136321191536581087255275493903497011795310565274368889440854746664727270780909808046997309405202816129582446065629165507696802 35614513

9785999535614491856857949608990228154099318962273521074758244856724526195100482672561527017230093439430290688053019264553530429770999782891887127297756731225080121569089

8921046206103032654195355883084346331184232432667935024698940574739104932355738728624978680751440498738143435118385255825858965084861976534830516577465354849251847062358

46361128961231063475613487829196208080290215418895671695470578412636065128050970044865339456921267610746489021826015136002176404207593430425154960471216560638 26907266881

0603281420474122008886673494151799724644431487852302819566773009243452527803547237133249811211144752719605172852639093240007410085041013534953404377507086826929095896450

56777697550518169976547742490749871376897080542223103073699862144213444970480833388629036192462099417006595275523499456084088835277119951256411788748775542690 69537890166

5837170505976068817790849116185702187460703920041121012738663425845845123645938488104904988917124685739269622180648113410347799910285334504336352935599469991079748386073

0938322241856554365049764945838570957547589241810795493776431689626073436458913015274465652114578614155770992104933437132765485502078197577561301902631627312439722426741

4660169404353621126647620668955718147149979122203905936045317129530955540122855108112121026572969756546972317828275358932526338331110696576581556117359486795475942419029

5582012445750907537337046035362261549710395443112933407783239085701809046135796288485527163402699650631291909942696959662255894979756287212877571985424196734753015874634

7073305073784579614535454252064423087824084140807846279673687388788958520447092790810107121198779827835924233933557556193713249979593876098414779898213818 26503432240130

6385745448854935897070486251427872429652100656649787070156838838645399590365080171114364104266278696724141998326547334111284879330934395808667890754340702026793784774345

5352656651292901343372346289780107820115522188723937312016093709704068632553677092619190039894188411459261537011666865895636731505221633041077984886772114941600466211065

8605248504107954794838142514049963682497073968941442666717487367463168517045635775424231051505809864764641739106287459904737924888107274308142629524890931985 49576289832

221

417190896199838410018154664437808237533284022544160414895993190873008472472855748818376975818953479390880198458062894211236113335463254004046653571731667323865207845030

0166195770809027132411891654620198070895546091872385437261138749662997666582523147148188003237950038066701185989421959948218404892260458854876888023375940142157915429528

7123568227193941659500062439118359342678164048354162384967924208087709963757338733972713857005101721639010281400619964029555551092051482372788939586225063582005762355

0460670033354514888308803206990796123184492379733390033115058829483802068767102409164859588883944324656344667574633584953224141651880328254882089918241029406831822065117

12210563581095080526760824300375506483382815036583507743940163080243748097984829566487277021341700558911396621800935230435395488825590266709000657118853359523301307008

190428286557679076216907339408560706158893619301348780259524996703159143624421997785198300437984441454865516590828204141139376731810408771959657893608900600929857187848

4543162180245386116837858586234510782849182169145864330972476771131495903082734618524789221595763617619415803413031319184160775944101215743941774531300010795056442252510

930673752322368854324279121353055759262100311221186203829750275381784254173117337551711968165240928436766301402843312610329234523189183702230344454974141886397398650837

28606861694362513165598064294041596095582815279474489407960067040156066354233056608882241092462255818735880428279346966420627411245975215220276131826209673609273703418

06804296209480151901121564632335461994960497408344435453142115006726825166877950667254182169648683330823372445854924129216731909818022915171702763953907659608450159458

5976073782236371820491172490303721728475214768189382828585241866401407091409917788371395692161707697901006019230526629784219510099218401232206291400604821781949015619479

0264663131287502908324448371554662484940386152485353273663548381024000726950375367248133315649581319529832958248945699470154098386358837510527445038754237416350534548439

0591298010160337024065091419654406408930992874563503612249158486023751328733731306177764383349114167974279563535826432656509343897438769712540489099543976103094470122301

7049596779161345460240920721494649710448748038255096475579233784085015704696540926099375292975065756324454786401781820868997603550120460528987523722840965641540104304874

66276071995929003518139082264656278006917499381889757550412455262811876953735750525012728015922292778267737194613719216516400180711103260665465764572568399041112655103

98984776200494525732076336611879466854680756555778816168336505980479975289388593446218272748276975735348116703851100088120600268451219316134987292279735435145251620662

46755043513009958875998659114434873302760578327386061081960325678335362521398537214632713714373366852594289784092798477878664746644957883589668082045147767271907031262

07710771780875944659287949105828127513178451193902192043621739992710417481747355300551496807137461376682610245822978890268640706208716918408059066840210711197550731636

71232104299798331921659946411876739047383582397271020669138686758222340768371402512878602433601375456956121012111866856952757609838762420131806009730015109487704186475

46034719095600164459132581601108870034240951086006596682466186137003404540305650156032108961119564327940113323206244129685271897339002930438752682641325237281181873418

7266831707982236681984925517311140484292263600497298366464717470320358925111903145526378203697482737644474657963470343627745210955607492099970685966310831881197052675407

608418520990745264126830901434767342985506555504958367106871903849244388501716241895924159706438955179197612480105118326621083951809191931686471500762285491546332100294

9231384348040570977701398274451934839221971306082370165463372568013935034810121226605114976878294628586232064347412112626633527821577473245248312413542866044140191905637

14456193361673409966378271496009780576875416953445440599479396181489149477288393444938623785445710715026668290496140379849969979904177314534985205251546802967974626481 96

68702286164231214776292412429658312763661501321599568755630321383495785041950236693928204408381491115068656842809603044852969738253800241726769870945805588238766750880 46

99912381136464981780323271388627539998143064746124047754171778696333861396561788596483176351902365389428609879813243105964107502022049060393598709159547794213809864179 1

32509391514302291823591945291150294344931341236215198182158312032029486003940276690126632207066257065165460052747522401019923873469029975061163166192818509767792260931 00

51354456493611645965660284911284552622478572687470611594626032730492603873190984272702230227936371756711926858364718578151551335203402009425573257024135674979298326066 16

89237477237909231564483699398221906223959835129277487404921884865148618006764683647476349043546852270441815032947346688059802599704514719086726975420909271463724793987 2

42908001465960306126644166022286040507213318150829460988850709805662279815499892792431405323990348617380802149334102633155051103722338875148162043988162961449937118414 73

06543972563580373206053741937671572155162520262879124347517695662745040518338716088598437046467204976925718626306817461111789650712733894134312640042190022842688463221 92

49602699285376809615893194890230425833901286085290213585525352287293692772573112483980587096220893068466462636341647431789867000474061566857637857107584274947429648579 66

76254876979594106944926811657656925791063739128091743329342766008234744512264681724479134541128328974457551465078695658990528246653499093871511116967951536432826151198 27

89866889710931016959104178450248828735231923844862422636297349757684392822631120200047132187201701783710975756685537139382475853381189805680552456175892532901241044606 13

86262529720144955188459248610761575320195237921069318224655177285864226604778693839844215191391884947712040124803222614617113859479756568591174574724441827556436675503 18

61737105312840615915488212497193717302126743920082041117549854683639841356774608838167338674171004703304512237269963296753350355683742559327885052848439709953415176590 32

93840285069219964260910899068429037128452934549490793498403703950159436563446313995129824581333531389648303954685377742386758238799595073127916663391213576229300823813 74

99508804241436770634118679457902022252519770235995448842923046328754923496722087300742261853262394415846865082615318656057785769972925335180744046382897304361213124874 16

64955773205830319852649228400338212296198294000357218896092227607689217373213756128171401278947680860184617347613358304779955557011084658991846526471602906243268230972 13

79199995026002007677928428128010218904655036864494068516374694740097967052287174665334653476683285299305839175291925684229461333033502992661474903553099705929443017566 33

43834322304415434370347646640492739502658109126488241517806384955847321292598486391433780405356376028060686127816888921524835645436191658459051120653701944847924254620 55

87901558333354325586591018391532755634325304791374070246658885855173264155785108271162140919115020187616175817025131170079441408634863148319055296155411318677747603095 53

99998608079384374976227303703716974916229200218300013533918129991829604023592948162240375734996478981565262268246922466182266003346563155444069190919446133592294764761 75

30984014456964949854178741721231360779557004623157574070164761282734096389797697740198761601941576015291091092531227618383478679518524193706079791655907057515180512954 28

31018535917318636529202913084205780739267574115631345641060048148590659777277558923397304760176118610946668393831705136767659806086454653275384417193324821035003461438 70

06425749998217425218212189204246983459694601714541080967173547847902896490057093695658507360279962669616846431118237198194995401755550793724878019693376504513474123702 32

785817170527287540676808677865733191918065414650702346930410760260438076498398418776243353883813581831396332297830451927354200012443477014391410202805835137615248683465400922414855558907372029219460949678309380472522542715397156446139320717562051057479473825563034044998409055189312252581517064469135994549410789766052839379950219126026120477205145883687727742039393492746617423169661272448881420286391420227812419353329742169450946795452067395697738280628009153419557209296208702178162395731539858049405991309645978436746168803327624713133218716363946718093667473440335755526219772548442024999633931748616681168580024161019357181587639139371591531076342504338406774911006239888597954611355539755839653053924251133851519729571507256719493159144543886748094127009297425772211701976780777554115314644415888813546404610034367546339543813657379417552022983704633781240452763115958787429154142041620662489126162385007770392863484726233343506441746226554888964328960847169212331084463333505337147173330331901721153077481815975318740320652065466630383402472404436419295851566207720195731935194885916298153300551052799525400109233465679859706454051429106571050430262847937593593880541200846812071657259589682779436299311192290462874989338151616124180754183887659727170003193889665336565273596570472983705556510026802792385167950336520665309178435468005322211811784568324368004432665902546285945991038547579652091234389503251059583028451450194530989232771984892807878455467496436275646169662618364866620367155784981383986825287619563857360852041933202576410865853082078346093735426744174587917981650977606748034235879437881661119989595666794464838215477158445223575130963086132598323044566468192097250293449035786589888840520552887864065938982709410986621527371617524492691264722285626744317906706605132503314572067834404637951426017333495920626461812732873779400130153571573662376126852832110391120166194811558779550403940865122360419497579357189797408115563734672074274241577377404441091238485558451967364835048886830993138982344212485495620233919819006035489851804406713580331408724132658155808563355292356505562434062217386358710059109166690201106008519210620615217298898708361337945882584198929728135937846386408146209732157488764585454529056485693476254092899106691922560246425452910014982009451475388690585078215749901642383258266123330308423601713313301974026436592810516974600611297413499543611415077890155231616362758211607345219385511272300896003399371087363400847445858281169291728401471592927127197382253553963987645924738369326112803742994013875808176175069360473088259160765499660285494158397942913044178937413498128512994331756758244076734176951024244331173208054198122503155476482578701650865766702309785271213432609608282808472227355161279802179432486389890602925191544808852944924913238575321489641284456505833651534398394353706322369086818474891634082930268567197857966041601215228834098644950300613637984752788790195341774384444672748004363482761808233991610087405112235165967744517830819216702751251435494699668726687370827849191991682025903711477659826068464297168288715196482705657937945663048499534258282710020287457514253134878869888080123918252754748629359202579750028182197510099462917837851103466716091576507689424057643488232454106255171455768523557408154552326601776743209479015645205466875325651052514797632011122660267524833213955089931268326349301368519242840309426023879053232047876793848815781799120758839989511824201625492439937529250292683380891296512472328149902698230237888614435318987992150720017872694765932161051852405076863648912351751671937073774216243588562940623594770431200526606256976250968421781211488829880026616044059222329331624176122908743379022287804561701357723750619521603426862806290537864968871393385712562416964079324475831369885918272999275782949295751304825043666028532371402054964473380738245775582570927007591535821362248787395

9806447536509228338732197378945098948812224316665010698739616672989920964420596817569619231839866191793408474257868154615941458938640602296132950038120038389767450208633

8557826679881065690369990815678277851637829345993619433669806529792215221536662888399402680386218387841389549979200722893711695077570617240023448728986838088894693258218

6337823435691207402895687188566870960612738621934987326322409606596069917602005453603816589660214387177128755309837099020713308471230391797557448381005068328095118932927

2191231654940906640214568359874463216265575739792837308702860612939772368385814919939258415742549146335154820414128505256116414384738621579485025909591694016701922227152

0515924463847867368440341019760225485058596203745202103401958672160817127019264607046079599287131210748035118825068233530449698126552095670808845419410225351991313683529

1159722281977965191751091412574906675271987992843990372741064588671698118350910120568349817677310095484699100664217037510129640279526680792620134649082657083731273088770

3498538168830180415910735087802789781444525165407708127488473796550331829893602610515170900920110222071001669479948606498841609925577821033292542220238243161693794552445

0771166127819572028994905923717876307137916207732805190043595063027837205242860716863197568319944953596546175823793319549221835571406382172621701189906243630164683498799

4306732489748402990062675636866393449866870295746329559276358227417390476736646832709872540082565827407937301342125015787224693593360232027880213344754711692547244278823

4788572020478481096682495732594650693818393289944085629329548452346954732470720957681550003136286818305873597665524456292333709800392024465397081980880975167759000829452

3393382538737975166366848481990619171892305302932752820522887389977798577574644330667384284668342381977822394151522436389874249517010656678530267666437085462406145075I0

9823826508232921923116946593605521624370127430023949246000846441913713453739088171054329719962592178595836890022753546934419927056354499446464846356921147395454234209303

5095619125962942760332314028381564195812399216843570592611552186436729390811404988429954013503045826166856151191924704248067778748838713189871896738261924739749089221696

5648998157671704289017449666205968008681990919566483987127996000606600983366508501317267050667807381053625332404361561098011084767554948774942365851637194652793284979905

7701845104909170153356861363244389487965903436759034956002166426558152444392827872271735945948307893872024847342032229005213360684605312929409749759889320134905015546639

9788069991700873715889175956776894727061811503019647289132576784816190919388459773052888173913796341910139122828818689576669158175066401906642575911857638875482934362992

1171912710549773853730155778381018844418606578305924104527243196692276439468819023029366036893929143527900678345492052288961178860518754083104918089177592609625711828832

7086436346727818162747255714850253575093560194453370570429727931651832436970736387485609272821695707559352179828293176304039885438950572501794655341196643840611832817122

5805809313865366901632341550423553948803977100712507041056787741620258599084007176820989414486746229922762892025508580816217315083897553884053494279190567344876604830970

7910866592252931017597478538255714719464529629087845195651740959478943963376856887841336334073539072743766372205280224459160534573754066183716958052421718003218602285837

5322597888350188042317887568940231975197437444613352597457897400554662442432497593440495376268236401505734726953980110100256582513119575389158493821251296799677253621276

4607639106722691844111059166671823174812066194772280535025793861898710732931143196219558359007573254549265440534485876237997048696398212906230465260237154569746398218504

0240616064721224738628536914542275828158930325786719200381523130703012314500162038355855970846362887285686618295880381514125979242712228006580721753759956097530281681324

31908826758112139786458977991567077123341300614050720737874775422713347772318791387780604116028338928073229462986109438994804268776309041382820082493276437844569466668559

69135097280929656029683784248319063766489758940229747652337370707590295732296764107444779028542205710833186416062834832840393767133814809415308100383634620986740923 14162

5772592601642413107683838536096774393896453881219871847087835760284657585006662643013183563775983439423256319567388921786474251153914648306105617618522661484986211 7352993

0039419626797835115432471979210099902359950101850452263362136629541758790411552911630059509298870937205111995320951919761191111178565685631458237425273363348744262 17894373

4255594438827210995525834404807815336301231817250476112098586213911950381276857684222706022880857922802278992701354502689828981286708469480858690773687310488241352 092533

7799281517828072247304329560570662234561899656929407990430103180055883851549560037101536288339329630838861183757251344292962374362569028623990818089667478407215416 482153

6466985251118356093769538838247792682054035562293103398234617217749928761143107112618698176716510102132817484322686049289996213744264891787470800521778991459793683 257690

8254470499557365460738332954455037605455616938462993526955598254814369522745135159635012844381657623878219028344778419434849167543322089886572510721638012574559205 006261

3832353331001746356326967882997952231221335929559877147842562152819660095824803979607418806864814622218467350238464996209482900237216747151301761618348664856900958 0445

2712924136107744855016454016588210094695318516709496532028368556343942755258676230939902626468802521002348398810810313959156722167521036403116827698020447084682751 021600

7652912859618123289239199898376154654015288476402389564008009111707771686632584715188865213418100963097892468142777674442490984819462072099118606178378827256060277 489202

5077565496092241537218489819139994930435616869362159642917710952569509706990063231006108564855544823176816949189203353825938397977955201780600261504538466022348058 428066

8080054097272424887098899181403017210837408519716844655506868668259576213176185347414437764098116897462020371211318615031805348163709928051005793939581839605381572 799053

1735646204725646756465733752324960442866627542283341194771011586148225132905747233696454593507786302813970350269335586720254206532019113645684278522271130499308474 015508

3205534502207111518292503782458415159542385729092055909315552709371573043650713919706627072083660506535925780753879966242782962602719086367858420342617942729278387 207422

7039258647899698885201728543732963891417985649549871231317421078111745847834714810113022006691881171390633223774639144278713501339101361465475823547313216387797859 422925

9092873266039806170514510518935261384635649007582348632915202586551031103413814049101573516178860757643964618834101756544414874774396958712593182068392921811680883 559712

7226537119526746391354728890901025277742990668168005319870623247556963164794199431877289989589073717447913822106916830314430210883271873693722363715982485112777376 58543

9522066498763027698123461345369762104739754844398066496855150018242872413962930020784239040770171753087400621103886981275161331107671042265609534206562796480309195 917122

2256860508660697491958719528511720186301457223111255958058030059590451780162091560033927158135619396152450582940942383067522310487602756833193139537061594056900825 467163

4076885187380620283937649414169522478976448274159243938907385870336831106474829591656363194075978797738936856825365656798119405558294689075620974863970515862806160 495538

0199298790107269885264068694896100332327284203406188454212897943218868970727382736277850137270114963870884664078607942655552555485664825367262388553019570089909441 814119

6819278225214743673023757726479063021362700929435193537520244824650445028177473530313055878404289052956536166955757460304468400201302583472857878603479642966228563 383909

385040595996338205201313459775949545959152824913675826557737086309853431031656991864774935235587698561676402569436794965082659516493618563906776134086344949610923085 28159

575947324469299784378750695779652340662703489343992200393314202215964047530798772485471928990319033113606753874099226587614582952454629274478253060713708610093255 56109

106290159370323457846510942541565848633675971714103682469068213645950225938235868980342052145842415621097712594198860517418460981052180233091349159305532364021180135 3993

827390760960188127085700221614994202352285301197851514825409266951856567653410404445485488177660629153334239355883040977080382629431894438507702390408475017104093236 8683

097904497849323892155985002143587007713428715847006930247167663122853028391120296158483322876218731270254402750986977752418671972580398456783452867233726268194259137 689

223732798696369957194750828057429990801609296385648875774366813059933003010651656716864331160038178433180947698249242660826392564722108563028212285843591291142036032 7206

018252323794629310925410251245417005166491749697850176586800128632544573787252932671277451624334360127340397859302221599617751736148666793967656332219514934298376790 3749

308251701618528493383442425041832303026410055781883185442893640874032039660088923438710023409268523884967322844566873657042343156698938113117085498055633424110903902 9402

069878836686500964163691705281565858356477475504883191164843064598996632703398001097062714315487174370481121700620918608416245963219627581891687594715082363689276171 7485

163145845180705435637970723278957450538758447100756455874737245671260758755826316241638301758948123727346583284264334984211990679033276995061878866730634490328278376 485

909168680653984403931713825696659592365734822356876095046002069573673695394353734489287894541429944924229654199206870717179908202751123228830206374093324308284076802 3659

962230745072395482913251001462315228385669963646481619930610803501193409855128773081595450154979098326100700084351632209140971316683905093077706782579384915921320992 865

980751664776274042102287069580316910197690665849294116301449041755241528407985584192045942224722408579452489966149963129567459931787449783413501974760024855830935569 788

153697313632144525114082928418128045024919929598456172978864983652967173740653502757578465341870784213098057357585098708923218338602766809687867445876739370425061052 9455

934480000337948441186910343848981992721780045698408825618002774024697155696343535370581771324966543170795487952577664211206856943407407369416521104530170777514495029 66420

150856734185613308793690799085988819541774261880314414174869352930128628687697963497164412421773800196909748627996089460936425306791041745935712831904029831131550593 038

611206192754003474299601297698456728568007574786825685265588055044650282472340621226723098765095246795551167576075518973671081866487339135554730387177148259924982090 655

636246368887442816354759738020092703372797235762058520194887311757364152085688799839625539506720457656370866768684961673992899051663954734806468841632161269623214043 0043

034978793765895525591261273349443137493187558515220350488771542061283232155425010369584201177706058113108574067216884473924121518906742976799528434604620851042298929 559

015388617177785962596590245374799644805737542590339557173690179397516001998758369909403534602006006114570812972728649244155588597502427490101997527856958345334494325 0025

780204344440868289075077439617367055383761578786385387000953573335902594668119751237389838726653687955430018415044807205276494457025799468680349949294168674710474523 6313

650471152698278105520599626500224544073428713991949802538333850588539364994173363696431899803695321146317461717020388070866349064786340422458469135590424245140281425 9720

943368039404264695762203519760525374669196864684057485227322214112634682007326809912835968043371248986512484713338659995815570362412843119237138052069855463052239628 6001

69326092476187523212570099594164545017597913048315852269009244055311865815319784593140513549679750197159130563640796787427438869743181215963320242453695090810854010748 67

45322336694887417447584560189776395844902174934597104770315419794721755903104895515071303375092264289474366150114617112854048983628782321775540335581513089008600231119 08

9283171979461527339639814755795610481654721822820928241262244086617316118295314627011962136619959410879358356432093296418935628950752183416094956286605476082023394390369

382944107069073784215937110084355080993495125848055614260279488811735778231410921563097755633435689280906240147043040680967454142850010531291101440719318106005561952937 5

99440981612645437443677397892845582361680657305686818893290555248377738869788338482126900338552557729632948503625724161794560688062505674398343584068858652798471320568 20

33268015840118612253710729945929721983140398249546430136414120937944647846329673080412403157916714681071721546579759508437906546392689441643670202671733332142868727930 00

67525680895947246054007739214366237747036693706479892806834363066623573549188363067408969053454196254921859594829635299144250681242195784939762493626997668432011717830 94

7897645342821592110544192539567389068025874295234702462536272058624452991614257874992015478349251604342385349243843410303807273770776570147437343580798451121498902138772

611307493125184849712899174590950039321906256807723932545455469176735002114278251591537139224752151026195751812555899192237756055625186257778715204024235643008015440647 3

78686471774548533756851330395773050554298410274520848824563801811743244115088666941720292251387140525933292189039302348495217837323533446532626937774732504109205508276 27

01360107626880573492834106150143211791258410093281226749115294969194414057983354038200794920527262073123858332785887564779056721104165441661470761288361000624384130531050

14008101079875557731522504246358724208134651707819681326647305265266870097539010235484005431910303058505073284856662189230736101609798730109604578786272596971821497774 59

31912172420855203832230977437336272600910791708541590654006954047576946452693527389589089465660935532169427129426114018933681755821233660928809368683610841295931368976 68

25463416180797337993114349199450619767414038367395910433250837896097654632163442968447504718087753879660115946910584698569346315467151967310540189434725327351012330555 66

49244625308979855259889634444538294144883825709671236053389198293136490349913321228421966087365757136943286363383874965694154477107061380343673939543299455488946044328 62

11704228102975764901084613036258098852044456528892538656183554374669050757948211069811160943622728171946884422390136435558225224330140937115484011366213884108247980900 5

42972492869087708639433835697580891344855483753767771965895936587515432750294934011636286284193306048171093923998791900884203487294343721494612397017034303370791698316 2

45766050613624590549183588052452030731298424258801870960784918163583376321417765526484660862674494777513116337460985326156771682160131478190445605770892030801852154088 12

68882246108542068433312797584809219544443889667113144461789314741293665128198790259329194565273687344836398893389984361211680689865797567488516548886337690035375291987 75

76930810573551739514437952727038004470490072857309252626316730990740006849045997587871320935333481479980729780306859252749215403250806206296793680290963657119655454749 832

65755760946724722924089206137105621700979339927932066567094589212083904984604758648011445523278136028145344579543873365991854029550601001878962582320644671450963980891 99

67566146598237014128736646880385940326652224087508860528841067197999140854487007293022017220260304786380710886172631415313923748994778191781040775452555369360545903637 81

6192863942002266964803975868226345375812853550796206064636202276341001562539119946325786788360870252437252628303301021044894326226275322073667652910916281998887191616786

697698617210689500902364059291757218594584763068921247043650275363283506500433461831897030508283915358506052517223442293318962943725776163152226873950058135915637995009 0

7045720072609698883738753698242621863149512139883571673563880630056290325475145199661817267782078962727991656377480299102250947244094198014686901886252851004233665906643

076501671002367873518980417508647603805560882711984788639116966057125758111614332203162398619539960810644891151299038321892449671151819857985027704469785184162807329531 8

752217073754278078367450606084368877693049830230414364689139837008269396406069456288164169529166544373908475728196961464119571580663688131248487829600519236538166969144 1

316437821280803774582231914250665372731866015855576310005458819146086341512013406605686183053991092842220972227727666420070998758231590762943912951563496720838098972472 3

042038732783508601474116048285204200743959607793967665745457415734381437629295610953115848209200019968349227462223320349229787977209325937345893185303632200218523329226

604326327773869929525440037460654804476294982724042291465600529045986614905305330341184232774713475287632417746860051035196802595048933541617738765023893191084066521380 4

667466529643577196052289272879058233362671718010478707415097865327441555228155097901531432569914109132995250276991641218479890353417341802888078594370474700374687166980 7

291369878105191348231743199718175697324716934112404219323832375815834075050325121024232721599976225595360817163965955459215200626349383277938717450987669553428787897746 6

443638551706502650448577147587895166390526126187671738754045938775924493697246870221984680515191826034341463515337517354398346584036650078008251337619581112539605958941 7

188047217874636046685077559560876646149799759072572546693095018122660597563383204512084638644649947755674711048825818624518114821602417311351155337639418016198608293258 4

828502972156412493543549370147221837149093135274651614042298840347258735848047100304971036786199690396640031890270141021874714397393047966712714696745248582196159044358 8

575047471161083558276186011598867992523277670077913488062706784305823037684432240837285577758508216273267775521575385493139188944331137097187976549309906304370838121079 7

273160414688741044273294030727774363782884423977594872346291732896432364895042303033952547723285292251820978632941279227761069976483964461549803039103687476367007442072 0

134868042097834465670780850085124892609127081572378968179538971471665316354179274132493745551049067708830538291046609861801335492736471578823175170229529255747672943807 1

843235282789387873058507172169878408709360848912758296731845035233009100824500894600016837289698553478156770898606637024379918187127137483594244296364320947275727110418 0

443346013051070221778247291988409544429155924729679314766416864679909456377046036999887007961285734670508717635799267264190775864829795801509614971798643932311870590230 97

451683435712523358744257165025130783843967812489541028789968672155583518182197672923726750882719132592890457053921699623157341359810162606343841974111595598712194855707 9

154040991260843153449472936182597146663520940304301794361263079707780953877079498286453666763263533422068945343030626974857296018808465649020974995525667340133802823078 8

629806818705205412520401199904304289199124015463064896075523480019294548752880557065055452354879178955987440250913007416419180939972946822103570189811867672154904495028 4

462599683767825168872707929533499355139851172380491169556661244188049351821195423145457932954973290115497632797004257252928851676055670695788818916668926496278266068428 1

788855135682201059864864426897310404206420388620214159313343335650607976483728611747854101318195388209632637397818675015670201035161382762235549907816708176258006326519 0

907230973113261264539480612746157639746972903819915758063074587538517334833468607608896462270121404016579559790815513643179269714327811600039509295305301566405385440144 6

81956741689143950050601298995325206242564025699973540563356851170626312937882096557630557832675616162922170394518589959392779546333735205016889846436488820731461399285600

15764619060882700521838829449052835018564065043364175350139349857051006350044232775532805166325015553600168558606321618016788228898592777398070823443012187649829882195007

64879374527324977571646794376825978653807770909315826869893218567541022181137066289150507191692405571751537297985467954771694404608728583402007168255878503502580689802790

43094618467258018625571769991436669256776624788825636710239050892975798248952227094184674434146644119119762486308867522691737956084643416367635588308512954865371125074490

03732288271992572715199660002166693895605038279519066233710710296452611253548201808162340593161238338327872154509090544271980320064422532360012498934484363675937191423220

77851596265768452530758485537358308416515247778499835560996791452905537128993380417358033223313804348260191619102805347598662385415120389560611327005564966892813167555120

97996763366535540470907939888668953068578101733026605688536895602118077216225892191992431173048925224971255312282117588225282650923582291225204138370082863879599408753310

42292025537883192590178881759789430772711316048915678508678373881223628875855526611865733674460226117363328802255620686149584672266053779357525558360934098916382963659980

00730784500136994588212051974271662951889363661788724519387988391498507463670116146235590918091464878247698323775978709634559315457068005528207062946431076384817183641200

42844883222416345306417776492806003167897001913734414599529081813008273571202445417814606123726714401875382252745351515524245007907975368799187115671028453031873265631120

81918735814205042307746297225403696352335748306562048561086409342738331833463432735127159264239390049127973028883783463744236046441965815843840531910829032323939150063700

52537770106594292190766393180810878765575969070784073173239732355063441358556746002928122819448262638691858136721604139546331879072561693081540996943760021476848236668950

95833864458433913973419577395445894737996499396501973180187582154388185804924401538657071887678789060589434397057243968067662330777502154247770823779044126904120760661710

75145832906568124019088806520592144597223687602261730245554637405620748081399377467009412522215327344884170636815244358256118696626136383402928664497006603799675017937630

16763918069437678433862149089623561820105640614012377885098835667084731143716888913942684794853876466509841171954337028921584835072586019765152460415435606762746411781030

79580551105852750942447296557129458895445020468225672010620986206771821667486885596778133367304894138883039656612191893305871404778455132876728030143220992705290210617710

39212773759052612468035783233612231672451314301832827898769529904655069884850334348398335992274164810683359631797050404801716357581169611089887526505039455450818904578200

33398880552733617405890660762567887602344899655818219507130679827984374701471305670367804002790888262609687517984330621616836497577339489618134418824168865022899678081620

79756256922749808740208876881035943692992039572656130506812883876144119591862400223644524800394799942440582531726824651352094895966587269366348840150992837598464647534240

03056155906510544916912421860787811768003899307609069048350672795121403003404594882920845356727163295012007112146837654494140706925962143328567874574683801853930454624310

36985598732642070733736209526825330022465956542110218831635555322102232583541986992666491353192963187823490149158700147749921910894650128017261865842411995774784463808880

17923589672365496158227535354996984130293949320567962137577698966542096156183935754851061018736631402673250619956581438409588435459271042752474485532015362900288790273630

71511709761157510447444857500232585814856078898512835095561221244135323878162333181656119292576209991851687924286242308017058600765585346450992122138629193200629166671040

05344413253099405031484201600328999231910328401724803603264117395877373643931580547563446116674421959053416946656368004997460891763261639369972680056711919008111164600029

60099906297666495080108508005158670385819718130552317324630173528769303497898533046076126706915198052104187921693861999113136828410258438748308631022755665240828141236688

89519506244729375224036690315921818643234026919293237688715177080767495238989921492457417629918580433486296060893631106258100141380630601231494362793068733268768771147445

49611824196603717301173226721154889414471580764346364476457590703340858879379388551175194674235340453941225240645707121446590466567348092426153884175026364996764039796640

39506453603305846710659160869493642767063841875250763968961560317311929268633832386744633518113308307439130354337227907141030795282271658841103444348578821809080282082 95

54428122003801952660218635952461056660631495761270251582303335224947078904661050788518613260070870531252100924188140333110980343254472187535643394322040465954026928504 48

85564614251052654795847216630572994563578827156077280214821750447870011247793657070567309892113872193059290580864978398631943466325792242834020275207962010766746046940 71

70560953513333993760749271301117650602224078447808214938396398812005477893800566578035990431148710163727703521449472844806598021024629628637432933375005342109840248586 09

77159460346265070892757684803201836190549885228928095376821315150435582517203728601695960958764251395013822098404961222624228173404340280893997262257739330366102986819 92

21337757916373560345378075501813255961569355010328329942384997475152433810011519501213178050138796465628491543324891943371832694701926816767596061591878897636526032086 51

26585226424524119959018981878845082876944376766318492238487992413740076722940680731052803953540236603520998420550430592120388275565593048083911659730624501772525278790 79

88546850842585551738383833851994344288915912253644118669644171242400135887960721910613494230833029789663443088271107467000523629974326102318027142266222618750572543969 07

73814742635221552448324008043756966990710294726405178030151618910268009263587698184180130345266471055199507316064326755040487453177281647974984931681635188811325461499 64

03183140120849999754505654405665114835838717438107104444681995736346286893002713717643069604147832273275678903050809576914347830867035401616202818491144123200843999282 13

18184843322881342551248886686654485270842304284009883138554901003794026484476236363753646511055008102940660991524814792631730877440642070953919991675563931676247589083 224

27072948295443281512295290316475060981071517094932166168130222008489990733519284840900143323698869379175997792387280564485863635604716964436602044525970486822151441978 15

91232274757877216398657527609840899889373777508937404065845611540453448982635679449628642247071613265874995955834400802449400525647531127582945824555238339938861602167 0

95409039509622984369753605794438167782815663517171901560867850102297106693787721490988919368454386725973007974938110342945358118921302910485720499673562211673336635007 64

32627405882954486157569614320478963305917252500296559546804876536262815497576765802778755902367873481625042453157027418353906499723714308623953642833379852588093656358 08

78758721141673582000237858579171465416121120260342725573842215518015874630577677939529367891253297770050822625008193741645114168473736572522677789087457996823734527732 73

60629946429241699671550862292806080316787715019020416166302049350758837769066186746162964701677056344183089676266188744032970517788145243401251222679410412177617218288 85

08159721642083847933339829566994034959379072820339780087960497013780702940145706183227529603485302837192226100593956449912415092778754613668231461289469981672588247118 98

54476146641359747240400116726464033832999040150263527121857991381875183821542253047992153889028461653792947236379633347931270846642722737043541076853791213190343311924 50

6345207673334380916812903926710929875717714802822273307085859022591392905258974002437571021699553265761333518518766386027619970039805239305274289337090169102367520745177

0169640472375386382876543190430290357981930446828632045430189142160750516996685123364451883139431581404652068503559767528406209686484001463298802638325495627213258275734

4853558300022255133185962288649772494481966641528190407028797109505677755838364707508929280129921465508984652700726965716889740132432879571982172311902810990922494210691

1519427044773587520266021778729973938043291783216346721288728433697903169348592455772175986332169229101312996493456569456831267284809584292509355156153586820337367220136

1285171957991790678887948977874155795078582804005198795143793102409735137542445229106658730078654625141882080807307192689839135049253775437442026570165148549039037849153

3578352391950918422941007958179462613046216881844121746806220722871046251493876491783338925853594154399135800585902429854085572504489429103113066841061052521529436405894

2822561951509029885349670118520896464332041879321533366847500909379474586244050094419795259305808470573044171422807785657037127947580934562908770479883469716932355169605

9155129039465464919469769565801044772122115297178854242063014493599903647048816869639454598739566495684468008279740648593976288861542063449595204778764796022224814045187

1122057621282895120964242624397691077791875989150916967488496901404178146248821899204721539789701004100445191637463548493777672404896305617608574901906641992085649882441

6659259136411497972110570920048346356219112592053159495207728572853502277178691134317095074741774046112597710544066392888757183933236000244502603875999517421359497976494

0400041440939686093193286423323138073107260523470222699550297533641333336376838307699122239114777055859977842874256964525973045897989161844009118754738104698043805595170

0629630329433750112437691659207229530151254321394054433778916278191406215516820884736345341979998879516117261028410632336985345662271408982502069128670444116902582047965

7650680608338935449086211438738256599464349788032327217582926945169986312673587510954845587846314075917201962433708521996779288308204170836282188671042940242600584400043

7735875331070418881422192092460714913350296369058466448832031947410173461128786735179422094145466041853403015518155623214316574733266610798980310906817008268873210193645

9561785851734505472858980078728721154172567402441979028843225315410192140135091238671110323213731459405115614706721289593263819675803769072313032161582473040701388589334

6366335976771547070197732495488145171495615889159727040316443495121859747041467171509731132947384808502107073004895212374842154038998185951322490144185729193570943752415

9215545692963115014493847033948930762435538342354395078579177058758873286872636137723131795763188119174939973645829559955961684714478441518985430774145594300916272777064

0067845262221886063381067248472690244026426741339072193530058424406225946425394836856547840534349052967430589748649564389293525069687282557307388653479795697379637939416

3125122113572366124201402646831987523491375325916515806193872666193916051049359265271321692209622463969924533949416814876975945022756931601737297825225932113922797 26446

9907870797211292701007289316414132897554051129860713004542449721998255923017335593991966625886284890280161029774147281472179960743046863683943583762096637059217800358151

6991294767315483262434722529800380095958755554513635248529233660366613345215784920268506151949203452902146178514203242331042284863520896879742184540038734941728320117627

3782264796397846777136587351119302070722256003750749407810394633895199845441663143229731608084404982813543030383363163531454052991483164256012510682085656900160302972916

5846789183221058699489100407801076924778257280672186586644935759237706601999726065952554332733642503894798336601431993073084809345161508804807646366675290866716936206249

287398148879904365333871639691167273697027312653742840860973486972932552788541993019041684282321395857966024873754065439260849531863413469468678923583360680339445576 1856

4870113259642775582026319256809971589448934540735451669323844921499118554933828244577076688230525469796128224404159966892371592950939237321195478945074080677444890038062

443457522461155572389422683859305152775497654543180834902387291984674869316260887179215124829247615893514149141589042351050735349679694874918633443047936252036510556 7215

6988823952034980523015312238521251326166449473704612481860990143956546372710175562161122110472247926506088187921878564564770201918708174098274263885178517823195293419048

1931571564040017826008047464154536425857968822131471202195068707370393121533322394296471014338817639918115074215554226048219902450082052031551588031076765688121985750384

5120447360279692388489439850407766939191917803851311790463726457872800566499501595762530276734247490355778730320694669762067937109531408787466090719090054787150227573861

5622840311999793601481740181407268559346424708186513726761279734277641240894070241225057591283320448767508382482335490062243196257292826480566009677509285325730388834182

42504410194438374908292890770441518151343279012631862709344102805833319718393808451124787757790528799614248096853758097666763701569484348743174757489914638891633504 33836

273988511029559099726899559047151129179455591269835942930673857430486989898559443261989642534349217117176194986881381153736011925283763481221877710943925932205737095 6269

81646452645930525413081768047684917996709459097562709945746416687312998517771315588620765543315102630236084922353201840024644269498222009388561981417423529421101204 48887

865176204772310072355773711756964540267373869878293238488465868548243072513224599718195176378206516770173496390729119732315211045083889636900343634564977138841805680 2984

140532309783687878873323574584371677859623193118212996544264227460331165621899580738570914074817090777072060125825537255988182554000170967909097413385517915050346241 3627

96294337527980392121612449422857348055409299617422186755267066387154019716495925804198245727233943587273849129806250522990823041441796420186323933597564085626472114 0987

1027568423284710544204769273722795869343255162372870613062489483176830059503162735392722215559603719126092705632090016884464223997459907628360386145156011467908671952274

42253415373563043636807658209294481681575624407583542094450414818369400724787199371608074714370480527241227205762001482655673842585276152042257561677566344890835515 90403

475597055278114985130250874121655616058542729230289933165473549907915612178664717813433928249941590501409236320169840868059967723646311800323091723144906596018394433 5732

46799472136366714309332268725922769959786634219848604764038331215159824633481575389136213747050626776094939156543444966503071575601905256149343412398650086334976877 25820

142616035876421886575309174051824174917841215303222383004188066393854558891787620068788140487669276059762638850841876717239068821513753446907420527968759386296574986 5441

776294251870300911496135284438920514500715511087309466495949907089979305234012957349386688178592724423081521590660649960755027237608127238705851213727455288861773544 5449

5938515895687751951802687798564825202662409444861882672705420747504335367998458468021181612451191791640838822097788641827568105850767756572864848283603702493287158198060

43555879980375757476331720000544959849872516688565706303352876068093081590181410593721378560788103151292531750411050960975165425371030855174854899280792792165082670 24775

246374998378504723411487224038878779685621658918415735659396870303193507502981382895299683035730430607120754662998058479510773229041914306816287029509007188141342145 8284

156116327645897977943185244670333572201518300806773009843428145985559436573897199032628610071674691150902659464279237556249374235121744508031213499874102105040262541 1576

3114123064033738402302484473936132777143177832648722787200003132437991158454107320083254717655335778841973881119878308116128253343500137910973264580456753562692848345510

2531756976137831443682524778543069370631432550964076224942709697276210616798163074586477313621029169131901935053917363387720959307728802113849522530852335642009147582113

2150814163455937327663816462099641504181427926147848561122509697441807399401218649576170877429853908394199011888587733637311313017101357779033475620443952626076779765685

3850415178002862202601739831535789490454442716570559649205222318835447428311193469603711941218609396474369683521630084113092122137612361931555091187753464456042937379215

1668962024254716803778182746385907968207356409342994334271792080288752211254331790114149116004796389603318772204714551925930589486933504992233576520706393366578610808592

0057759573577060563469345760388491080506695516093810694366212875882733161322864831431471767211570461923561465003774053872176274111366017823585584517310029820778999364681

7768759805771969044293265641492888950616174327395453482331666399791748498402747835405359120022260943990531207076601966727432146673132505991961537491912061092648781953777

9061425351892234661396095319606252617842571586992437826609161717464971634720477389613148671942948249029198941916758308889233973117415554172680947533102737797997098175650

4505473602276786210697540450592614388377815161792537901060640229167380269625734343046453004211042527662303055207247573930679272639371318872288012695855490424866322830702

27740155528034220557317260915929275132872044337772363815466002242627227955242640479069128534664743956703901536664482511862340278040253780886661135356644106913769723882365

4053705720326485133071180018862177768059795321806543675321022250428000439940618518128895361407337239506631151707000571386315302132936855380184898696963028510893012021795

0647072487750320999483675687172470029055814569840514467469450718871737636802873473556196853175307566120156930570344309876149723068952866444156407483458808986525661664379

720289586844220392181943171512756411776147563714059368640001035880263891259692381706227637167628748062838160227594105114626922888091294330277664959472497384473093376327

4600371084359078599766718005586870287301832296672925665119592610059415810036508929062603999789107646931019522717446451994436169991555641564121510871438208088680752297850

8148022862341353184392056663971152460890481318445192314929106328154027922489378228251545768271624596117639566886461742395371586574462664399615547890516373252182578333253

5644589892905951926058659798671344827447826266678984191962736059352022149668157043655690416708257527445881757281160956148185722436954647505083028443075317077923557132934

8761178390813029105991835522622374686711575705937749093797579381952473316322662359826956998047343344026168796547513042934616242661346074732526957031148814696916429336907

1948154548179082929107206942973187597197310154261993356461532836182287015155903310706146530421700668825337970132344950607141683526860988131227220540903094664606618585799

9914153978144847741564082258903540644906463510615433719400401386160350714559736014278623451486573479621797846757021898995133336443819291905300857739950452349349571896846

1271137688957597933234953320895381453984677028512410913999962409428615356154952015641889962125930051264420968659725289941843503668188048075291059723360083654823570191986

855092603500487657378829516292374183271323676865849464000596705067783453610036744259491885819559592690251239311072595121211563382415896067374800718324687784130780969382

4148291518956042755017542065174420881340145436070713556026763499597575960041036160961213773621820223563980101455924936015689714897933365854991863497304103495007905509710

373294892197640588699532018966493350820431004885230594298486801785556564538971529638687198239389278862831305388987044163387485323665505625430238286131768314743993446156

0931076538494758464893162151583588989339195673294433479039090964500201525452974223609334873774857090601864807051699125755933251820304412057338911692494979374444181721018

0048695274815824860757712217241382985252976703526885042133034637032059011127692708423122473740399034467618957001025917858966147045611886905543180001357411454538480916238

4560193982145769801540367447309332421416472755521908773969174173735064145951846078512401813774545883762985179066094251799695036587235132911540694118558004057561078043579

1910515438952930178607056885781172174213915509532072119708984152215425316479193046160498117600999404341319099118921551365122611550181311073519406748964186094028486928305

5022119924343866309661229983761658981274730669004071331315325819303202814967455702892711980502308342949072461054910957995507893660263469846566281880554901043878989574409

3141529641433776902605064364098326821763362870988262723974302300550638516752892264837509508861372198333534609848906855685902444678886336439604378182364931607506979525 36

6177704480786285210468209326826682897220715910980081977801926495253830472463607958939173700369328966358022050659802853387029968092228675427129133869994026633577360863754

0472021149927333995596386713941415995063555038162271317992876143298924595866321022805072720176632829028139513624639259879408411977424214784974887928534813292261758042969

5360568496416333588361246477604717663033985377267173732323243519297973342376460670072590569784778225901022471861849551087004140155276349224305850649791746999412247016670

0310104432762653099301528420684246859523591053096968105843118551037608085368103331709534908134883131172235937743874146218392650171560903279402818993561269449639671433207

8290473191666678085182551977172880062773545399159278903401078628896366115708075792637125157532125643458797675822798605621785390463443878260224769831644730911677313769865

4394413974813448003818298103754950588539835429146322753291226062391782931996213986918817711118424419627718789923057350447245775383119433851793221285766035212168779011404

7765897677843569635136532915149309638039104754501169967587802799979553895585500590455332979356370264077033348112055967910966088046544581991175696733538179402029774208446

7140554762538001665796195719912620080781668202885915862485723615599401625547770791411160067640782608077107894734372899115676130685073224963159123163419758846276472881920

2367627163751947669533254204908916102491648373349659172708001471152710129089029612110404724620656228209632836266708889728464849194505485241475581339237736269212766280901

0703960329946262725094714117691214291335397513015143177467168584029059686222170801110366607146302062064220739673674027544451115318680357371197061263214355234685554438245

6532551949622309244222627616181076353271218486711038748633241670789046885223329211115019790098723766740155479167534474858911628126868673604222994356076826978331735176394

1137568187311853109391473316134714642957480258866120984333362644789232779921718938110490257508983329575231138511638411810192449913293008778472536273658801679273239119566

7737734602923116714725275438773239540964407417449308810335690168994473265062935681240746859168925465092110914231643396434965353999052260304711487117501951086036214 3788

7928407449825270332425169177953432393380537534154286334492005727579681918742184272188588466626602813391591226508703295562974312100608476462382406120209740885851097134502

4453456269674845217493795199836601359599598044210553393057994635412156592603739545481307090026816164735807530907005794698595121857669282043359313336580210439358016107908

2794266448782035280157498477771875366638868714692849223359797020185921637526370647072392328071177497552365362417062631546327005902663040247398045335302040939313049739713

0791718151488632385160351409187151727259632060397751818987737942983354896214929883065168797261732334295186029197912354209146617618580812065785097554051812624547853587142

349872282450762802185554164393735572873413177079533182641069580231812678272926217247904786733132302602879014768543358099932443723491884995859948625830676000122047336344

668680030217744283089567321206573109092985212685308293535203316260961238719270474910316941151648388474797456771234335574429812684461432753371060377023811587306886 2889693

9413236300606050428996520045106037486769613649172511721417104539723698376574825092862531991761037960505070047452751987069243830797208133651074580862533987045295036577394

7943751943255366001421055464414822436061646770791716585117656108592356346094854976447796211655113187009699029140731514839039089918159185783326502779539578418251970561524

6751810745633045708295944288915066671592976041280335474515510043994933991135740036810821452010037166333769521213375323959064451506523337907475042857815969527569618178470

4238178420315992417112157281753138255289908317222708031933401849974624661506864137178679359480593272851964335736880274143158690076520872345466373639831869120209656207541

3488741155043517945705202192086628621570465012959513127937440724676204192266556744533344472968171487354493873384801665428264237833848317565438333617440873218792199714309

7193907561528997991933481684566486989431576014380286263353313618572379316723660636754943800525296713997403509940712193373758571204555949602844456404613060336222636216293

4122457615116541938791684813280962469524445695462125087911893539832219637899994870575517487718861051045258709120015502718111214008330339459997728658704523419166730406855

7004717286117263358849682710717450035389033631066658091122161122795352059735631542387862792211740027929927660272309100878896448671977510644852854236760680678328702716021

4912208907383598679167790798465468476544328863327545926899764713611821919363719709430918976095893307419509153578998159456268174031091186213611238703266328745925123801722

1859237596420397178011973301354548630311562876453973330103535199368908917165821184472025394047093178330601239641672709312163693791933239184259773052761479229302123013163

6529561376233305284546377449667838557241630555328610532755207843894044247233087001494007564853949389708563666247235115549684263707422419853407218843317118086247851099998

1762322580581202049072702367515599603855584667283973473259596127104496948996928070408723556135501883486098273344942119279511596389142170133713625405959158400657637103 3621

8594354090721495079719264247416878866135096201313031939816564431842319103674142051255686332809855207709323995574220458372892438309481108423300876415366308472416897637519

4193998480863927695317901643727802977688806162490841933764103645096126040651273694733432136475166867454187542353324904525140012619910255049422060899086534891218519778520

8035382979351647361636394852849756284971488562703642543761525303485679142181383415467656303629359432715688885113964534175501135552342266095177381781803893864430908305399

2738653198839237082514434976695795125406640558213249534760824464237959520467403716910402286506016440118821281688727839234273692926062064096409195961459043145172341616179

1510706177671741511297009743626357169179809791310760755444007274823165853639170769125919005551128507328081677051347490741450119502481084276777357730810360845003755565026

8658270894906640961146299690429226983808434968138914924798862248716712812408926279700650937412914280120188192206542159389736338193225912707130384894216293191100490714922

5362821862035617644685446995943076419072713387818263384790269051413488524088341597040931667176458485165390460010963472932317024526860807864918007702454260533859200916633

1507927787324832590160442171566874940579151896771159131892750178044518249937438743299329143554374680946834026083464252681707351360267844117117547680302578284327412712955

5092671085740230474696002644571189301805811218925757250024179106647302011294693754953338392710767838158558088756706132999649915893949904087497782355039210513630164671634

236

0862269365394034567695186527752685603128680881568916991604601367935600028878486501738703611861366168233700637624901718703548391653008880657523737679906815547888893864623

3804336788144738626369751444635331513645033652509877954130939941467601122228501278273455755159561984487267288862169113912786444182650107159343331816055288098093137576021

9544842366891814048761296983574036801175518913300572269947591922872439694710724497704047329675133848537289891985144879126933995627276286301571782705735523845019366528869

4250301571288649098993055897745148064974007108137602067660610028335398320724359456720594945121684402530561416115047237679687125269315631930981608232979504258981667480087

8152648677364144935695842879538795111120900413882435069998882091565554032892502288051416967879299266268622467052549066749536250132697003182451011407351929815270911168287

6316152545336231324226804522288961497091739711353525544012360861881545414708532046722994693907148818860332682826172282696478516984097556132809109049299420589020997586802

7011829714381130616650165606940509417447084136593172946036832314886783783401584666526277938110347185652734290112646968995135220438138835925408450875742934048304805257026

3674681999971113924994308238094814731925760115285382473572083149105271608169922281418675329911795524477487920246982478357701790581768433766677768902177649062193699589654

6765996942872180109781369213674462209747830040927181905137635612325486127214522261680518029325681831093141396659245310344236884339706735287266383000454195146442303262301

9071897598561247023586500542075982524898199075031653803249502601693723058314817314752430435942498914879189062802634091227267353344853777985327688970476167261585288351406

0352527088519929217133070578576387493937455594009676153752177828011626903772652898962034412615988106321682532064438164061291711721200955674738391672229623555746124390155

9905448832262644162568712687048500344921141575761431548788382262449382571907205282243565403066864339495278663919782619662128890293170809150693354760936306950387796483806

5009708771258420744211499716985561589989747876513750578536272453652178066289777507327157034985477471678902956639583511119977254308821083008387197030016360375482320311811

0345196341997195708016263754256069696618343629726907066223061431318636181161133168418495161296479946354081551662886453122010561796238101443846201413252468510264137934116

6216666044355543396726083900293342498560592304772543016048596898781615324252348894799274995680405750878596158465639968827705058248080375262444099228426558107196531396214

7422223415350770031361866522902424242733975223220119730089596891049854054474276975638059626226908788476436765519375681951996304422809024719659779814112299761130996689484

0654703043061615428405289846055561052774316709454797654256999443256151512704117768402472629905184687393844031749092277867137465048775654003526182336135822096915951653100

3029947026121379832699551547943004528250404116178992299479111764121739926937741658202028350242611557953577101928695026460543592411800668078233417498334223525119403957869

0357868099795735556646348184109235356638053216250587339612730165179209152696307741603539343614876508656958944166875931028197227084213006069890327681248136434088291450693

5350078426900283389692890036766306519621256911370825149526413073002057234260061434794784184662076337424740196523490639302966223377308206402287040880954039448926023755930

2757838186727111955590362643818036944102698956099702240268518929057056341157634566345353091783644912706551465214527451609570926960198193514825042308309332402085693823257

3732465561978380507982367839148964413212119032583371930512612143512054346721380249172084457240675607838911836144206172196093241887871539065311934562423143050595975813896

8001459327268036990315314858981784218414086270354132340571406372423344162305201146005372433545440858047849152738356053700832984194419408785772894289429890556411184890127

237

9881742427130941732502224649989776184995844482431963338771360641700505758811206260189035461258593451545618175684097314733842014951893758158996012087525756276033295003 0118

318809564291086792993649140874263226672138684915224129903291462932026823734909566257903206428045338516755725663359643282983690679715448949144144284457366131214716525 7729

283228387225191227818503331845753752311813889104687301102053329343303322817674447909206656325018838874991783124527795687803251857087877108213218175422991370299903463 4082

4319822001818143016950158675647723184551735160193539741180681625549863346929742793638368312286209015008476329602715420554092347219774875557373727125358437929973367550413

539009626075460177047832009209000043703047720623969311236199692306945192122807512806261090339608085511993936257664560584547489298456610516437763230204762933488331366 4553

345734804735715674449977347178219815739262943566148533256352573800753734245856962732264432925391218548350084718726153761193599211755449446875172209534021714967332300 85430

312773430084421703922356580523746997811952384744493338373857748511427462252203934675721232785066105269132797730634628873726222419584671667202215168082910005267022364 1512

652274077600461979496685044241492903303752615324755653009315314557741560785488843720415714060087651280761331140002151760928982489862945062647986397278120873344792984 7854

531512329334051406847255746928486263150354770925719144201422185887802572791283311779822123368077931168758654777139994623954398600178217140445115877933764582521759199 1088

192383005166331028283723613412721407224623795391293388364187931553299328948798748615386139152307468917410066261860777226791348713632214751656850844199178069486195460 1934

089370819232141926382775337591945703264502363043475687173452958399553670973947311374513943328197791122269397254591249383798231266070963822259670190083814532862904610606

586856320978015085422334848110590617385229862052817896049500732570427222020393613638247958310354325985507262140340985962778601721689559875030328828176804094685209388 6403

363652364944285765333810979533420258752306609947377791748340996405620837330431676710875929826666843546700959970485895374841511522145022499454415283865780292853017658 5629

101388144172669383790207050034191012138679134635465228748140715338202901919235146721268382751000173948051792235759103106294117826715838186378195464884312297363020759 0729

496131322642355108491026499847418870181274039872030679358312315482878780386867207634549849519911344509912442473105052272527668320660348538056734851263693194665299251 6290

262646589416341396091509721872364027550026970108838683249414212571204886964565829636160986536859883788390280207060702963996208929169242011756462921271784144386609444 8415

307132753827418051247560470084561419607860495448592558130716152717681871096104170286462445106386992799031329802393832292307860024611121256253749299206962360554973977 9337

090550915061599580746264769307061465473365729538801084659307737092643932709617335897987551332985173533580576198203756071739649512102605682421535394322065787806543336 8166

837918392543102962997862558313815084290234604146428506331820780266740857504296549353954494865185275647088143513231959734978991714151693732568338933162833896451848870322

639893055689451839191243082932515654023675385004309455227522986219363499930799560689684466187459894748823413664085188532193673114375894636570214222303717414812012726282

910573318578392273347952606800413122404444690695700343265791095617342284655138302877708170928004370327526445576200902948987017264718228932761788234679959538966801140 2866

870526336706006304261299460849499563827559906026477765219702537583064118146128754387609857828996342210595022534150439826096187609835216523165433169772144125177003803 9021

598137974891320292927755438711703391163224807524657249729623124765093517943567483811431528641333029089123777146612469044864551164926799346341556211882281756423024051 6948

238

9544428168314140490438057886059010737006718298499365040749470278557386272032710842602732695690064120155580946913710129842552905449576450645756003740314945879082105473559

1136399067278064814591917064338706971477366524778443386302556983881025898793095019713128407089187196967493940026571940572215929586883457866981031818359493810271931161525

1530174090403194517238322459633052678626421000745736336797264614352971498884605529190782295721345692646383479217594057805130367348879544947334464560679667691278267990494

2003628806990026035221665252664880972246721212946167822822474271783410535858490938180843820769671226221556492524464101160066383911818308730856354226721501721889134911144

3407423167201858015440968394172184552924703066633174396992032099913723079392087063326814950270241836323739355756594835586434275852715303647534674601181623121808611137993

2483545148228986306253693327937473726404693126737565340199730090761426212286501158568944820803714283612048583161747503907712876046503361236135224312142049114096204585829

2255435749009027171143100562027796642732820368408835142189973676612851541741701550559669295433553384988687023249020610644580716922863343391855394434659741831033154532910

2591303606462266687977945573490454674882327531737599593723227310371044521133115338289304247739724195727440116541848431556489404892135805570855762755849553488919138564379

16383424089396022097880195875047614164578733843443198087351575166749682003791537961029734944432109476073270046363343661259071179260382965776504898339968200528464234206854

4946993038712496466424858116044200046669339857416855517298369829263584910447179338446832504338447175872526993668623375707985863799511764743787742210295932621738817179921

125649607665490503647530112846059719986422397278433919677403895823191755732599419379008549282598066076789498548433335533052044297814686422621546390705667804793891317765

1922049935761663882196322357224138758048818728755477834305533714162429159181440724910183373607258613130585839379636913731605046386537876161997656835278960391654122119712

3163706463843508750588046575531967200804810632083118215379561380098353559526093637000645317080644202888377266908268009424750615773653069536999464734442641799088072365856

91623899636517578076237318613662803000067759525456983035935020931034010665488238760590630966715258031902701805651077417965996417788950664060278847170680779275557035102223

7147306795006509607538053426398202615407127213785603227432886168024173389459790505032137974846614903095301740230095495752617958896983609703142914084804583842017705933308

7278988292106539860854978417702268001994317231256072796693509378461673808145347108132937634521964744163193311786906499824823727616205615024443944723233791069608396885603

2674365944761324366862391058343526372587026552727235468109736136753799885434022478297321958647384707984985141728538675277923065840917432060501099102238929818938645721604

1689492340208559404805979888719907538994483624575918179587264785482436871784280511816570103599948961675645814411774359994155741564054198094077706078181787327808839235166

52729811729470451824894886940253978497040401257850170852522948003264485539829333954102504934105444614356130453712369616822024270875468032257772246764538690691735846329099

6597892708572413606852947228418998881119769492577756734731492045418824993538607544853832734931602494458301840052011005971211224881899260140903390584301410505598071884441

54763356093389295582703356383918920724411566241363467937554167389089309186860803126378923091291660755009898084043087717386876849306238533335091504106000383060163948853687

92106123894105743940346062401637185442521771675451639760025505022643961152599429430869349869074629783759970161295030843803660600589226585293056378866958466784875726002

5329183930718547261012014353181230082628245390756526384816628430671241409153532301737357772231705454533185733039863611629092807965140007625802958683252113035625213499854

239

006783329057981002626637678051720624754016353702521682187355287204019963596188736069347306728409608128864989228165452185240832827912818493863635272203008598275445998998999

583511157436878788812704855717381485740307803629420485942064433415790169338959681533585277508781574397192432227798831706054634005330969611599543732039412995517719740924993

72819386910424719168074575580541317281683365537965275951040258293766006937948476305024368669308749861291151557902990891475114714361655097781089168915938904322861163098008

96160154365423970713173398762556138393349278906057471453816915692648820151026214721832503409165624542935331173283968374175550697887724604398556261085337374028770997288047

61149157785765104752908911381780654692220721713254159467978055595740544953255877928432324750482025729610721193054272034454311190184326515998329511924254995688662924512206

15544354851877843376022845731855255302038578067996423334739432832550797681431749352903653552357083362272954029760362245967870224679610872900653691581103297724112719688718

46317120153108720228291216785136832868288998410063083059699732951240183437928007586687788984960437727540275895229293668539226751399282371615964473732982701750908375680022

744666915911114977994466711356910889243791993094247213080730819842625931443479657908567008256288588361144633070690191063606859951853870417106238568043241122994069976977655

21718948993497188045038643217598286433134023273173503448755279378348641310419949659552577066904567184355021562018967279397342682162568608592248811316664734142989101388757

127570483051459436636099246610627201124409872399971042075654391506863102013575984601467302651199034298650639676000696687958282892433978259058748567826269263303468372332321

24066157760295156537226106822903836613368341504999895934280932018654247036073590765608162195997593438201724618076958178347221271503991239373330805981643494623136717495999

9463042117638181478301910210213344735692656258805710164468798455661720375874281490998433930465239312200035644248650280200221038723815085543608061085953817427853246549792311

0151018126674138466294626740034072909243067756491785793427751652950984600098628219865193501486314131133823408186418101959888722959358560343722423672396015146278965654853

35331740074198424260136066735529840754402573177714954021927548762566366320979513483892323984730934282729939009549226286852582803762571040814041407092153377982471219334648

2520718387853744500723852593605676595762204021945192479291241307585464859181278455595125339485377327439546532520168622505372850013045372400046474447907459782510294447904

75972689949375374692808933115543550514205161236368344100734984299470708653487282616826119499544459888850359607914367111963913932091120095413351288550899249339285379477665

61641592545275885347906803485930421014317785771172451137418462432155337224056541214942322346734100328640923722757147317038093058466611410528665349292157043843719387582544

89184989446589748921123980435592536491908658906691739080886750091323305426654820771573640252081624830165987303608659808379615767364117736273446697615666539921348293456423

99129280785997987881520429221519091416907854973618725169309999132270006749972335155765795143664747023748766149644049261348608299760978362604927823173889497922468520977759

95080498826972392495765987223064695118767799160549567269969085152582265295227385885439302173427475574391874411376633994128594831234384848812794576013671006676165974395889

63255453067081684294512114091221200910866698998915001020556924848523722554213107166139198282765742981882917518337208417523869676828059102315199253128014453772216474366825

9508608886364434672080407995745610429010196560880839309828716061604912163604586908622289737564557413574307159108936724233166447733282968241883149217164949725140119493690

5670952961527043291961756410101851405960839542210112530043203244772904509568672868692837978994453347325407832005428354880451308874136319369558168287460790465669459004074

4288418738123256769967164692679986955886052080637298383211238624681202881704348155814062949883062634593344998884038652605737422307385786640002377415312885909455125753539

3369408694442939407522182847100107766580951275670201477540829825943655390077790618030037104030191092185293284782411655189092287029012404160045214931709357753608163420856

5523320144384538895834220684171388239953227063563872426113302172360887531969248201179065222750848154653606346843208352519510833216068431733436584680591301574087870182295

9876582283002042528355669320450198198817145826117159840001172323248462236802337849839057195842083394188330250004100260037883422114836730547496096779242970449998379947904

4054349710896265896766913002849909603860630462400933379790920357551625516664005711218771723903001503960954051845816999386430449804010399161286593474495582760668348248909

3373386292669896469705317415608922966624289143819273723567260303050110341597015039075941159915617925116562289244176755772026392710897852605994713153357004590483012453586

0224570771605821233218529587582203051937289020017734320619428734214752378860830070299796531556810301128799258939183387796470067520270368877240584066436919090287438763388

2097145801017495101346458402812780113168139897806509007407674642209638998045332620765149608259774522758423904134502684618616814579533717594622683030636661453659920280300

8432528514981788177127257386753550285133836792305674324368696202727569049569472142242467988436041192263169155678827648422239119627403671459897414454318006168862933763562

39752548161092018062894420650865088651743884451744029361570891066530518191344083524173853908952947331269090002288147617359240547275574100872211860248070655273478546467081

00332528804948728188464766451387194846470027398366369678691108722490689445254499301361359823021009664966265824979074179330260447961646789612176304735470941059054767787436

2769811146481959467653953313260216045188056852012381853835993525090558673048169589393126688887107245163758078691852980464437598493901498640886729121561514693505446800392

7753771628002844461708872834613320160278469351417103718981359285654440473889353364342252995356306714864357582266150708472242121395749058812364726080795665391821078069759

1919627299613768250527190168013501825936503904314892374222182997294359105047665101984354967715836390356050902744094547376200086625518953798973986955249442094528368937291

6225586445881857232200509734021124202427013381380975063870736878622413461626607614186589036757567280494685139294924694749767044828627850379939428327879122033297139754384

3644722779419524830053313308326594126816548143183672418519071645371183945618853767186114634510098763556103968824034693227431638685638936692078262878666463162305865623208

0344670332241489658442908620119179775183607898117847087626296153194003478154640506345659858453959336783920471778161961151978159915334832397561116212221045289683087138354

5998806585780135485937490426395602017295786811549407988789989527859449531291247582481713710885909691407061933036180030332389191321684024237117855941479381751226153735529

2820484620119087835578241076795898728264838018836306051625874581323807170212707031160159931956653211055908684463723011219393528829932843569560659719893014841894192469651

4195047131003620913846840871436786883238124871873805822177969166872677052869491723296929757412937157650310486149826499639425425153552278926558176593281228135199049986283

3891769509864987093885286522464162414980091336048094161672069334242500172533359024122452069662742838060791570974610193234327442284279030092197167819796597905954912721055

3864472407600831000588081819072447870343657454279475046660211686153282079367042283157677410978706565288995809215120790091062488938646517486033663048808838585836547541959

06352901696079598667197919151547267539998847762188847851073660556792237455158009612734637037295470996414489544035707005097959712497079149894057504201650307392100837573941

3281657808571980285113042479613451450427776366605487057390164979963880067492763356999017421424708604276336870153889542554486051966156011457074311012678360618976337654085

9558441673969989899171468666484090241900149311117346206295823077870579486764638556758721099513646830997771609454655720168122853937767374099428303951557494531692065820371

4520458277535783379827116157535547547595988090288825001513269030621837535588152280080499462199263139514759007671504441201028640422323467572146522255433374645407695544296

3733651832944081148165531123178885685344936256510923382232508875197006402178562624050439203931155127424198127866118045752037903135226382150021077213050240662418300286247

7655911114130894749764170442763287774736669514152962797298473622901963622315436319141841799696896280359277506155139875578536726635378140081715318318793314798031663073541

8382498547746143475821220350330394913462672508437397313313461349718786522154847952113280655897451002032487297922392568927537490270425986685614904053752845046626144264259

9122957699498459561688661293427415216860453695085827710553806842470639686077813289931062855112879954394336700991915208886144555645744227925330947510327863999828608664778

2469776693646959682093306304832523027286162884091854021075857505952335491745175635055894316749129117086207384696004878978263910905625739599487492495979011090191614659080

2074849626393582792650583647676708383011968855505051861579809721852980782027546790777352845948855486420957184809577355026418379662201056062017672410164759618231771444198

8100961027947776081962562468208545749937593917557725550439016442709090994033368202101818918809431446872511944883976072610642894763770508381680928472870453085471610727866

6310122036688758290646249653294421504042608960795596028483768115833106563981887154102229185636375546808614676806062617591254751326265896334165706265117268170549301094065

8630266922042298948302376443236714219463649003200189102951055373197420393993038094287078665529147692881385645878149664779808233148266402166554846713406240185971412175424

4097871277182877853413438373825809953777455566839030970006149284077302470091720199524620624539198585923713450742399985171822495428895132343503182334483788319295955533487

9010992117189922536529429365336258259539094655296358137449729369974651187533853748709421770808174570174225160409043884612157412445285020237983903699911389696734778804970

4208886393128438915798686149953572063769482149209306281251312280804996626255322428383991735202566745252299084099463586468341113042084589880324142878241486082621057749

7533003770602151682554216858882552052891371938772498690820253933629404730475050099704407094693589196990053347830446358119649104831638160680743239747518873774504853932080

1189209217620032541285900192128507800878010809612186997215672787878360378342850502233591047386100379003368195821534795312420332192379118697973810932850103678278806081274

5282883103918315704414803727152691119589138273502062661678813673893005894349871926275421758678377586169291156419695497805005027174414821422531771664564897625594375826764

3094912605529285775355652731149295608791108215961940175702492044626207794680537769554416379023844428627076001335882937229058956135310992448792377397001263801039062103629

7100540090233266205286855129238893654007663966398392572450824496898926245992439438450876557890902818985683433451099619638409116437670596054841952535056878723052067916205

7973980086958750046561196154504016297677700965470185284334977644679496028037224229231209258155238064515073175826397848971659561076294495873794568519845906093691349732270

2428238293584834762009727957320397390824404902652459373962572954491177296085988591097983191619192371557431477730561327586503018671287922333999409221062679094625850811692

8279669238172379855127629935238608831771468972857155910479691135290085367375489955124718689039574466000484795401447802882232082425670184511475458385946399776041121232182

242

945120720779081768233151618455421959874974455890151199270623605189634073539348756409531560340706935958257608166769558749209824048066527562455192322003251452941755396468

271902352195763796378975941750521586754201928536559666201450456338552783571256712051282275196092953909245784188554012712828606220318403041625230457334991986203368394185 4

919174431281417027468348053312363654640800952279867998081678614387670229917642042193577030302889240633823337118831666892347256653147154261973692488185677444236830676285 80

803555776670852493943272675918271305802820474498683109238845183965692912826914135802285087920263815575944588577839176304153824777450273630047967685689590574290775328240 31

838874195328472505758236998984185573609737934792621075648001276066005684856289529627036503340728499245707682166060043085192144993994615927401867906702492509188084254975 3

825000573641152765627882068316837456136116466105945296098764787617810657325699441300696040441275748480590815431525219147593078104140991804367706639566726344353562456193 5

640751356546727167909411879488480868092309383287382250042851308615564673823134489119765303305903494885070904364085511708968159681885355596836776708287592695900122268130 6

340790521808314175342883875013165292187663333574314371010163477966588836188698834998328981164007025965502047611990688459306411880137433528093795109830522058657551902553 3

132473935518421572608823101683181764097241796642821705539235733182359972105591467580990960157125453625914657358345108084976970807172669437183081249664139285293242058148

352262744693758585695716870345022377577777835981585809545667455462729794730797569320508562640503528302556722866775444828408620998138979303709253264259640385349961705463 86

083723031489481001371999013197012628010849698683279805995250749377542359999787837446577837123753575785267771063814356136789079473724792426181599280191554236358409224109

269519498675837014008069125945944555925467047320392800470111600155954170202879503592920177039686011345630444923339774354557165457271495173534529046221587766932103972565 4

053386823913058096600212123248311387049072616349881777525063398803735283044137885040535291419573448626221480332949544888908545079248054268374194066918855516757216622110 9

208997318526676782935266132904276171207173433002345258919537355747509268743152936342844964077178567399584381489421046285181038496434277355192443854011056003640918609065 9

830023373684591499584014449901293693725893952889834811556455810610307945458047566465048935767852752111834079945987442056755069846162829748169274340319574481121269219891 3

243550319837054664442496096363374848655871436934240008426830694664726867821605430760555551571130105499636942141201452860567154928145035636088579354204498831255719599558 7

078583365330551210839284998411268479057129722024655013870820524474927234191950360303934603476770851534725433807691354302103233118270994105254373163717918996081338444036

735092091110631733765874016000186973042529842024488852703317251469249747539319823503525226176160943848097053512457087513186892673700507442774120709790407346312262052900 0

639189290990331964353335333778277303380095643552023728118934142222131638412466218762656292621316537474409523054540169190591210298332548738443149969008179176662444562571 05

500140603668001332495809064102834877164193564714304095505763803857385220987161679591048522208070839544381665295350087746068113602730824885662613928667728371703032378323 3

467164064186510599162481767070634565314676407449265899040024929433820347665482730341522657904235101310929756783048481363297639763227467327299098939274496385721441290848 7

064480704661630671832626973018955050522617503670443765606386089757480919952639440353604654399823773561902465026931295891402221059887340881632225625868617445324411589493 03

555099774824741515073734119475733373556527627418519164245146201049878262945368950884462321176358003511841835926759864297128139853841446909691719079951674046781893587502 7

243

4435035511807996777624987062847592913272616884488849391411287453205724835916236067416163446388013123511612633214157350735873789089864603319101047904951088053276243634380

1953205313641343554593647041984399732731928183025760085767530258321696714646834358840039142037072426708260176162533902989867152675622198130603529594844646404039688560064

8176610903305607906503876896844673169485434389183689353793341689876461040506024109359855532664312999759390263679696925137693692615910228511721654148892157262235977366770

1463984585521047047638897225490221795578347385363008193343789887481568027697599949512125441570271376490277515078779949109561489574262273987940332212825753245784561955 27

1497177374892311071758004485261087533971440933423641670007394747605338766380254295863552936913034769868990613336245908473538052999169523740650576570393490154739065565289

2580819446962840122139245511198760380740566099410523116436019256847220532851072587588370887878354074617797601531730495005829381247505250302170023610957367090784022351162

2259723813014440798479181330321054432443110271979100591373809348377586361399771943672006559245699381470276191846661211804296171866528310695776609243771615983151245936172

8010390163655204661937092525125363139655920117782142768019428047658653485815874703119926770979133504911072051652432462312538315744875129541560355275026367961454610934441

6853578102731389969995468673435181034432552595169300233005059257279978159048202235890526047920348250750421717345546642531252852594784406842133378979724559983814529024913

4127724342450971795118644615062815283928650237212926436840813268942316931518881895385268180047631030777803899244121518719858272225449899967787514587916023970766604413330

5940017725196828827618138902540142141154032221880752214931387303943894838634880488587257312960544022675401194443442712395569023790715007138610641985838887956880548633517

2808444644724813277238595610521301309101510626429327631624722285985257102637159462999401322655051880446573989800377544218462997919330400605176161879860265557607689149567

9624033020920353382300641985751212805490876812649998662968202160083935774256923900145096765518968300114985803950066246338616802255066870759127483197530001455406015411980

7841134948294656806017074182194482694202914591931179729442535235139331011218688316780023266376919519389893049565596364430499826362332222131737958633752729015980267624382

0849089436428312496796216250016352996689304462584580416724907109041427954474322776405586064497993774815979061292220293061983352618004261179665496758054829018106894737221

6120265716263043635302282129337245083943534370967854324380583128895052786635657171288036985284548714950798562466555779317050790289986859714364593077973507014015227544766

9341982392638989294535343190218016938758502877868797020461682319735199280766975865127606468238391696014867111500960384593882005106152664625627270863739947150257718107230

8962043626415525571521279045835429590591012051850860519983358295202764452425123577351536145132912213357834119671870758566063500029766458721899656846809835422555679769978

6152958316206574320737210996844061946085275213997207746028379618294067782659809969835866089743865103656242556250842354354563015121971111673283365507058324791727356514276

9410984986357066618802528434536119774234900016041091359806582532510477723487375858846669785802999922419736650411551196281600473215765070051662898453996379194144719612753

6849638484184078353921949516076075298476708413827460440301770757999666767568612536105140039171681725678050138978371865837968976150172098480602272151208763671116355293561

9613040209273964185286936047265139668755630400875358568683131412868609228255512422695956799303502490113667709364034990374845874198910890189457051985781247844035757867139

3197085549893809997065410579169020959889749873844727261372435180656164993163897351119703310393978040928094799734337726502122497143409823782365196589288627922327799039418

205068569837671166237721514412022766337949173743734228317972794011937453904905797144617183602567322140552194621862859145438965623402489453798119559649687070730218607805 1

319309467852848444202128432499307154135644230793862725265852084922694843993948530535272706823358639486081705774075169738521202106289594177160792541306991346081438246328 6

693523126590734303668095385956084501673932290965428288540978637787221825907274341956466116595934087134481202995796040005764146848563841920840258326855212379549624891158 6

260094009876541358587061925113653719481486085710770837602097237465955311440307339494234844861525237226662531720908162269400021122759183425529828168997196876101438519190 1

289880742428052835555332523257154858756144775201194680111550944054296573101936215959187582194822074756153078334803008549943308739821340270700031128588792796739627366092 0

712697115138195377155464106337455584919631691755255599189224089797833124273545384179602784595989647060095284180866467711094641598431500095974947022075958749369609348923 5

131553087952267531922887519326905699269590112428008437808975802237272111281427715809215786190314382328222158431197363867972272768495813632814701827652361699870383165480 5

714617520177978010749005200986222538671343789879848810445030271773860682106718234856661032815984091857149084770741257737215296236289514281939493449231002475529468768814 2

282688263381992107050031482296907812706850236096795245615631762537844390868388545471766250548688614538858940190650918410158852088336961298773167275269197562386427609513 6

945838426185218338957051864132632025229313913483808212879643388136298455390423731285738559006232871979159091218171033492088287365725960067510345169173034840666477312472 85

364986970322550580285121031458139716550054021999125846712475229623304794762041748357339651293388462598605466902068727431079892009360086299625852649556926342244349075889 8

712075405572391778889837474006283110359893639753683143163791553508445994998151790357191664595797264053563579622852232320999562529072956596862566647611821743688936552658 4

209735880483823563270384629174104274634032764044247217919633892333004523529200287573013563038211172897133169436372206175085815202908497246369055667628921842626517819765 1

953852464303642562032129591608958533598154070650245252165708822643889695532003307123860177199426979987427119660353048528358184414608549134506444317130866665673247447943 2

280547339137606275818904283656586598954664885617098590233571893647110221915544801416081086328652657202503734779586980289697568759656655171572997817414912551945083377 94

974466980602686431815234229243167763265101550537467770920415647646247512496723011428848395397060605607246531422732188958383550139850224798061638234945366128994040935591 7

809265868236064198494786394927392551467596218556484340287163998424165640792243204992151935270942749255097209837640554799507636962370089615853142082854978064406575694610 7

401242830116770917540880488062665750487449700100644817281703371857168769905205043126912675894235464242632192681612607135255937799846876848766637463737084830913023031875 9

755192524399178262640279966163670943691125300862843502978667148387735570095408509510925426723708716285087204991001466660693435254396813242277505204120843117783620864259

137414037018939058491308853076718033777597798154504060045084281692653954942424173439682579794296332233121321810780129321979360275038885263104587257887904993301724937169 9

290336354529074965146405609127548752881574874850466564788157133643242701577120650882647257091154529355415106455103509471072017880017924641359721384299487104597755279842 7

450697742656414883406830080923054646260389483267223960406246465945742025221084288128266889675278980243468265578962656266379745368910929346892092484109702357131627533023 8

902076986731432584276094818388124585758734014097068671895618131122947002197844824158154450981988915294242890693466056260795656643943393427998358823571167575116329255621

245

4994709518352233124581091315865577359175847036941484881503863264098660524416089944214237244673185768540526164596080359046160459477472320562878910886992342019575526455549

1518628810370985562898184920845362817607417557822466638790726046775548345174434818904968805637650325035359262713745149886240548295624736933344962330902739113708105840712

3726554039444066616546669812794016117438031442545836113138700180051064222504742126749539970759097152710314165536334773200549635491466719499256643771113519647122484421338

3344829914740191692970978273304820965702365744166216404138597571994011075505481383567509753677020862148022476905759312845796120520106065272215959974558956392927416101354

4777148660272222807878919031049330486064234888941265365091980468047266792907970770229050202576713844368458156348290022266587224538924207586781264807425984310372314483507

3934916328568708798935330898303553843839500797078015089931667212153557850476824480668120600816092524900550656068820053943975294042977799777390954812182338828159978628439

3448937918615556384584794259298949054384503736976473332256852424021601959044724016399444925891545222094738197285735608141629442012082969234213675190569210533735923778160

0665668763641463665790435737824436651668510493519577657907919335490054245374836258264105168437864450295918358983964405685936888263686371050276878385312777705665995603001

57235823714715472190567142601433687966468436491413949995727221505804289921810098950799344944214199021446892553928676917510398246758246373438052397350830280687187344606418

7629932031217040704830461291642631982086285078695172901899700143428968262441780781916264417195044507576668563242456745817551859136043599958574598825035538466741328204525

370108050902351148404819530588806416040823552351294128140845448810370856942626000271639230100860372142424842912649571956987321905242756339490162246454625934745670300567

2837508467249825279836734956177089836516547827889426186105183276920850360362178003391524933714844455014157911625050689091071382758022446505098609868832767797118325793918

7162167678835622419886753868393157565897768652016394528273886744065178756600689498217355748074443255775906927363784818051335709627018971520589097081198620052276793499133

0405828456720358566472410588945530875223646183843964025960112548528766088662848307636012870066672280267000361494230261254165504582961699316384370582967507053229269109161

1967493612737291631205863688479052509527351543806222193273002895992407939074438573669093529492586894401073422178024370771528161242531908822717327154338237401474543621843

3325680295994077130149833237045109699654532027453350702177037070611381910516358803074747708198082653195105524034346189080408772855887102619129910922959508822518192058518

87441986348625188245665450780329536348102634843303083724181136556278301939101618351740322384509791468752162387714423922232324557636410797470012553983247122601905304886666

4933322848632905380868351709643540440256866791168844434216940251772691672005423658795245864872919319419639837059108346595565457374554274722525638720491964846804561216346

7555780018391145805072029104091774619788296504861235552872697552000423820381832076425536409632089339245449675981523092151894730501978535100152995735305428112836484236594

7435959609595698016202753023959419953344628208264979236079421886804110602415874150857519458061568880834301854125378195459697414236778741870667215842752231935277070188776

2803032337402862660420730505237852035421042577244255914042700874907643524826938681071637669307303727237417175424585224773575827029599664985410231151013874323704799159178

7902994485501682558654515388125485742541460429120123228556148895327177172401226279944108268691399729874382556815811126238732626104264249191467303139324078996737861432904

1420844114674153516742689733703219069028747706020088422190321251655289117173856747773113661537343917519627079549216170937800543404578737896895794065840260922267066639747

06474619148511446524438074152055212868620200671723263684717967231549153359492453428928874859317664269936209673407497745053070568431410133263287775921305762314700874 37384

50762603058757749787324207140665136179949569456010819284319373673284179893518959542351970228934729769771049753586499567318504695098739662801395315247336067459574653 67342

25096595010987669623734140606839350348198382718211868440601761576560254701139537357267834526945596079177094497172723469473343678038722377574791686895552041362153801 42825

48673776952837038414279340044001305689783556227986071360705966044685334333224708199605172761522010806671065237950190709749374182163352993865189170077889883476360923 61880

52690600640828079713434978942759592887202610785155541123973730460975923178834688072538351059080021866044902897911896725936755213964729068579735073675718216974036670 69896

13457874506109712035014795653751641661531216473254416789277507245943628192382562336298103756532891328239232227250679417094691318566996226763008474933294923772482025 16805

50666316199034578005029621609509712783134975497484970807501469286177972053920877355735426344654046917507878362978303712221263449953760458706825466567329277539722860 36781

92473260758753750363960555755152470447904689279007417440809916215353523795515931682503917084115733894721483377058789369541534827316071703023202492909454665531205525 01263

22531741427372940893582323130414035967091049257183835201953577511103030189373873667295688347590028046690150280415692206279410789768280962696610121374881195563119677 79804

68124930647837405316274760628345868471645943824367753427608269557653234427605339971298708052089898561442065934722157535135731353150964325686326997601181676887233090 47847

38782610883002718765026082629252331944769099404167002640065555971369918146697911356778065745105729647119638264380698960233881135307214985853687086285871789268796297 80160

89416393630120941635523027773429963015263435775125041351983411873620583345473118538074572684337206905220856255010506100937942856140474184559211793392727085972511528 80056

94028053614114492402092462087948741836224854453701673479359020150069908947111950095477169960645156934098095760872306116798593054474249458559276375426550790985078262 27452

41428056419619579470161814101885939670292884088175071326949126451479245871388347220957012545376287115461358447101311323201495490944640147600030237632857171395365471 49001

35558696330692581126404792005317280921179128700967881389373295949068769162309178222864353334059339679160242893274844466315594574856113204517830646491662241813246295 76750

91859029883332306551450236294043474054925561176422160938847117341895740719985035273669869338669851702573938066023027910628085352549353166194585293885401347619818297 90192

70269975539762709721332077521428883136382794037795481043639684621695249482298443229689692085335553085317409539710027448732528352757362479458012780445503610606455857 80357

36262525563606477349056863832460058826457299672867064706881971880489959182095387698672412610581231337188328153873053240635171604883731863483194487855245340213105960 54326

97873627899027362358152686677286484137632175406689989734882611860180029360022362615884959038938183834781502164731089138369537380683164369908798085930128373528762206 0053

62275872876794657916805763581432409253055023886548294925725127609771043084142413271492230145550249153801165157010725991966088910334458778020184201986872557983485892 79411

57916548984180796559816529244002860008928330899598461251541347364124755370565807249607337289686395655103449758583001718801392934081593465774074916873140199038284277 12262

33324460588756739838593500769513118556316845738386555122929408030684220362567245918113860635048015522616706356496428673234596566937992435872932911668849839364206979 70390

19159319455970361259262706370837171360797222924483897365994926322185943095293445517054009459274870328435199388140267085915289496359507636380732347053462309324415095 75691

85048089195717391916531000241571429356686909706755384850261080407434064574263428325221102071034503745383407217192728609307970908786402740375603419620326095180233319446 60

47043934800540635869102941831438198076626369229201519626745478890548730085334220881597403289253567824780457234485556638842993651785938154287147347054077625040798071086 83

25712720965952470280931298490597903061967508059944421798850698316109638043185757349320897027921443393134282900983890292760099810349716753400553502665754851358206981718 9

43173652187372727038665243420592696839958587716580753629304917458210275330126702362227330521370927475754927540322486653632392842887880718119434477544394315746337377421 90

51446263840148384522306013263650278845147170479058318058348940856949424994415155483863342377204069960193358031375344976028444995154109011381560664132329431053540356633 49

82500900534136214959747529802823984619728367006210584613977815827467657982601784729657646395894187742496331695884228391191590565640228193496801758163840139429208142088 20

45469029946376520599881978317544801271199655622131732442716080219316644607184506702451604612011797638272392113483394387989629058401796863609432553006506288897323925136 16

32702390752395982653489399468065805948768646275141094064993115341987321729914312591009777188686945571244449358286138125976643755113427984573762023435625689831225304205 90

21490709996741603215467075388165029658639915531513429055133316532483088503470195490556744841010321876589956758394558238286883108141862835473194755252747115754025434804 66

17748038786859837871569491342785308294728873541204319023209059529543286106613269760762669266113521146625276984127734085241913826858280955075837578751829441953816699647 59

33805810974419704088768832525737438561825911089750196431479372572070809405896055397098285800445563078599861108297845198039988230941625098680263516282280756081650704834 89

64416836183659463297691973326645044332702652964407332608359487129720563680362999226922055550219361313094392925616898258938095311543481328951849165428725462863519781030 23

73003490379917930768861220453265131810138168987919567846678661443310580143812599127914158876671702906759990712229281727452785443191763775186488544690552141829475460755 37

34560608556346420396176670957528745494901204660351964636873577292974282347505496786544592736062760898924697824189071366910300926678191303055919511695693178319754079624 03

38421104664465240458186863932609646353033471129213154369571442206723727019032161283163660683535359140279885260953147441976705764010907530604721386570676654997265613995 62

59081850853030455592840761412465221809965435307163185074886478933137159804019105801024255417135661896120601106976120338712176953627748147024046287959447965689291666561 51

62911773694618494617768316635945285117164140087961096558671942116381654578935594374741659019104026506996537608910884905409088076624223624452531328521785687211710750728 7

25802437027495835664624563513977205964778734767213709697787437222228544415051625814759001600103498734216428737941520917280874385287068745299675850634623615656838065684 6

85866591288392993987491928045097599357621915303453396402416281637564573379859690128292741879627625038064030579982239393509589219952785102916463804783293620919278040771 50

41873006891785738178253793126532269564298480575057043385936993433934560449623259372433043576671477116642620566716193777375770182049361597820655177596044557427140159585 06

24208614320221027947003286440974985311949339722052560172438067839808063019803303871401082373702021780239991965948474208100416246313999893872678129698371396533436978063 94

66764286002415282487378563941292793538360770760830075008546883668496834473813800019480999463927939725480926175571232307228799147294996278293181211799953119139336829142 17

08003917108392039962532465712426711480762187323888663027326326051022748558768582482996273785630375020526294883169608949012916313726348858049819248754553524887326239439 36

75045016547689340206821458565566171050775119437380589413425696041313947581919706682302632423409024450540958834108768898805836001905800856199491033238845013313960419 45468

8283590614806279170274255056983630386819040807698607465088442367761705462260845439722219403554202652033104455269007046882774582145696636674469994288417347114807023479517

9430778361527574001167542381049782265169967932701790991325359692564131021203176759602636795510869259891936051529316116399937906052216218262573447363342026307505726425225

5255006962503721338053244844653771497170577712173855502140364109117838972141797635473176486099732370208765716623286273640666665252444844237047431260270194052932539203 88

12456116783922607147800119571847504556116152503844822082691866745295001945479554742670311953388463367504753411924305172890558306396064273009317899033971343931584059161 00

85863938221382022717081924757782100150391638486660170908139091359533406190146269042409568052624010705647776618407365199659831201591486191247910453282084900376962357904 52

04928147481448465817268742916011125678009811772691222090702378514866112437144591984668576309473751209424335200446505329739125167083018255429713022606746609800526039196 27

55798389509069243190047375640183607454934859101794475577162863150554887610286729181867586476644478659627827294039993209905335496913847574304220358026682005018352485651 51

70342099431107260374750821643495885414321045735574198011829400651638459077831309473009929939541827181805997731995522537656235226168798804828472031495890625696894424278 40

76171974785212111708675294460367053355570333619526994064352330819095743708046560784123500619341564951004973317383620400422734437891578965349586115919181358972444956070 70

32232278280157914855808866326700924042342031392716468963013705622185503990346229467917697833297015241275735845801308975952586398550215463677050900703329079755832652569 733

09925199423352342674324526287834348780393209901478697413172811254964459042777973691266877075337938105295979219560094519647245214566447811294088842597399502289315673204 89

00359565314881818133524884486986948006125364742775500050420404621034425555880866214425237932464586130867091526143968878163367734969512409007739167264140941242164561853 64

62085838052194808988737746342851400043979235067024246706670697307923553229765566845704762902563225978319618333972491546952552514351434797307250589853935034414301037276 93

30828701075552612213237948915324248501547845700977456836739570646245323167834271605995132533503844764618045298891008925761423645689209371216233577791901016985274650743 31

33940386055838356051252991146475251049740083923818804095246656978219067500774512734041374694859890324303842177375760809258836398494285249166123974213403066399683994573 15

31459218956185429736941639291274586602149193256062908354788945387058909910287768426344909242758195382277154455075508282078488360093885750407133806431446435106635772542 00

10721512821487380127832552194772666696435196399513880976645324332723557684264153138097822980463213153774625235404290603580170561927446884847759849178086745549832965851 95

34616200108125422426766724465919850037372467000945314183285138022443386726425015956759211414749624518491204720646756094059335988791797905900136748853864506869620656149 90

83496822578568113455648395727288903768564734858702787470924437841701540033745698274823246922177670738059985075586166871378718683096809676558370216611140772207852210640 26

82698887307289605168437835203674022500330129422087997280733552033208498251879476366174290552837591285641649741372819365114332541321596654479750889663784405834138448245 57

33859312236750877569174580777066637614313837033604334586746501571999949357092631498313539476010997460000005376581449457332674586106330493215029738393935442737099386415 060

48173133810760582530943941987857532365723323047959249575058023465548576017407830040764894676825864095103880437439926955341506926095520996684963621960975419902566892730 01

83039884115526354875841019186258565195894991754367013775262962123556260890759814472451063453168437420392696341251225494255324466039223854149218025474882876575364066956 65

6854499049481491528050382535285564672004425228943146357459705605413211134776183459143500680409751657893967117854851652292067783571075255139531452823122246077432648578021

4710696712582490549532520029801187432980385609820847498781051624257385362174946834560794401386804272597372177995976134849268369259049454472854534855547672855092626966333

5154395319199262090222901179165035405320535415573239628658778850100121050944836436935545656529870251042733731127036864327018539304622986390184984977908527046638100741060

6419934676834879214901823792623153577676004525398309488349953129731567381103576713809506026898810326060604431764355029953307435121783536374226307308941189098334119638055

1522024968443939425305314272823845197724460166040399583547279137257994876905630996389204706293760505652182054213457997735548935383680336528207608125148943624690017425014

8369489103249786394191146005402306216185569908610246731548646098802027810734737377704918834398152960696061717154770915580453914721379448863070275106259100795373999513948

6174490302297892181523395588210157649640209459194090016061165874111119807343351033819020401202992932403144329877164721941875892567634592321990922901557239337288860713694

0617761645093456715228049982747509278302635993198163991685562604495679935194832038447039650098985380112586133337947755213032889161978793373276844741328316215382135 0502

232982479519855842530644062732756704048542157953717334383013657922716976581525422725319026917215835634790655014339322118454577433731865452718559410210622948825204347308

78381222298316872357783733734451559948232923392695789294479982400949392664270739432327471320097160347570744107285030796303907572702838050209591617550700050166279242931

8124505238672399685851931779590388404675565133255758214943153517959498790175939954018699586162066971538933325756001809207795785012715852501324370267983815321683151058802

7374558939822516507965568067405529335804164586928523165289250624410345072547517466969576647599120655684798317788386244416469541919134116638313595094269840789705549465807

2945983183865462177521063114551043363353661775730503740634292089127750223620941830938203794204943018325648815146217473149531227249665194033266694654817482534172522912474

9705114616160054281880540354198947234572559079014198518298148145990271405143266071026229634919481259342145337898724583097604861888669585130390729173392077225689609836520

9127931148217974750644655976134038757592240223960347357084918337811214989258295323354443785318333610174372171270755616191438053272660244948668475034380752518392245923957

1783023471419022350350380794112727748728950400230804740722455600124762073159921250457886191338649903391268514724330910608559436605624848676475330103363659857748963154641

3465281763741261581100781736701249544796541160122490093946034993309994023927494381350400802944899168793936676108675503546923865708941470394616477845589307011895036016968

4300781588665329135620556891697251578127543955416215195210530013133622187193814571974435467846367988318741333305512922100157473047822274735436233806082009833664536855868

9256331574692024315168312340672977314170507985368300211465536273578120633869732468790648595875816722867648874376546504490483805651802297321117954056543794461504202156340

6645608062174908344929657888829537930350472608621682320154984933606588503695960666607613634125466771426576696693825333353829686677801855476376375874137561497408516836293864

4445437282528212759657334188069780403863756243213593533382769143640318722051944488269989986780437905031726519533324288476168006516298029907502324788344342406567828812886

0769637406496025630715656767505203705308391361662839216418450982844674305122844423983346413015302739217504201192661457275082813903767336357639254460529716017653757738989

46913977067171842123029668128720497136453255289930864012330523226054081611387889253194878617232763152748020476997010841422239979291101028956919632942803327708353525389O3

51431858828631937429265422212927628526004225453979981005955366783999892392918091503877361400928153081940786066474842382259576352851784924147444843684342520668215659669L

97697288057366703621563551249944361226689330580394804546277509788735672716123758257138391394108724319551951605337651555589253551826979714756066893173531053191521229835九1

73023026758392741649314224393897443100874491196222448073715852494755275828413716873388568451397193571743513766510489523902901706309271226468406692783483998862859594239六7

93645641735587184019901875725463460676207062627896232203480537183635179469108776385199110783793669022647896142652819899498944095836469265144316565821247179207892033514O5

93667828494017798965297981505475435803177585622558690610123023401093612955353576358432997463079288408167023366788970128145958847428199428749866143771185970104607861183L2

22856749469131900448256430282620072487875253943979012522035179906701088644441733329427038504777629594843149909989126253488800224701126853866059437662227036508267221232之

23515725503765038540952531017575868733835311996936024166003965092807274380713754704848445686848919687212310981990987307479532054140103959736196967523052442164056190052六

74275993979137268686278535430551791257504715766018492371822393484939196929539449988380196572662503651757494033119659579411721257623713831651147962605579178440278188122五5

38340896398270497890725743044303341179108109059950541712207737379747750303681258203099584411948679998579401117139332423952627036199270637724170339781113332382715682773四2

0147905472616543401932267144218019610533705026333742731904551879487134852498626682222111111318914455762210422898347390349981259778110809013186425670889103674298030491363六

514213249820339892382623311457754007631638251268666848503158411111411558864267907213460409221705074598247207024552431403520119565313924583310091425363495878979074393713六

59709552725556666007024202839100745946246631307454467511305993777512041192805564972915123491505532781022986430608405376440989174443107699871727600381516344286065218306九2

10091799279417093193629424774580681733533597017598932860314853201568769791565212418967004030959172775708160301869459714079982443633328754319221407359969526268303856595八4

52089565549778101327453943408643865269541406604205250151308378086465729974925690376170930028161622503187000302737144860512516407239007008823823906811473995804355590351之4

12218232982743667778903853858623918147815883532581378931913051647239015905150073292582746204289143926675349521549429240927856129414258693729514689314270544422700093241六1

09334447817626188849164416925608135907757944738620601592404012063374998954252911634095244853142318247458686792135102696628751705210646078470497466615451881415103516734九8

15783088805062902524727849132883585857196877031633009755390490488456644897468888248422504227410769069154782415861985184219579094591392694553934970741708260129913613729三3

19908996124476112702770438892717012348861763196368502467208266987608481975265151178468397433083172604878540303329427864436091148976287974130203367539268931859458016183之

91794401009583249805875064583664127695292859826577033300623458265495553231665323056373735121952849214896392942381059559822709275997303299473750568744987281293470260662四4

77615834666170491626975717975872429291141879750748782171533419974526805573225600314170463422031897578207730237386246978504165097975844527164585220435513975928752950895四6

52280662696943449901488020041811864203977422040427026695544603299092549594352025279648873458005434584684945297535391583792911305703761773663375795239771087393379547332L1

85487906192685422400839536103687789915221045210650200380051808347709316151054412972685089966422824644897642323194767560243809769463100168877605725679893692800865024874四

67608245459575013283810001229747305653999137661127606785834512958030384050253041663117398222113792207474393966300340704960764328821987339877333802859793609821535465591 00

72431709715706097105968688690664790679515081011519705136357516361120759637386375738584999837864530579030443943012950410974378372747321582107022267675703966141860877443 96

90962477761429826812572551820738210629142917928982577970023307492988858235399319356394252680619486420670833245104681706704055421341876516419219776188680295892187243673 91

29792967021700260470795407599880696538294706826294750079920917805450108721318367103021403412399398867410140472913176444209080110919803524432113336535828527182843262367 25

03700241162594822559744880831760673701548562891186654465504563052137790450573281851201979665409430270049744632546121224141128329366437940321984479276656109271707635594 01

22055359024467307073784068108916048402263161316537882261306234764932281552291952234092518396017149557062533930191906745971899007765435853945373057085876733937752255618 87

59086266557260714813602684304809463378108948705332533469315286522818485015039938003366387897338934112884345773523321999762575438761947829060841049231708697927266856501 77

17845357601444068717168690095280680343189335630427097277870650886333372970310015932022324797004101814946764613369688447909453611490174462989749323003758025319176955052 41

62500655284261143730762265420826822134594675370336621842181656644347757209630013688515145967947203601213939946326114541444682752648661586756681602323921704745125701348 65

90616430860058855687920847833606246304195846417470830336606533005342062042836863188883242668160351752140074640269007587610894794635150844959617000501827708966824263274 75

52939134464826875600962762224250729627218837874695585396808698731268923748126981350128702594852987093853722712995055630159371662858218865916052074038057260330451319721 67

92914771867563052935572768823902922662197305804873114011753081338921701186180117337251436635308756573420894170480721933598874797364268619879410285421295294104365480616 66

46656095350681267986083772342685220615877477450454408597241235728136293950248772132291814474603528240905004010177366269864170218167035181897467051204279605436662794521 74

94148564864034233759590549060136096930916841062916367946892321891209127401951706180387212284307087960773113725441306054172989950548378827727046663864191137798869740632 77

67999608103275565628709067701614858118516715255720673109259432660248555938871841243042256167746151083897288342422589148505084729051760618997775830070656525208474082088 42

17337391076739811488077333220165889114100515854223884063672865672508971288503845294031629188371445378786613054000917050111154718543635583103320721198133485863115342360 19

20293718304352616908440195500481450658937687152384712418582070564413530487420515614512086668096886553647307014517115558399645835678003490949039327445144117799163796310 33

82544506126729662982076892834738348275637617138396079392508938287519388908374248283953342256640130058887766579178359774040670750177108719391145784682542500121104011567 81

38129566257255109176460365967780579961308628200115012516792484947604849884203733393954346866859023545752309540738153041167591919640339500823221212203194858219244347552 93

37530173935181814669205300676835498326089762673603011784585548752703073220035324122391029670664816719559254532213478249740250027027480599816837621411818338760834879258 09

81381516616041464208075202053745495801305135529753878317955606609534527502899952843052599586314977990312599592685238675997576441360225760470651198719323526264910813019 35

91599676247754200546833291360809332318453091032664269570273636886158684198635589869662161263694234626987065295164603659088977630943695328921497180259712631079198634234 33

68337985427815924375361059523136767058842514726866925962255623338825444915338951480780353160096236272362629321093838121344292596168977607129610965328581263856352840691 87

07269095087199084875880597680615434384983307862237329950538595446528725800917384821639524761866919428440920203252858635973427365210841779240653376994870919040065 36284002

65799102780858816942811241986780321267661190821206876943902568983185502950735813628332588132934978756119965706477032335460135933015737186998527595278840155231304 46664670

07044570167374730294477842583797133957981023419274301141663356310472002062303466720043473633620918607406379387410837798372659102262266281668368174614508881058679 40209169

62370702672270785567024669662152359232489065565411423216530012306683158137095011751649747417707710478671144231270308259497028065230972652259535026093760320464181 0401282

57029715290996630179728749671607818738643460436560312600913080199055913449698243059810448121422323919883233051748761776106038024224869336078269834879905318712019 36565718

81137988285470003661803376461641808005621106566571354457380355721627020669870665961163026692813351285972342273474043550450301843661570597586025918972971762937820 75851521

43663005844137543530152873638639375592024940199122961478312053390204021524957162375177139207404812163206956168783744067514276611819357040926225442812572446747935 66990224

00016162793569997737936222932889951096671881472547244744233245083281611358850626177813475226377416689306796189069817385426207116834520208661722554021513152030142 61536352

92417624887240193843284703145335685532116346032411910969004980661636537048300440101181291865610897469806955769191358515559383379158906798187857369687334916653170 29348327

44826234967893671344072677226840840390785044733709169016194834174928447685766055823894997626657072609591728102611237088220424048967417981761059712165243418987697 32540518

35763906869675246430494598140399601983368186282170586073372014699356972850002318515413576999413002897984546387206092599165675004257455772138558216066238188184326 085590864

88423057729024583754833271946592218906082255771930310244284508802380411824587454595405991187938986652434677760671624111861001010409073491303067136969073215943484 81297454

53214650616117015870792378267767522436635663919196042731264289021440187347592854707425670344996917675233598478138865565705899983318601234615036475938171158603856 97047892

49935904441279760418898209130348330214953067826196903024060825099184094962411171475019366825471966844473398515303878500511069790200649735354559385757078833367688 92461107

93462714441980272903059667094654269466800036557248250538537265700346455298437548560576643654846895919870325550908590937420980486241309926743237286356117618097163 68736588

52587542928998684070667409131052333116913901759061177908405550410409730126750877167686004326404717317308748944579536828065166828841880976387687751677225401507003 93369379

88371358231367550158528752403375539868709478397561579463085259914621207238560952292201024194226436550096437281621236592956492120341931085580480579252070560970733 11003610

87257336551244363974172687751532206542256393391424987919220299243040151353261830425163982175998799403271770630652960196594660331660919322521397851742027560453282 25580091

10705570160851977806571014316301211880912558064986030948049146676107720745502634506956158153283070441964462644401978953041673808332923846545153725133316840331525 63896589

47110087988118295323646800461339453767014912177042821942828505066221884630508804097835701153266549162552495263850962757967494757716034923473596280176155626743906 23303470

04453811940952869755093858067970266114568484848463089914943151524881780244169740998906039084394418502530473572469373056161853793406882946014254022114233731397240 857449458

62377619322551855689110364637684070516003606140643570811847797770405995641200104601413000890788089377595295745047655039053183599146854554075702541944535881728230 21718408

73401560659477066982172966386591321277165192516662912421600260824045564181828242071555930414128479212381485951448399656715057646027136104073454888170722718008329 68140120

39662237524091762593050119645743753899286461890218741075016941551730748796555794341221340186194571111414514347679110508738584379543228146239251032152517806102200298506 11

71511522924684406233670927089226453241078178623493002036486152993070136846970506636958762130387191849673626286128773135307721316622088806901178452231994936532272443177 47

90440805210124268282757760576852072110323533635517332228465853855790725929976584978688690119345292746422752555850487824174159627743927269508630037391972324328096423320 98

58067469745511572367038795593244575845312600239564518634241370010986150265195096501276333828196736597643559400008983705072192268375625342457964215208141538140352834232 7

45745282188653998454027744122254522942265414501024628838852570646391990412738586374723771970181661501559306552580402449092982449535013277809532564263426263432798995168 66

92296919787694623076382134287918534016638265861389810556650674010634208285394781800204536422696790163479917796814269701962431418370179332392082069639114568653435751789 37

02466426959525960065432609160427090327733412487937622089785456943184741943451062039746655514720561706101946122286668159690483839429043099283167735414442937221925763921 77

79234223773397148488181910169982018278621131812845363263984386215655096776531988541783185586741582390043053451170737372867280188233545785306959967788067417964300979384 13

25405448123808319030586540851532278542313542837423537587216882363220550706527064952513569636752121046320641843223935637795209989654547200169683510743077019398841978707 09

47694987486420085547074572670602171274095669266543143833776902401314278999775677872417957135722306063230523856351476313255726446759773377698628417509403393812169214267 35

63864623543402066943063507513940274428897658237003075649301065734518698212694757885041191961361046093431644074153374395772435885256502313780947543195773054518785207209 86

38611622730433453295875474855127338327972219185035047871897884249287106896182108994171586776120838015161888651454001285859716463013644965160551493822792834449083767432 81

98921142910431061110868692165527920798201649329374581342150765511878489276738254879351178323516112085517841083762117528150078518547781860767942864148432332331910972668 50

03293412891404585820443155761077878576257742383848993157237395983811060781246778635855689965273688452409425287045964359207605793466843241453255963562487488211791208183 97

03478464175492914224861988169835836819129231902407171295700076874633450585460536189741365291148748788266701276631750412100720315781088924376640367730911755309040928171 19

61362105392341387334225937678768526257135122434134824492437863318852768275374310490354455242143571369313856402904000656993625368177991878558718447307705825297435057486 64

12708465442603847274453318352984893683389897808289018623074484047084490852849403039434295546385274085475901526881600037477252281178116115742371398195074067417534318414 63

50544342438357451924861158088003960668343940190898737819244040021983228452076515479211369255074787416583660242218546219464307751521861355815198875404635517614095905437 38

57005250135809390485967216202160698441615690787716840681274143766140911181396310855991516614810795445153249599213668254996327123735625684147454143123120604881956549350 28

23579799437677571835665900690204179933690917282410490623132712442494160142851609628077906360383358414199270915422768425817749922199305725803259512377120129944248407012 2

79686794447341806792582657934958914767888378915440932631656700608894731093256105619880308605486520877456454524748535315405011168124284749579043721594155394367029071257 78

65368975422894964011618502443713140843463035435806477212711435043136254843866092436155003196510855005090715833704150255688910245840041819335202521244984571676792556822 18

02326639623157096458907968699494137343262118149547389785624882172010558278984533333136548489450888785675190404595926531952158119244898113529022134550383565982719369493 1

254

7656578186760047007731691816455034194766774380181590333099302566595666806010250926530115150160622468616389719309021432141704002191457726858407865209722168098063403409743

0869072988223097519519831002986126704890817788276877579611783302232901846001907160874768423152240507743607793306997142082661884079404692580632742472750415657425467147233

099626039572865385905528880059112257746897621928238904065552137197494522321276833845155145768466112776688207883475885860064886573576316527556999411910834661445618670681

274946339286427366151155891224086859707892700750339421987583635973430041069559226761035533059931059176312793511629714079606501253293619401366506307941570050211188043581

49453379132085873254843046365963575445347023689576407547704415571493176867930227775311988218483730191178628930694011158935995127482991205306812955192982493827214779554 10

129443773451756425117165262161669991858356468746493579240977245513328023629366757099076978525142266411911935854345013740960793760050865528342280657264780220420613079516 9

963953324761662289154668469409691451525256341542697672709654289236696840319253509490673845090202972431809123127018255195138346002937882176207767661470179105698709532770 6

600308416181059206587486560051483952184824992543254856132508585197436417072553803196406928071563221322173345451876615575526390318339396568420029460701119129003740322343 9

767010625918954911081661777562192162312113709031397199510930006010300715333345290159788935879859588478418005253796008798347110926578754895419075386106622018995978954059

343787394706384236767750218336928872866052783454452202405805711362960075974665197771111263096946950342384651053143366709511076068629058782907882208713346420036439036633 2

98098850085133781496316618857710806631255804432420769864751221658236162083468148414083152527539605062706669652630190259384874403412660635747377192472521516522939406695 52

4534224812999183265659442375912055988200472864204206707422287275907409713982157237963214549916729673080286486448406832219033684902698992971020092041186578702517859183575

055270334768877563858546273960750997614740057219289818296672620120311458260815855382692225101382561102542944306796243640089978100544006780213427670764254992593671010228 4

67486622594117529405166755581862694175059399711537675399665098330161573697092270055730969595564927238518257578755341788875263978855596462447492326407482162854233803633 59

4937489952680601675414219788350902373105768750328191825905212533184931830610070220267858052785156302432419555393856571133062252246234147971144793257899373626522022284579

923034601177104157440941246819729500291141398676155035259981948073523915452858111822298181564944792756355332235062904962376021888577294081893514505639433389353397765545 6

340866555282658136000816463334525449484505955566705163107708872505206626022560573629214451674721138860169162930054205124206415555272401945587359509752625533729093899907 5

288085242342552687896232612745355757808171530910245963535539481476277196916419842611182758862589058043661548756206816374340800476001644198438029373414682867717616041428

541260642564158093743961591292427136313673177612445899338591777306113329884665524574828349252311945870699891238106008382022702659856257633391903367294000126429069966838 4

238179436127469866593189619045322232936031663296014943687182809327430491523893225625519111473007531481288072064268422269811361855201544798087601072876094502692494540614

879525977809863600696691013778141234900206869198389264153767225517687495152040148830312041130202278064596458948058713779967688734939187037982143916720645908696518972256 9

169850999031020309665736330187721579107878142644572561741158418587769963563529157661151716117556191214637721380113652263627127850352145333065842404271934657080561805642 8

6269988319327273724685080806372534586774314356471343299136108458711456701768060241563987452403858336379398356339349445950307144276942139216593351516100280682600186217108

02758990916345462460226985867174542310977297306019504557502121786672926863366512580710505001747213369207269807192779063301291903342918220130117130206861246373615361 76394

805561122679035959983458207368030321560508366401669154640260872605066519849075749621296331192046086424701059966275204067990852111188739395262221717183945674358436625 0259

40756778745520688317702101822639289293331994618955621433939875377741823349077638559935400871935321557810634349264692166801973069587793471722254480791181111963926392 76480

01235177257192747883057839711366906064552543319190322289193609549718432491009045406627292385027405633838548590188268149435843645843980262611116171765842816097957652 50678

76117951163239926351700326195341509297500553404886640965835191659794919350863488219261598144843692437545565112204194805526823075332476974088472778743545235927210880 40526

25835368689199319844609897160844674263673591680055284786340626912717217317175600616771714634755616198078843903113584777164260510474745766361438320854993672197457399 797866

52277503539806189140888838590932137437527203362302578779561047292856386085130915778464960008736333920310489778169199045483721157693261472210169373395677086569137611 10869

15327835405568948605071082295424809180558080958528406673528781480386538021466467571438965475808604345129553935513095869321108629931110605839939424965760106574952402 64494

63655244424073059903652849089666480404679455176056890276317171918768727725748903365671778563832165305692129115050326412815732707501138355197893094089107488034261090 8827

41413711940912309437167869613636072477657104623486150406068547046457718789166038214014347509730536910311084407969550456237753811982755215952136501877563397073543958 01204

71966019512882510545033173051621634109051819226052554631226435532259295747572882001626270808236042444590358136199045960449167540375537272061819889995147771614932760 79799

93540532317930374352758499542681718721374730002593356153619211112926163891621846956695620335649705965093323716885518784203330418075030665055606251741605052623316640 91925

22382588709552189028812957505217116559791713082534046084330797746547688166691968144768973284839179217767764271032151745244795073588076328905416731603181192610241700 38177

56596118256541525180967987602616343017278370327961732925081347047654857656059010727673521385459827689298733297583299446853656599192720272312841968966632593594772667 22350

01137195026467308449262860959852620522409521822359200314706982697792667225042365592919249205435035444239094088057620105046530926977313494108572799763801130492797398 65584

19898876583320159334396104687507963520178472987317304408427266965840609618054654656319302505014959888401925085596318809233280147303879129195795825101292043776533474 10891

80758072471272402761666296862622223166070448752292147150714615960773351238272166915455272913078876136704003344771052077005942899727117736591924299132120809706489631 2558

84391194426424834555020727461522672099256445652834679899490660341741368476155773134407346980037980421412267132037246532107321735737604919320627556467665490391302868 99780

15277912027247510532924595527398642066245295718008680915733555397019651293100483231470413504942935965118265724049820443139756703147053709850613146155991545967908038 20633

27112705397643894613068335246691567644480584790531856264957839354546836297097508864072557823669299065081270643207674253904368857138109407458556596741811348102672978 01287

65977058162672845756159328126606457532869835675416943733517186915449429803953280956253296717647427841921710549638534142832198621486189526791483040045423024372446249 42769

58818851304787948051509240221847272687432602896204859856831807437521486290992133913992180695380744363471106230210202399080166428312197903931088989028681277449139877 81601

23696349353837904833738861649739886424085640388600121735603712566302824193284413937152603550255365006846913799512533015708816931761980595766096113281453624863814374 70813

7609973392793407138710218356044379465595763210859726380561131738627799618662828106580006360605669651605002754632000642838339900470686106215897801359187080238837576895579

1117132702187191386124460922850946621788009123566467142528458131688323343861702634545310635732686171417326825210992719583249078323219289804851229823033793785926972269319

5895033354126067763684519202799011294351329008525896049061348176846124481858634526732494412395033702428955285667455763776548313039154490724174469903334896301032696085126

4053688782221214621521942382509078188940264353675156891044885683292128360258157480099845832054876530840824256135089359722960318588388580838358185664412153867087676012 4831

0104630844744321880144790933674746884478564148174459091245398103230088592063710156358756516430959611273964015861769781314087950737142883177604366898196264134750069719405

6519504550516774239794019689986252789008523744580732886706974177396357254506098542345678985204250732286070942026394841828669256606166544407204577568393932312265407812478

3166688018030184225045200535518686848505584530852538497542612057943053533080747750826496088529445572785003441395589379335051184030225298706162915059642259996005856489572

3365317986913669699442786657891252568462604147981108656855717390721330789331085259033333113718587728703492627902715673291686627166749081993182583166832828500157570780 1611

9316922193151475493775515980465409283999109493742010371708560860588178544900570410413604043513762426899815268060925540112346532950434918053747735616666710469298309567892

0316482063931739221248405512034780632111316813373224063216455415588237846091942738088502838312362266549744300558099898299574258432353764298631465635055283560477090757473

2826367704396309792346562979493445664139608514643713039321367742124709044527215406879215426306425972602920189946552981151426126049007638671417302357272768390415597234506

6669386646058829201247114417883178223482153389187605836327618181943322769555311258190484751746290562001346899644071619823205434718461102051131555093022682510749199014960

8178562605108859036584745150376384915134000329516399106219240557283008103521761979616832213981692408357639556211657126112192905087163255258558649616638254193591482181876

1959232920569955063764581826855752211527870118029943354674153627620774978540541333031363354236824101084647637490638852798414900764646976485400947963589549754614481376369

7059163569983681198752505479306935320757076678014844770142471624190816682249007420711186488154772891718653596776539579933503342728214605416964960098470697958559264304287

0363664713071314782330611576419913222420646099898830762685836055527409904784676107604241784215062851755735299964786255295428367429870664579433758010140740211618614484329

7657442634285287047785563083096314352787830419450197029465757777328167468580874539316039372533158992805794346314087358608617788263349277461511849116551306818467136773488

2334108513640394793920887688633633946138235834479408156961091429387734713893423773619109646056424447477908207604966027135616895410644483213659808293890972961891211834291

4906163896386106937520895346883983344467189821243478072387407457697554507436846747135024858818399665568196344528811941833172636825050611864900394125520574571203603557802

5141904352671837219213848299058032246958424323158984432510396544353505354322921674704077861468485976255744615351188003143056995492784716745449726976128393325183819722232

8360707522781292813010656941262948730634268837338181742170608647548276394242391402753218042951903411635170469807423351556057857562450999253201787499636640473477038985587

3065076038709977318431281098978988208543559550943253902371895216820233442455725753078792633985509016455942373396625223351648750589556942172972448959988250892321120347958

9415465460303787861759157166139886932687374968473054965329378214756481057938082853005324470805065692942234001095934829461453907889066162640215013073533003319207456372637

70770999399922886212243248802062634850888530360107234368901360642758142528398785949179979611219637975765192452186709608809213711197750008781593043072934488393095757415 92

413752859777972918934538505080383198677459002518657917237080857416429715380788406071306868036198241971577476389507253468404569192759531937223702229015580065607604738547 3

599044779967487499697694271376686955331951253377640985870966838632639261649456086841403745684207194059507017430354691821509004664939985517413893851975731215682616228622 3

188109672974760601302833119371611408747270676255856777511995666748615196491297019331808499410961813929649278936090212535443327375064260624299412032736255824417498345094 7

309453436615907284163193683075719798068231535737155571816122156787936425013887117023275555779302266785803199930810830576307652332050740013939095807901637717629259283764 8

74790177274125678190555562180504876746991140839977919376542320623374717324703369763357925891515260315614033321272849194418437150696552087542450589956787961303311646283 99

63464604220901061057794 58151...

This is not the end

Made in the USA
Coppell, TX
19 February 2020